Operator Theory
Advances and Applications
Vol. 105

Editor:
I. Gohberg

Chebyshev Splines and Kolmogorov Inequalities

Sergey Bagdasarov

Springer Basel AG

Author:

Sergey K. Bagdasarov
Department of Mathematics
The Ohio State University
231 West 18th Avenue
Columbus, OH 43210-1174
USA
e-mail: skbgdsrv@math.ohio-state.edu

1991 Mathematics Subject Classification 41A17; 41A44, 65D25, 26A16, 26A15, 26D10, 58C30

A CIP catalogue record for this book is available from the
Library of Congress, Washington D.C., USA

Deutsche Bibliothek Cataloging-in-Publication Data
Bagdasarov, Sergey K.:
Chebyshev splines and Kolmogorov inequalities / Sergey
Bagdasarov. – Basel ; Boston ; Berlin : Birkhäuser, 1998
  (Operator theory ; Vol. 105)
    ISBN 978-3-0348-9781-5   ISBN 978-3-0348-8808-0 (eBook)
    DOI 10.1007/978-3-0348-8808-0

© 1998 Springer Basel AG
Originally published by Birkhäuser Verlag Basel Switzerland in1998
Softcover reprint of the hardcover 1st edition 1998

Printed on acid-free paper produced from chlorine-free pulp. TCF ∞
Cover design: Heinz Hiltbrunner, Basel

ISBN 978-3-0348-9781-5

9 8 7 6 5 4 3 2 1

Table of Contents

Appendix B
Kolmogorov Problems in $W^1 H^\omega(\mathbb{R}_+)$ and $W^1 H^\omega(\mathbb{R})$

Preface

Since the introduction of the functional classes $H^\omega(\mathbb{I})$ and $W^r H^\omega(\mathbb{I})$ and their periodic analogs $\widetilde{H^\omega}(\mathbb{T})$ and $\widetilde{W^r H^\omega}(\mathbb{T})$, defined by a concave majorant ω of functions and their r^{th} derivatives, many researchers have contributed to the area of extremal problems and approximation of these classes by algebraic or trigonometric polynomials, splines and other finite dimensional subspaces.

In many extremal problems in the Sobolev class $W^r_\infty(\mathbb{I})$ and its periodic analog $\widetilde{W^r_\infty}(\mathbb{T})$ an exceptional role belongs to *the polynomial perfect splines of degree* r, i.e. the functions whose r^{th} derivative takes on the values -1 and 1 on the neighboring intervals. For example, these functions turn out to be extremal in such problems of approximation theory as *the best approximation of classes* $W^r_\infty(\mathbb{I})$ *and* $\widetilde{W^r_\infty}(\mathbb{T})$ *by finite-dimensional subspaces* and the problem of *sharp Kolmogorov inequalities for intermediate derivatives of functions from* W^r_∞. Therefore, no advance in the exact and complete solution of problems in the nonperiodic classes $W^r H^\omega$ could be expected without finding *analogs of polynomial perfect splines in* $W^r H^\omega$.

We pursue three main goals in this book: (1) to introduce the notion and give the formulae for the perfect ω-splines in $W^r H^\omega$; (2) to describe various extremal properties of perfect ω-splines by emphasizing the new phenomena and the old features inherited from polynomial perfect splines; and (3) to show examples of applications of the general theory of perfect splines in examples related to the computation of N-widths of classes $W^r H^\omega(\mathbb{I})$ and our solution of one of the most celebrated problems of real analysis – the Kolmogorov problem of sharp inequalities for intermediate derivatives in the Hölder classes $W^r H^\alpha(\mathbb{R}_+)$ and $W^r H^\alpha(\mathbb{R})$.

The organization of the book runs as follows. Chapter 0 is introductory. In Chapter 1 we list such auxiliary results as *the Borsuk theorem, the Chebyshev theorem* and many other technical facts employed in our proofs.

In Chapter 2 we introduce the notion of *a simple kernel* Ψ and *the rearrangement* $\Re(\Psi; \cdot)$ *of the simple kernel*. The Korneichuk lemma describes the extremal functions and the numerical value of the maximum in the problem

$$\int\limits_a^b h(t)\psi(t)\,dt \to \sup, \qquad h \in H^\omega[a,b] \,:\, h(a) = 0, \qquad (*)$$

if ψ is a derivative of a simple kernel Ψ. Then, we present the major group of facts embedded in the foundation of the theory of extremal problems in $W^r H^\omega$: the structural and limiting properties of extremal functions of problems $(*)$.

Chapter 3 is reserved for the introduction and description of basic properties of generating Fredholm kernels.

A review of classical Chebyshev perfect splines in $W^{r+1}_\infty[0,1]$ is given in Chapter 4. Analyzing two proofs of extremality of Chebyshev splines in Kolmogorov-

Landau inequalities, we show possible pitfalls and technical obstacles arising in the process of solving extremal problems in $W^r H^\omega$.

In Chapter 5 we obtain the numerical differentiation formulae and derive sufficient conditions for a function $f \in W^r H^\omega[0,1]$ to be extremal in the problem

$$f^{(m)}(0) \to \sup, \qquad f \in W^r H^\omega(\mathbb{I}), \quad \|f\|_{C(\mathbb{I})} \leq B. \qquad (\mathbb{K} - \mathbb{L})$$

Chapter 6 is the core of the book. We prove the major result of this book, Theorem 6.0.1, which describes the family of Chebyshev ω-splines $\{\mathcal{Z}_n\}_{n \geq r}$ of the Kolmogorov–Landau problem on the finite interval.

Chapter 7 offers a detailed analysis of special features of properly rescaled Chebyshev ω-splines. Relying on these properties in our implementation of the limiting procedure to the sequence $\{\mathcal{Z}_n\}_{n \geq r}$, in Chapter 8 we construct extremal functions in the Kolmogorov–Landau problem in Hölder classes $W^r H^\alpha(\mathbb{R}_+)$. We also find the sharp constant in the multiplicative inequalities for the norms of intermediate derivatives in terms of the Chebyshev ω-spline on $\mathbb{R}_+$.

In Chapter 9 we characterize extremal functions and rearrangements of the problem

$$\int_{a_1}^{a_2} h(t)\psi(t)\,dt \to \sup, \qquad h \in H_0^\omega[a_1, a_2], \qquad (**)$$

where $a_1 < 0 < a_2$, and the kernel ψ has a finite number or a countable monotonely ordered set of points of sign changes on $[a_1, a_2]$, for $-\infty \leq a_1 < a_2 \leq +\infty$.

Using the results of Chapter 9, in Chapter 10 we characterize extremal functions in the Kolmogorov case of sharp inequalities for intermediate derivatives of functions from $W^r H^\alpha(\mathbb{R})$.

As an illustration of general results in the theory of Kolmogorov–Landau inequalities in functional classes $W^r H^\omega$, we give a complete description of extremal functions in the problem $(\mathbb{K} - \mathbb{L})$ for $r = 1$ and $r = 2$.

Like J. Hadamard and E. Landau did in the case $r = m = 1$, $\omega(t) = t$, $I = \mathbb{R}$ or $\mathbb{R}_+$ of the problem $(\mathbb{K} - \mathbb{L})$, in Chapter 11 we offer the corresponding numerical differentiation formulae in $(\mathbb{K} - \mathbb{L})$ for $r = 1$ and *all concave modulii of continuity* ω.

In Chapter 12 we find the full solution of the problem $(\mathbb{K} - \mathbb{L})$ for $r = 2$ and $I = \mathbb{R}_+$ (the Matorin – Stechkin problem for $\omega(t) = t$) and $I = \mathbb{R}$ (the Kolmogorov – Stechkin case for $\omega(t) = t$) with numerical differentiation formulae for f' and f''. Generalizing Stechkin's result for $\alpha = 1$, we show that in the Hölder classes $W^2 H^\alpha(\mathbb{R}_+)$ the optimal numerical differentiation formulae are of the form

$$f^{(k)}(x) \approx \frac{(-1)^k}{d \cdot h^k}\{af(x) - (a+b)f(x + c_1 h) + bf(x + c_2 h)\}, \qquad k = 1, 2$$

for some constants $a(\alpha, k), b(\alpha, k), d(\alpha, k) \geq 0$, and $c_2(\alpha, k) \geq c_1(\alpha, k) > 0$, independent of the step h.

S. Karlin showed *the uniqueness* (up to the change of orientation) of the polynomial Chebyshev spline of degree r with $n+2$ alternance points and $n-r$ knots. In Chapter 13 we emphasize the following interesting feature of classes $W^r H^\omega$ in the case of nonlinear ω: *different extremal problems in $W^r H^\omega$ have different sets of Chebyshev ω-splines*. In partucular, in Chapters 6 and 10 we construct two different families of Chebyshev ω-splines maximizing intermediate derivatives at the origin on the intervals $[0,\Gamma]$ and $[-\Gamma,\Gamma]$, respectively. In addition, in Chapter 13 we characterize the Chebyshev ω-splines related to *the problem of n-widths of classes $W^r H^\omega[0,1]$*.

Chapter 14 describes the structure of the Chebyshev ω-polynomial deviating most from the linear space of polynomials of degree r. These functions provide direct generalizations of the well-known *Chebyshev polynomials*.

In Chapter 15 we give a solution of the problem of n-widths of classes $W^1 H^\omega[-1,1]$ for a wide variety of concave modulii of continuity.

Lower bounds for the classes $W^r H^\omega[n]$ are obtained in Chapter 16.

In two chapters of the Appendix we list results on the structure of extremal functions of the problem

$$\|f^{(m)}\|_{\mathrm{L}_\infty(\mathbb{R}_+)} \to \sup, \qquad f \in W^r H^\omega(\mathbb{R}_+), \quad \|f\|_{\mathrm{L}_p(\mathbb{R}_+)} \leq B, \qquad (\mathbb{K})$$

for $1 \leq p < \infty$.

The numerical differentiation formulae and sufficient and necessary conditions of extremality of a function $f \in W^r H^\omega[0,d]$ in the Kolmogorov problem $(\mathbb{K})$ are given in Chapter A. In Chapter B of the Appendix we offer a detailed description of the structure of extremal functions of the Kolmogorov problem $(\mathbb{K})$ in $W^1 H^\omega(\mathbb{R})$ and $W^1 H^\omega(\mathbb{R}_+)$.

The monograph divides into Chapters, Sections and Subsections. Definitions or Notations, Theorems, Lemmas or Propositions are labeled following the system in this example: "Theorem K.L.M" would be Theorem M in Chapter K, Section K.L. We also label the formula tags as follows. The two-entry tag (X.Y) is used to identify the Yth formula in Section C.X of the current Chapter C. However, whenever the reference is made outside the current Chapter C, we use the three-entry tags (B.X.Y) to specify the Yth formula in Section B.X of Chapter B.

I especially wish to thank my adviser at the Ohio State University, Professor Boris Mityagin, for his intellectual support and expert assistance in the preparation of this book. Moreover, I am greatly indebted to Professor Mityagin for suggesting the main topic of my research, *Kolmogorov inequalities for intermediate derivatives,* and for his patient reading and numerous helpful recommendations and comments aimed at the improvement of this book and my other research papers.

I also express my deep gratitude to my adviser in Moscow State University Vladimir Tihomirov for his wise scientific guidance of my research throughout my years in Moscow State University and for posing extremal problems in Approximation Theory, on which I still continue to work.

Chapter 0

Introduction

0.1. History of the Kolmogorov–Landau problem

0.1.1. General setting

Let I be either the entire line $\mathbb{R}$ or the half-line $\mathbb{R}_+$. Let also $p, s, q \in [1, +\infty)$, and $r, m \in \mathbb{N} : m < r$.

DEFINITION 0.1.1. We shall say that *a function f belongs to $W^r_{p,s}(I)$*, if $f^{(r-1)}$ is absolutely continuous on any interval $[\sigma, \xi] \in I$, and both norms $\|f\|_{\mathbb{L}_p(I)}$ and $\|f^{(r)}\|_{\mathbb{L}_s(I)}$ are finite.

The first results concerning *inequalities for derivatives* of functions f from $W^r_{p,s}(I)$ *in the multiplicative form*

$$\|f^{(m)}\|_{\mathbb{L}_q(I)} \leq K \|f\|^{\alpha}_{\mathbb{L}_p(I)} \|f^{(r)}\|^{\beta}_{\mathbb{L}_s(I)}, \tag{1.1}$$

are due to E. Landau [54] and J. Hadamard [31] who constructed extremal functions in the sharp inequalities (1.1) in the case $m = 1$, $r = 2$, $p = q = s = \infty$ for $I = \mathbb{R}_+$ and $I = \mathbb{R}$, respectively:,

$$\|f'\|_{\mathbb{L}_\infty(\mathbb{R}_+)} \leq 2\|f\|^{\frac{1}{2}}_{\mathbb{L}_\infty(\mathbb{R}_+)} \|f''\|^{\frac{1}{2}}_{\mathbb{L}_\infty(\mathbb{R}_+)}, \quad \|f'\|_{\mathbb{L}_\infty(\mathbb{R})} \leq \sqrt{2}\|f\|^{\frac{1}{2}}_{\mathbb{L}_\infty(\mathbb{R})} \|f''\|^{\frac{1}{2}}_{\mathbb{L}_\infty(\mathbb{R})}.$$

V. N. Gabushin [27] describes the exponents α and β in the inequalities (1.1): if $\dfrac{r-m}{p} + \dfrac{m}{s} \geq \dfrac{r}{q}$, then α and β can be determined uniquely, namely,

$$\alpha = \frac{r - m - s^{-1} + q^{-1}}{r - s^{-1} + p^{-1}}, \qquad \beta = \frac{m - q^{-1} + p^{-1}}{r - s^{-1} + p^{-1}}. \tag{1.2}$$

0.1.2. Cases of the complete solution of the Kolmogorov problem

In the late 1930's G. E. Shilov [16] found sharp inequalities (1.1) in the case $p = q = s = \infty$, $I = \mathbb{R}$, $2 \leq r \leq 5$, and formulated the following hypothesis that proved to be true.

THEOREM 0.1.1. *The set of extremal functions in the inequality (1.1) for $p = q = s = \infty$, and $I = \mathbb{R}$ consists of periodic functions of the form*

$$f(t) = \gamma \phi_{\lambda,r}(t + \rho), \qquad \gamma, \rho \in \mathbb{R}, \quad \lambda \in \mathbb{R}_+, \tag{1.3}$$

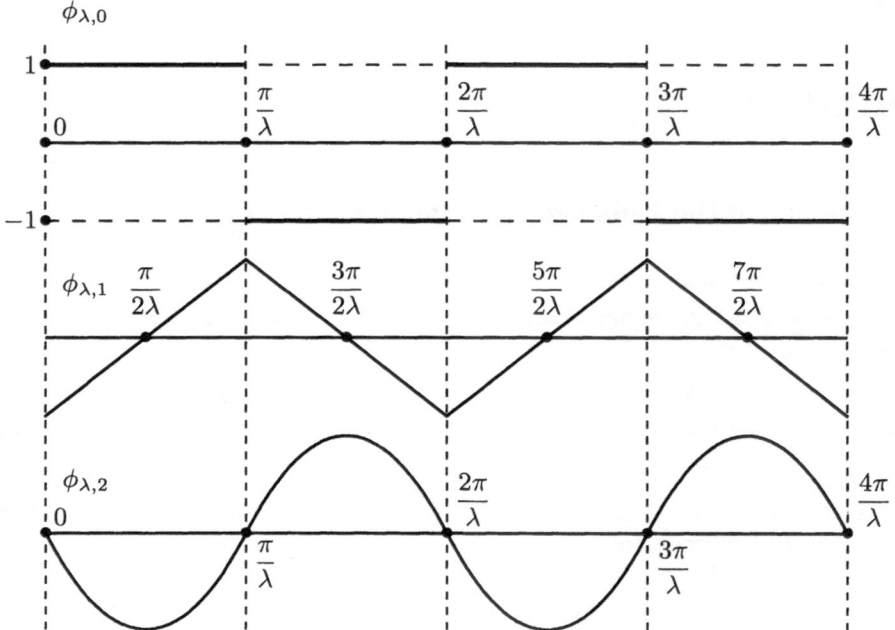

FIGURE 0.1.1.　Euler splines $\phi_{\lambda,r}$

where

$$\phi_{\lambda,r}(t) = \frac{4}{\pi\lambda^r} \sum_{\nu=0}^{\infty} \frac{\sin[(2\nu+1)\lambda t - (\pi r)/2]}{(2\nu+1)^{r+1}}. \tag{1.4}$$

REMARK 0.1.1.　In other words, $\phi_{\lambda,r}$ is a $2\pi/\lambda$-periodic function endowed with the property $\phi_{\lambda,r}^{(r)}(t) = \mathrm{sign\,sin}(\lambda t)$. Figure 0.1.1 illustrates the graphs of functions $\phi_{\lambda,r}$.

The functions in (1.3), (1.4) had occurred in the works of J. Favard [25], N. I. Akhieser and M. G. Krein [2] and even earlier in Euler's investigations; sometimes they are referred to as *the Euler splines*.

The full solution of the problem was given by A. N. Kolmogorov [41] in 1939, who confirmed the Shilov's conjecture and characterized the sharp constants in the inequality (1.1) in the case $p = q = s = \infty$, $I = \mathbb{R}$:,

$$\|f^{(m)}\|_{\mathbb{L}_\infty(\mathbb{R})} \le c_{rm}\|f\|_{\mathbb{L}_\infty(\mathbb{R})}^{1-\frac{m}{r}}\|f^{(r)}\|_{\mathbb{L}_\infty(\mathbb{R})}^{\frac{m}{r}}, \tag{1.5}$$

where $c_{rm} := K_{r-m}/K_m^{1-\frac{k}{r}}$, and $K_l := \frac{4}{\pi}\sum_{\nu=0}^{\infty}\frac{(-1)^{\nu(l+1)}}{(2\nu+1)^{l+1}}$, $\quad l \in \mathbb{Z}_+$, are known as *the Favard constants*. An elementary proof and a refinement of the Kolmogorov inequalities were suggested by A. S. Cavaretta [18], [20]. We review the Cavaretta's proof in Section 4.5.

The Kolmogorov's result led to the development of a new branch in the area of extremal problems and classical analysis – *the theory of sharp inequalities for intermediate derivatives on* $\mathbb{R}$ *and* $\mathbb{R}_+$. Since 1939, the complete solution of *the problem of sharp constants*

$$K = K_{p,q,s}^{r,m} = \sup_{f \in W_{p,s}^r(I),\ f \neq 0} \|f^{(m)}\|_{\mathrm{L}_q(I)} \cdot \|f\|_{\mathrm{L}_p(I)}^{-\alpha} \cdot \|f^{(r)}\|_{\mathrm{L}_s(I)}^{-\beta} \qquad (1.6)$$

in (1.1) for all $m, r : 0 < m < r$, and some fixed constants p, q, s, has been obtained in three cases for the entire line $\mathbb{R}$:

G. H. Hardy–J. Littlewood–G. Polya [32] : $p = q = s = 2$; **E. M. Stein** [82] : $p = q = s = 1$; **L. V. Taikov** [84] : $q = \infty,\quad p = s = 2$; and in two cases for the half-line $\mathbb{R}_+$: **N. P. Kupcov** [53] : $p = q = s = 2$; **V. N. Gabushin** [28] : $q = \infty,\ p = s = 2$.

Also, a number of authors have successfully pursued the problem of determining the exact constants in the inequality (1.1) in some partial cases. We mention the contributions to the area of extremal problems by B. Sz.-Nagy [83], H. Cartan [17], V. V. Arestov [3], V. N. Gabushin [26] and G. G. Magaril-Il'yaev [60]. A comprehensive survey of the Kolmogorov inequalities for various choices of p, q, s and r, m in (1.1) and a list of referrences can be found in the commentary by V. M. Tihomirov and G. G. Magaril-Il'yaev to the corresponding Kolmogorov's paper in [42].

0.2. Kolmogorov–Landau problem in the Sobolev class $W_\infty^{r+1}(I)$

0.2.1. Inequalities for derivatives of polynomials

A. A. Markov [62], V. A. Markov [63] and S. N. Bernstein [14] investigated properties of the algebraic polynomials P_n of degree n which yield the maximum modulus for the derivatives at a fixed point ξ of a finite interval $[a, b]$. However, neither the inequalities

$$\max_{x \in [-1,1]} |P_n^{(m)}(x)| \leq \frac{n^2(n^2-1)\dots(n^2-(m-1)^2)}{1 \cdot 3 \dots (2m-1)} \|P_n\|_{\mathrm{C}[-1,1]}, \qquad 0 < m < n, \tag{2.1}$$

of the Markov brothers (S. N. Bernstein [13] or R. J. Duffin and A. C. Schaeffer [77]) nor the Bernstein's refinement

$$|P_n^{(m)}(x)| \leq \left(\frac{m}{1-x^2}\right)^{m/2} \frac{n!}{(n-m)!} \|P_n\|_{\mathrm{C}[-1,1]}, \qquad 0 < m < n, \tag{2.2}$$

gave the exact constant or extremal functions in the sharp inequality

$$|P_n^{(m)}(x)| \leq C \|P_n\|_{\mathrm{C}[-1,1]}, \qquad x \in [-1, 1]. \tag{2.3}$$

P. L. Chebyshev [21], [22] and E. I. Zolotarev [91] set and successfully solved the
problem of finding the polynomial of degree n with *one* or *two* fixed leading coefficients, which deviates least from zero on $[0, 1]$. The definition, explicit expressions
and extremal properties of Zolotarev polynomials can be found in the Akhiezer's
monograph [1]. Finally, E. V. Voronovskaja [89] and V. A. Gusev [30] applied *the
functional method* to carry the problem of extremal functions in sharp inequalities
(2.3) to a complete solution. We refer the reader to Voronovskaja's monograph
[90] for a detailed discussion of various extremal problems for polynomials.

0.2.2. Sharp inequalities in the Sobolev class $W_\infty^r(I)$, $I = \mathbb{R} \vee \mathbb{R}_+ \vee [0, 1]$

When the calculation of the exact constant (1.6) is obstructed, the solution of the
problem is understood in the sense of *the qualitative characterization* of extremal
functions in the inequality (1.1). Due to the homogeneity of classes $W_{p,s}^r(I)$, it suf-
fices to restrict our attention to the classes of functions f with the norms $\|f^{(r)}\|_{\mathrm{L}_s(I)}$
bounded by 1. In partucular, let us introduce *the Sobolev classes*

$$W_\infty^n(I) := \{f \in W_{\infty,\infty}^n(I) \mid \|f^{(n)}\|_{\mathrm{L}_\infty(I)} \le 1\}, \qquad n \in \mathbb{N}. \qquad (2.4)$$

In Proposition 1.2.2 below we will show that in the case of infinite intervals $I = \mathbb{R}$
or $\mathbb{R}_+$ the extremal functions of the problem

$$|f^{(m)}(0)| \to \sup, \qquad f \in W_\infty^{r+1}(I), \quad \|f\|_{\mathrm{L}_\infty(I)} \le B, \qquad (2.5)$$

transform the sharp inequality (1.1) for $p = q = s = \infty$ into the equality. The
qualitative description of extremal functions in the problem (2.5) for $0 < m \le r$ in
the remaining cases $I = \mathbb{R}_+$ and $[0, 1]$ was given in the articles by V. M. Tihomirov,
A. S. Cavaretta and I. J. Schoenberg, and S. Karlin. We mention some of the results
and emphasize important points relevant to the content of this book.

The problem (2.5) for $r = 2$, $I = [0, 1]$ was solved simultaneously by M. Sato
[76] and A. Zvjagincev, A. Lepin [92]. The solution of the problem (2.5) in the
case $r = 2$ and $I = \mathbb{R}_+$ is due to A. P. Matorin [64] and S. B. Stechkin [80]. S. B.
Stechkin [81] also revealed the close connection between the problem of computing
the exact constant (1.6) and the approximation of differentiation operators by the
bounded linear operators (see also V. V. Arestov [4]). In particular, the bounded
linear operators (also known as *the optimal numerical differentiation formulae*),
which best approximate f' and f'' on the half-line, were shown in [80] to be

$$
\begin{aligned}
f'(x) &\approx \frac{1}{6h}\{-8f(x) + 9f(x + h) - f(x + 3h)\}, \\
f''(x) &\approx \frac{1}{3h^2}\{2f(x) - 3f(x + h) + f(x + 3h)\}.
\end{aligned}
\qquad (2.6)
$$

S. B. Stechkin [80] pointed out the following relation between *the multiplicative*
and *the additive forms* of the Kolmogorov inequalities in the problem (2.5):

LEMMA 0.2.1. *Let A, $B > 0$, and*

$$C = (r+1) \left(\frac{A}{r+1-m} \right)^{\frac{r+1-m}{r+1}} \left(\frac{B}{m} \right)^{\frac{m}{r+1}}.$$

The following assertions are equivalent in the case $I = \mathbb{R}$ or $\mathbb{R}_+$:

$$\|f^{(m)}\|_{\mathrm{L}_\infty(I)} \leq Ah^{-m}\|f\|_{\mathrm{L}_\infty(I)} + Bh^{r+1-m}\|f^{(r+1)}\|_{\mathrm{L}_\infty(I)}, \tag{2.7}$$

and

$$\|f^{(m)}\|_{\mathrm{L}_\infty(I)} \leq C\|f\|_{\mathrm{L}_\infty(I)}^{\frac{r+1-m}{r+1}} \|f^{(r+1)}\|_{\mathrm{L}_\infty(I)}^{\frac{m}{r+1}}. \tag{2.8}$$

The next major advance in the theory of extremal problems in W_∞^{r+1} was achieved in 1969 by V. M. Tihomirov [87], who constructed *the Chebyshev perfect polynomial splines* extremal in the problem of *n-widths* (or *n-diameters*) of the Sobolev classes $W_\infty^{r+1}[0,1]$. In Chapter 4 we review the construction of Chebyshev splines and outline the derivation of the solution of the problem (2.8) of sharp inequalities for $I = \mathbb{R}$ and $\mathbb{R}_+$ from the Tihomirov's result.

In 1970, A. S. Cavaretta and I. J. Schoenberg [19] characterized the solution of the problem (2.5) for $I = \mathbb{R}_+$. The extremal function in (2.5) is *the Chebyshev perfect spline* $T(x)$ uniquely characterized by the property of *equioscillation* between $-B$ and B.

In 1975, the Kolmogorov–Landau problem on the finite interval $[0,1]$ was treated by S. Karlin [36], [37] who constructed the family of *Zolotarev perfect splines* $\{\mathcal{Z}_B\}_{B>0}$. For each $B > 0$, the function $\mathcal{Z}_B$ of the norm B was shown to have $n = n(B) \geq 0$ knots and $n+r+1$ points of oscillation between $B = \|\mathcal{Z}_B\|_{\mathrm{C}[0,1]}$ and $-B$. In particular, for all sufficiently large B's the extremal functions in the problem (2.5) for $I = [0,1]$ are the classical *Zolotarev polynomials* ([59]).

REMARK 0.1.2.   The definition, explicit expressions and extremal properties of Zolotarev polynomials can be found in [1].

The problem

$$f^{(m)}(\xi) \to \sup, \qquad f \in W_\infty^{r+1}[0,1], \quad \|f\|_{\mathrm{C}[0,1]} \leq B, \tag{2.9}$$

for the interior point $\xi \in (0,1)$ was solved by A. Pinkus [66].

Finally, Yu. I. Lyubich [58] and R. R. Kallman, G.-C. Rota [35] extended the Kolmogorov–Landau problem to the problem of *inequalities between powers of linear operators*.

Notice that in all problems solved in the case $s = \infty$ in the inequality (1.1), it was sufficient to consider the functional class $f \in W_\infty^{r+1}(I)$, i.e. the set of functions subject to the constraint $\|f^{(r+1)}\|_{\mathrm{L}_\infty(I)} \leq 1$. This inequality is equivalent to the constraint $\omega\left(f^{(r)}; t\right) \leq t$, where $\omega(g;t)$ stands for the modulus of continuity of the continuous function $g(t)$

$$\omega(g;t) := \inf_{|x-y| \leq t} |g(x) - g(y)|, \qquad 0 \leq t \leq |I|. \tag{2.10}$$

In our generalizations, we consider the constraints of the form $\omega(f^{(r)}; t) \le \omega(t)$, for some fixed *concave modulus of continuity* ω. This discussion leads us to the following formulation of new problems and results in the theory of functional classes defined by *a common majorizing modulus of continuity*.

0.3. Functional classes $W^r H^\omega$ and $\widetilde{W^r H^\omega}$

0.3.1. Definitions

In *Jackson's inequalities* the errors of approximation of *an individual function* $f \in \mathbb{C}^r[a, b]$ by a specified finite dimensional subspace are expressed in terms of the modulus of continuity of the rth derivative of the function f (cf. [33], [48] and [50]). Instead, S. M. Nikol'skii suggested to consider *classes of functions* with a common majorizing *concave modulus of continuity* ω.

DEFINITION 0.3.1. *A function* $\omega(\cdot) : \mathbb{R}_+ \to \mathbb{R}_+$ *is called a concave modulus of continuity, if the following conditions are satisfied:*

$$
\begin{aligned}
&(i)\ \ \omega(0) = 0; \\
&(ii)\ \ \omega(t_1) \le \omega(t_2), \qquad t_2 > t_1 \ge 0; \\
&(iii)\ \ \omega\left(\alpha t_1 + (1 - \alpha)t_2\right) \ge \alpha\,\omega(t_1) + (1 - \alpha)\,\omega(t_2), \\
&\qquad \text{for all } \ \in (0, 1), \quad t_1, t_2 \in \mathbb{R}_+ : t_1 \ne t_2.
\end{aligned}
\tag{3.1}
$$

If the strict inequality persists in (3.1), (iii), then ω is *a strictly concave modulus of continuity.*

DEFINITION 0.3.2. *Let* $\omega(t)$ *be a concave modulus of continuity. The classes* $W^r H^\omega[a, b]$ *and* $\widetilde{W^r H^\omega}$ *are introduced as follows:*

$$
W^r H^\omega[a, b] := \{x \in \mathbb{C}^r[a, b] \ \Big| \ \omega(x^{(r)}; t) \le \omega(t),\ t \in [0, b - a]\}, \tag{3.2}
$$

$$
\widetilde{W^r H^\omega} := \{x \in W^r H^\omega(\mathbb{R}) \ \Big| \ x(t + 2\pi l) = x(t), \quad t \in [-\pi, \pi], \quad l \in \mathbb{Z}\}. \tag{3.3}
$$

In the case $r = 0$ we use the notation $H^\omega[a, b] := W^0 H^\omega[a, b]$.

The standard Sobolev class $W^{r+1}_\infty[a, b]$ can be viewed as a particular case of the class $W^r H^{\widetilde\omega}[a, b]$ with $\widetilde\omega(t) = t$. *The Hölder modulii of continuity* $\{\omega_\alpha(t) = t^\alpha\}_{0 < \alpha < 1}$ serve as an example of strictly concave modulii of continuity. In this case, we use the notation

$$
W^r H^\alpha[a, b] := W^r H^{\omega_\alpha}[a, b]. \tag{3.4}
$$

We also introduce the following convention for the subset of functions vanishing at the point $\tau \in [a, b]$:

$$
H^\omega_\tau[a, b] := \{f \in H^\omega[a, b] \ \Big| \ f(\tau) = 0\}. \tag{3.5}
$$

The classes $W^r H^\omega[a, b]$ and $\widetilde{W^r H^\omega}$ were introduced in 1946 by S. M. Nikol'skii [70], [71] in connection with approximation of functions by Fourier sums. Naturally, a number of problems arose concerning best characteristics of approximation of these classes by algebraic and trigonometric polynomials and other finite-dimensional subspaces. One such characteristic of a functional space is its *n-width* introduced by A. N. Kolmogorov [40].

DEFINITION 0.3.1. Let X be a normed space, and $C \subset X$ be a subset in X symmetric with respect to 0 ($x \in C$ if and only if $-x \in C$). Then *the n–width of the set C in X with respect to the norm* $\|\cdot\|_X$ is defined as follows:

$$d_n (C, X) = \inf_{L_n} \sup_{x \in C} \inf_{y \in L_n} \|x - y\|_X , \qquad (3.6)$$

where the infimum is taken over all n–dimensional subspaces L_n in X.

V. M. Tihomirov [87] computed the n-widths $d_n(W^r H^1[-1, 1], \mathbb{C}[-1, 1])$ in terms of the norms of the Chebyshev splines and characterized the corresponding optimal approximating subspaces in $\mathbb{C}[-1, 1]$.

The problem of n-widths of the periodic classes $\widetilde{W^r H^\omega}$ in the uniform and integral norm was treated in the works by N. P. Korneichuk [44]–[46], V. I. Ruban [75] and V. P. Motornyi, V. I. Ruban [69]. To formulate their results, we first define the analog of *the Euler ω-splines* in $W^r H^\omega$.

DEFINITION 0.3.2. Let $f_{n,0}(\omega; x)$ be the $2\pi/n$-periodic odd function defined on the interval $[0, \pi/n]$ by the formula

$$f_{n,0}(\omega; x) = \begin{cases} \dfrac{1}{2}\omega(2x), & 0 \le x \le \dfrac{\pi}{2n}; \\[2mm] \dfrac{1}{2}\omega\left(2(\dfrac{\pi}{n} - x)\right), & \dfrac{\pi}{2n} \le x \le \dfrac{\pi}{n}. \end{cases} \qquad (3.7)$$

For $r \in \mathbb{N}$, the function $f_{n,r}(x) = f_{n,r}(\omega, x)$ is the rth periodic integral of $f_{n,0}(\omega; x)$ introduced inductively as follows:

$$f_{n,r}(\omega; x) = \begin{cases} \displaystyle\int_{\pi/2n}^{x} f_{n,r-1}(\omega; t)\, dt, & r \text{ is odd;} \\[4mm] \displaystyle\int_{0}^{x} f_{n,r-1}(\omega; t)\, dt, & r \text{ is even.} \end{cases} \qquad (3.8)$$

Figure 0.3.1 illustrates the structure of Euler ω-splines.

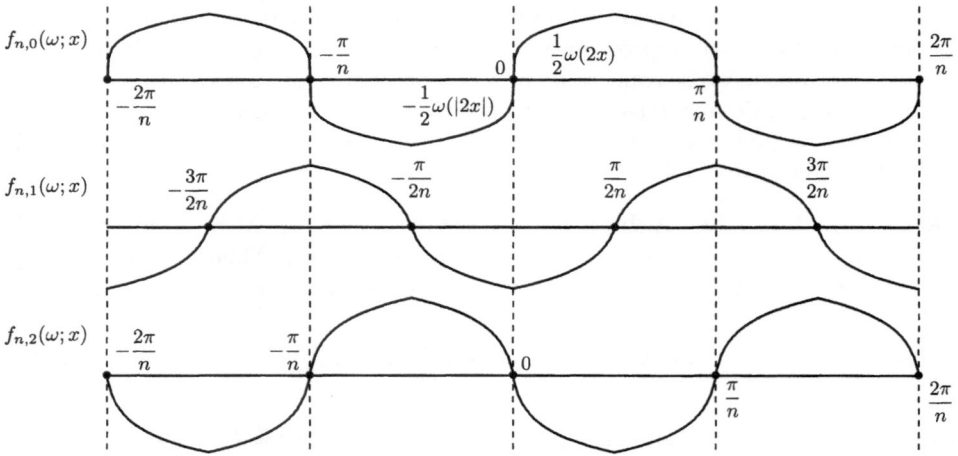

FIGURE 0.3.1. Euler ω-splines $f_{n,r}(\omega; x)$

THEOREM 0.3.1. *Let ω be a concave modulus of continuity, and $n \in \mathbb{N}$. Then,*

$$d_{2n}\left(\widetilde{W^r H^\omega}, \mathbb{C}[-\pi, \pi]\right) = d_{2n-1}\left(\widetilde{W^r H^\omega}, \mathbb{C}[-\pi, \pi]\right) =$$

$$= \|f_{n,r}(\omega; \cdot)\|_{\mathbb{C}[-\pi,\pi]} = \frac{1}{n^{r+1}} \int\limits_0^\pi \Phi_r(t)\omega'\,(t/n)\,dt; \quad (3.9)$$

$$d_{2n}\left(\widetilde{W^r H^\omega}, \mathbb{L}_1[-\pi, \pi]\right) = d_{2n-1}\left(\widetilde{W^r H^\omega}, \mathbb{L}_1[-\pi, \pi]\right) =$$

$$= \|f_{n,r}(\omega; \cdot)\|_{\mathbb{L}_1[-\pi,\pi]} = \frac{4}{n^{r+1}} \int\limits_0^\pi \Phi_{r+1}(t)\omega'\,(t/n)\,dt, \quad (3.10)$$

where $\Phi_k(x), k \in \mathbb{N}$, are defined on the interval $[0, \pi]$ by the recurrence relations

$$\begin{cases} \Phi_0(x) = \dfrac{1}{2}; \qquad 0 \le x \le \pi; \\[2mm] \Phi_k(x) = \dfrac{1}{2} \int\limits_0^{\pi-x} \Phi_{k-1}(t)\,dt, \qquad 0 \le x \le \pi. \end{cases} \quad (3.11)$$

An excellent summary of many other results in the theory of classes $W^r H^\omega$ up to 1985 and a rich bibliography are given by N. P. Korneichuk in his survey article [51] dedicated to the contribution of S. M. Nikol'skii to the Approximation Theory.

0.4. Sharp Kolmogorov–Landau inequalities in $W^r H^\omega(\mathbb{I})$, $\mathbb{I} = \mathbb{R} \vee \mathbb{R}_+ \vee [0,1]$

In the cycle of papers [5]–[12] we solved the problems

$$f^{(m)}(0) \to \sup, \qquad f \in W^r H^\omega(I), \quad \|f\|_{\mathrm{L}_\infty(I)} \le B, \tag{4.1}$$

for all concave modulii of continuity ω, $0 < m \le r$, $I = \mathbb{R}$, $\mathbb{R}_+$, $[0,1]$, and all positive B's. In [8] we described the solutions $\mathcal{Z}_B$ in the problem (4.1) on the interval $I = [0,1]$ for all $r \in \mathbb{N}$ and *all sufficiently large B's*. These extremal functions have $r+1$ points of alternance and *the full modulus of continuity* on the interval $[0,1]$: $\omega(\mathcal{Z}_B^{(r)}; t) = \omega(t)$ for $t \in [0,1]$. This analogy with the case of the linear modulus of continuity $\omega(t) = t$ makes it natural to call such functions *the Zolotarev ω-polynomials*.

In the process of solving the problem (4.1) for $0 < m < r$ and *all positive B's*, we ran into the necessity of a characterization of extremal functions in the problem

$$\int_a^b h(t)\psi(t)\,dt \to \sup, \qquad h \in H^\omega[a,b], \tag{4.2}$$

for the kernel ψ with the zero mean on $[a,b]$ and *a finite number* of sign changes on $[a,b]$. The detailed description of structural features and the limiting properties of extremal functions of the problem (4.2) can be found in our paper [6]. In Chapter 2 we list the main results from [6] employed in our constructions.

The solution of all variations of problems (4.1) for $m = r$ necessitates the description of the form and properties of extremal functions in the problem

$$\int_a^b h(t)\psi(t)\,dt \to \sup, \qquad h \in H_a^\omega[a,b], \tag{4.3}$$

for the kernel ψ with a nonzero mean on $[a,b]$ and *a finite number* of sign changes on $[a,b]$. In Chapter 2 we characterize the structure of extremal functions in the problem (4.3) (see [7] for details).

In [6] and [7] we also posed and solved problems (4.2) and (4.3) for kernels with *a finite or countable* set of points of sign change on *the finite, semi-infinite or infinite* intervals $[a,b]$. In Chapter 2 we formulate the corresponding results employed in the solution of the problem (4.1) on unbounded intervals.

Furthermore, in [7] we characterized the solution of the problem

$$\int_a^0 h(t)\psi_1(t)\,dt + \int_0^b h(t)\psi_2(t)\,dt \to \sup, \qquad h \in H_0^\omega[a,b]. \tag{4.4}$$

Extremal functions of the problem (4.4) for appropriate choices of kernels ψ_1 and ψ_2 feature as the rth derivatives of solutions of the problem (4.1) for $m = r$ and $I = \mathbb{R}$, and in the pointwise Kolmogorov–Landau problem. The relevant results from [7] appear in Chapter 9.

The description of the extremal functions in (4.2)–(4.4), called *the perfect ω-splines,* was followed by the solution of the problem of sharp Kolmogorov–Landau inequalities associated with the problem (4.1). It turns out that if $I = [0,1]$ and even $I = \mathbb{R}$ or $\mathbb{R}_+$, then sharp Kolmogorov–Landau inequalities have *the additive rather than the multiplicative form*: for all $f \in W^r H^\omega(I) : \|f\|_{\mathrm{L}_\infty(I)} \le B$,

$$|f^{(m)}(0)| \le C_B \cdot B + D_B(\omega'), \tag{4.5}$$

for some constant $C_B = C_B(r, m, \omega, I)$ and an integral functional $\omega' \mapsto D_B(\omega')$. The explicit expressions for C_B and $D_B(\omega')$ in the case of $I = [0,1]$ can be found in Chapters 5, 9 and 10.

The multiplicative nature of the inequalities (2.8) in $W_\infty^{r+1}(I)$, $I = \mathbb{R}_+$ or $\mathbb{R}$ is predetermined by the following *invariance of the class $W_\infty^{r+1}(I)$ with respect to dilations and rescaling*:

$$f(t) \in W_\infty^{r+1}(I) \iff \lambda^{r+1} f(t/\lambda) \in W_\infty^{r+1}(I), \qquad \forall \lambda > 0. \tag{4.6}$$

In general, functions from $W^r H^\omega(I)$ do not enjoy the invariance property (4.6), which explains the additive form of the inequalities (4.5). An exception from the rule are the Hölder classes $W^r H^\alpha(I)$, $I = \mathbb{R} \vee \mathbb{R}_+$. The analog of (4.6), the property

$$f(t) \in W^r H^\alpha(I) \iff \lambda^{r+\alpha} f(t/\lambda) \in W^r H^\alpha(I), \qquad \forall \lambda > 0, \tag{4.7}$$

implies the following multiplicative (in addition to the additive) nature of exact inequalities in $W^r H^\alpha(I)$ for $I = \mathbb{R}$ or $\mathbb{R}_+$:

$$\|f^{(m)}\|_{\mathrm{L}_\infty(I)} \le C_\alpha \|f\|_{\mathrm{L}_\infty(I)}^{1-\frac{m}{r+\alpha}}, \tag{4.8}$$

where, by Lemma 1.2.3 below, the constant $C_\alpha = C_\alpha(r, m, I)$ is now independent of the norm B.

As we have already mentioned, the construction of the Chebyshev perfect polynomial splines in the case of $\omega(t) = t$ is due to V. M. Tihomirov [87]. In Theorem 6.0.1 we construct the family of Chebyshev functions $\{\mathcal{Z}_n(t) = \mathcal{Z}_{n,r,m,\omega}(t)\}_{n \ge r}$ in $W^r H^\omega[0,1]$ extremal in the problem (4.1) for all nonlinear ω. Like the corresponding polynomial Chebyshev spline, the function $\mathcal{Z}_n$ has the complete set of $n + 2$ alternance points and $n - r$ knots on $[0,1]$.

Applying the limiting procedure to the rescaled and normalized Chebyshev functions $\mathcal{Z}_n$, we construct the set of extremal functions of the problem (4.1) in the Hölder classes $W^r H^\alpha(\mathbb{R}_+)$, i.e. for $I = \mathbb{R}_+$ and $\omega(t) = \omega_\alpha(t) = t^\alpha$, $0 < \alpha \le 1$.

This result generalizes the Cavaretta–Schoenberg solution of the problem (4.1) for $\omega(t) = t$ in [19]. In Chapter 10 we also present the solution of the problem (4.1) for $I = \mathbb{R}$ and $\omega(t) = t^\alpha$, $0 < \alpha \le 1$, generalizing the extremal functions and sharp inequalities in the original Kolmogorov case of $I = \mathbb{R}$ and $\omega(t) = t$.

The problem (4.1) for $I = \mathbb{R}_+$ and $I = \mathbb{R}$ for *all concave modulii of continuity* ω on $\mathbb{R}_+$ was solved in [9]–[11].

In [5] we give a construction of extremal *Zolotarev ω-splines* of the problem

$$f^{(m)}(\tau) \to \sup, \qquad f \in W^r H^\omega[\tau, 1], \quad \|f\|_{\mathrm{C}[0,1]} \le B, \qquad (4.9)$$

for $\tau = 0$ as well as complete solutions of *the extrapolation problem* (4.9) for $\tau < 0$. Finally, in [12] we describe the extremal functions of the problem

$$\|f^{(m)}\|_{\mathrm{L}_\infty(I)} \to \sup, \qquad f \in W^r H^\omega(I), \quad \|f\|_{\mathrm{L}_p(I)} \le B, \qquad (4.10)$$

for all $1 \le p < \infty$ and $0 \le m \le r$, $I = \mathbb{R}$, $\mathbb{R}_+$. In the case of the linear modulus of continuity $\omega(t) = t$, this problem was solved earlier by G. G. Magaril–Il'yaev [60]. We mention only one interesting result from [12] concerning the case $r = 1$. It turns out that in the case $r = 1$, $m = 1$, $B > 0$, $I = \mathbb{R}_+$, and a Hölder modulus $\omega_\alpha(t) = t^\alpha$, $0 < \alpha \le 1$, the collection of knots $\{\xi_i\}_{i=1}^\infty$ of the extremal ω-spline $x(t)$ constitutes *a geometric mesh*:

$$\frac{\xi_{i+1} - \xi_i}{\xi_i - \xi_{i-1}} = a(\alpha), \quad i \in \mathbb{N},$$

for some $a(\alpha) < 1$. Moreover, the extremal function $x(t)$ enjoys the following *property of self-similarity*: for all $i \in \mathbb{N}$

$$x\big|_{[\xi_i, \xi_{i+1}]}(t + \xi_i) = (-1)^i \left(\frac{\xi_{i+1} - \xi_i}{\xi_1} \right)^{1+\alpha} x\big|_{[0, \xi_1]} \left(\frac{\xi_1}{\xi_{i+1} - \xi_i} t \right), \quad 0 \le t \le \xi_{i+1} - \xi_i.$$

This phenomenon for $\omega(t) = t$ was first discovered by V. N. Gabushin [26] and later by G. G. Magaril-Il'yaev [59].

Chapter 1

Auxiliary Results

As the title suggests, in this chapter we list technical results which we employ in our constructions throughout the book.

1.1. General facts

The proof of Theorem 6.0.1 is based on the topological result known as *the Borsuk Antipodality Theorem* (see [15], [24])

THEOREM 1.1.1. *Let* $\mathbb{S}^n = \{\xi : \xi \in \mathbb{R}^{n+1} \mid \|\xi\| = r\}$, *where* $\|\cdot\|$ *is a norm in* $\mathbb{R}^{n+1}$, *and let* $\eta : \mathbb{S}^n \to \mathbb{R}^n$, $\eta(\xi) = \{\eta_1(\xi), \eta_2(\xi), \ldots, \eta_n(\xi)\}$, *be a continuous and odd* $(\eta(-\xi) = -\eta(\xi))$ *vector field on* $\mathbb{S}^n$. *Then, there exists a vector* $\bar{\xi} \in \mathbb{S}^n$ *such that* $\eta(\bar{\xi}) = 0$.

The Chebyshev Theorem describes characteristic properties of the best polynomial approximator of a continuous function. The proof of the following result is given in [23], [48], and in [65], where the constructive methods of the proof are suggested.

THEOREM 1.1.2. *Let* P_n *be the linear space of all polynomials of degree* n, *and* $f \in \mathbb{C}[a,b] \setminus P_n$. *Then,*

(a) *there exists a unique polynomial* $p_f(t) = \sum_{i=0}^{n} a_i(f)t^i$ *of the best approximation for* f *on the interval* $[a,b]$ *among the polynomials of degree* n, *i.e.,*

$$\|f - p_f\|_{\mathbb{C}[a,b]} = \min_{p \in P_n} \|f - p\|_{\mathbb{C}[a,b]};$$

(b) p_f *is the best approximator for* f *among the polynomials from* P_n, *if and only if there exist such points* $\{x_k\}_{k=1}^{n+2}$, $a \le x_1 < x_2 < \cdots < x_{n+2} \le b$, *that*

$$(f - p_f)(x_i) = (-1)^i \xi \|f - p_f\|_{\mathbb{C}[a,b]}, \qquad i = 1, \ldots, n+2, \qquad (1.1)$$

for a fixed $\xi = \xi(f) \in \{-1, 1\}$.

In this paper we adopt the following definition of *sign changes of an integrable function without zero intervals.*

DEFINITION 1.1.1. We shall say that *a function* $f \in \mathbb{L}_1[a,b]$ *changes its sign at the points* $\{\alpha_i\}_{i=1}^{n}$, $a = \alpha_0 < \alpha_1 < \cdots < \alpha_n < \alpha_{n+1} = b$, *if for a fixed* $\chi \in \{-1, 1\}$,

$$(-1)^i \chi \psi(t) > 0 \quad \text{for a.e. } t \in [\alpha_{i-1}, \alpha_i], \qquad i = 1, \ldots, n+1.$$

Then, $\mu(f; [a, b])$ stands for *the number of sign changes of the function f on the interval $[a, b]$*.

The following result on the relation between the number of sign changes of consecutive derivatives is a version of *the Rolle theorem*.

PROPOSITION 1.1.3. *Let $g \in \mathbb{L}_1[a, b]$ be a function with $\mu(g; [a, b]) < \infty$. The following relations hold between the number of sign changes of $g(t)$ and its integral $G(t) = \int_a^t g(u)\, du, \quad a \leq t \leq b$:*

(1) $\mu(G; [a, b]) \leq \mu(g; [a, b])$;

(2) $G(b) = 0 \quad and \quad \mu(g; [a, b]) > 0 \implies \mu(G; [a, b]) \leq \mu(g; [a, b]) - 1.$ $\qquad$ (1.2)

The element of the best approximation of a given function by a subspace in $\mathbb{C}[a, b]$ is described in *the duality theorem*.

THEOREM 1.1.4. *Let $\mathcal{F}$ be a linear subspace in $\mathbb{C}[a, b]$, and $f \in \mathbb{C}[a, b]$. A function $\varphi_0 \in \mathcal{F}$ is an element of the best approximation for f, i.e.,*

$$\|f - \varphi_0\|_{\mathbb{C}[a,b]} = \inf_{\varphi \in \mathcal{F}} \|f - \varphi\|_{\mathbb{C}[a,b]}, \qquad (1.3)$$

if and only if there exists a function g_0 with the bounded variation $\bigvee\limits_a^b g_0 < \infty$ such that

(i) $\quad \bigvee\limits_a^b g_0 = 1;$

(ii) $\quad \|f - \varphi_0\|_{\mathbb{C}[a,b]} = \int\limits_a^b f(t)\, dg_0(t);$ $\qquad (1.4)$

(iii) $\quad \int\limits_a^b \varphi(t)\, dg_0(t) = 0, \quad for\ all \quad \varphi \in \mathcal{F}.$

We frequently use the following estimates of the norms of functions $f \in \mathbb{C}^r[a, b]$ with at least $r + 1$ zeroes.

PROPOSITION 1.1.5. *Let $f \in \mathbb{C}^r[a, b]$. If f has $r + 1$ zeroes (counting multiplicities), then*

$$\|f^{(k)}\|_{\mathbb{C}[a,b]} \leq [b - a]^{r-k} \|f^{(r)}\|_{\mathbb{C}[a,b]}, \qquad k = 0, \ldots, r. \qquad (1.5)$$

Proof. By the Rolle theorem, the derivative $f^{(k)}(t)$ has a zero ξ_k on $[a, b]$ for each $k \in \{0, \ldots, r\}$. Then, $f^{(k)}(x) = \int\limits_{\xi_k}^x f^{(k+1)}(t)\, dt, \ a \leq x \leq b, \ k = 0, \ldots, r - 1$. Thus,

$$\|f^{(k)}\|_{\mathbb{C}[a,b]} \leq (b - a)\|f^{(k+1)}\|_{\mathbb{C}[a,b]}, \qquad (1.6)$$

implying (1.5). $\qquad\qquad\qquad\qquad\qquad\qquad\qquad\qquad\qquad\qquad\qquad\qquad\qquad\qquad \square$

In particular, if $f^{(r)}(t)$ belongs to $H^\omega[a,b]$, then

$$\|f^{(k)}\|_{\mathbb{C}[a,b]} \le [b-a]^{r-k}\omega(b-a). \tag{1.7}$$

The next theorem on *the width of the ball* is a tool for estimating the n-widths from below.

THEOREM 1.1.6. *Let Y_{n+1} be an $(n+1)$-dimensional subspace of a linear normed space X, and U_{n+1} is a closed unit ball in Y_{n+1}, i.e.,*

$$U_{n+1} = \{x \in X \mid x \in Y_{n+1},\ \|x\| \le 1\}. \tag{1.8}$$

Then

$$d_n(U_{n+1}, X) = 1.$$

For the proof, one can consult M. G. Krein, M. A. Krasnosel'skii, D. P. Mil'man [52] and V. M. Tihomirov [88], or the book by G. G. Lorentz ([56], p. 137). The following result is an elementary corollary of Theorem 1.1.6.

COROLLARY 1.1.7. *Let $n \in \mathbb{N}$, and*

$$X_n = \{(x_1, \ldots, x_n) \in \mathbb{R}^n \mid x_i = \pm 1,\quad i = 1, \ldots, n\}$$

be the set of vertices of an n–dimensional cube in $\mathbb{R}^n$. Then,

$$d_{n-1}(X_n, l^n_\infty) = 1. \tag{1.9}$$

DEFINITION 1.1.2. *Let $\tau_1 < \tau_2 < \cdots < \tau_n$. A function*

$$S(t) = \sum_{i=0}^{r} a_i t^i + \sum_{j=1}^{n} b_j (t - \tau_j)_+^r \tag{1.10}$$

with $(a_0, \ldots, a_r) \in \mathbb{R}^{r+1}$, $(b_1, \ldots, b_n) \in \mathbb{R}^n$ is called a polynomial spline of degree r with n knots $\{\tau_j\}_{j=1}^n$. The notation $\mathbb{S}^r[\tau_1, \ldots, \tau_n]$ is reserved for the set of all polynomial splines of degree r with the fixed collection of knots $\{\tau_i\}_{i=1}^n$.

Consider the problem of *the simple interpolation* by means of polynomial splines of degree r with the fixed set of n knots $\{\tau_i\}_{i=1}^n$:

$$S(x_j) = y_j, \qquad j = 1, \ldots, N, \tag{1.11}$$

where $N = N(r,n) := r + n + 1$, and $(y_1, \ldots, y_N) \in \mathbb{R}^N$. The points $\{x_j\}_{j=1}^N$ are called *the nodes of the interpolation* (1.11). *The Schoenberg–Whitney criterion* describes the necessary and sufficient conditions for the unique solvability of the system of equations (1.11) for coefficients $\{a_i\}_{i=0}^r$, $\{b_i\}_{i=1}^n$ and an arbitrary vector $(y_1, \ldots, y_N) \in \mathbb{R}^N$.

PROPOSITION 1.1.7. *Let $\{x_i\}_{i=1}^{n+r+1}$ and $\{\tau_i\}_{i=1}^n$ be two collections of points arranged in increasing order. The interpolation problem (1.11) has a unique solution $S \in \mathbb{S}^r[\tau_1, \ldots, \tau_n]$ if and only if the following inequalities hold:*

$$x_i < \tau_i < x_{r+1+i}, \qquad i = 1, \ldots, n. \tag{1.12}$$

1.2. General properties of functional classes $W^r H^\omega[a, b]$

We rely on the following properties of *concave modulii of continuity* ω.

PROPOSITION 1.2.1. *Let ω be a concave modulus of continuity on $\mathbb{R}_+$. Then,*
     (a) at any point $x > 0$, ω has one-sided derivatives

$$\omega'_-(x) = \lim_{h \to 0+} \frac{\omega(x) - \omega(x - h)}{h}, \qquad \omega'_+(x) = \lim_{h \to 0+} \frac{\omega(x + h) - \omega(x)}{h}; \qquad (2.1)$$

     (b) each of the functions ω'_+ and ω'_- does not increase on $(0, +\infty)$, and

$$\omega'_+(x) \leq \omega'_-(x), \qquad x > 0.$$

(c) ω is an absolutely continuous function on $(0, +\infty)$.

In this paper we make the following choice from the equivalence class of summable functions, defining the nonincreasing derivative ω' *everywhere* on $\mathbb{R}_+$.

DEFINITION 1.2.2. *Let ω be a concave modulus of continuity on $\mathbb{R}_+$, and the one-sided derivatives ω'_- and ω'_+ be defined in (2.1). We put*

$$\omega'(u) := \frac{1}{2} \left[\omega'_+(u) + \omega'_-(u) \right], \qquad u > 0. \qquad (2.2)$$

DEFINITION 1.2.3. *Let $I = \mathbb{R}$ or $\mathbb{R}_+$, and $B > 0$. Then,*

$$W^r H^\omega[I; B] = \{ f \in W^r H^\omega(I) \mid \|f\|_{\mathbb{L}_\infty(I)} \leq B \}. \qquad (2.3)$$

In Chapters 6–8 and 10 we describe extremal functions in the problem

$$f^{(m)}(0) \to \sup, \qquad f \in W^r H^\omega[I; B], \qquad (2.4)$$

for $\omega(t) = t^\alpha$, $0 < \alpha \leq 1$. The following proposition reduces the solution of the problem

$$\|f^{(m)}\|_{\mathbb{L}_\infty(I)} \to \sup, \qquad f \in W^r H^\omega[I; B] \qquad (2.5)$$

to the description of extremal functions of the problem (2.4).

PROPOSITION 1.2.2. *Let $I = \mathbb{R}$ or $\mathbb{R}_+$. The extremal functions of the problem (2.4) are also extremal in the problem (2.5).*

Proof. Fix $\tau \in I$ and a function $f \in W^r H^\omega[I; B]$. Put

$$g_{f,\tau,I}(x) = \begin{cases} f(x + \tau), & x \in \mathbb{R}_+, \quad \text{if} \quad I = \mathbb{R}_+; \\ \frac{1}{2}[(-1)^m f(\tau - x) + f(x - \tau)], & x \in \mathbb{R}, \quad \text{if} \quad I = \mathbb{R}. \end{cases} \qquad (2.6)$$

Then, $g_{f,\tau,I} \in W^r H^\omega(I)$, and

$$g^{(m)}_{f,\tau,I}(0) = f^{(m)}(\tau), \qquad \|g_{f,\tau,I}\|_{\mathbb{L}_\infty(I)} \leq \|f\|_{\mathbb{L}_\infty(I)} \leq B. \qquad (2.7)$$

$\square$

From our argument in Proposition 1.2.2 it follows that extremal functions of the problem

$$f^{(m)}(0) \to \sup, \qquad f \in W^r H^\omega[-\Gamma, \Gamma], \quad \|f\|_{L_\infty[-\Gamma,\Gamma]} \le B, \quad \Gamma \in (0, +\infty], \quad (2.8)$$

enjoy the following type of symmetry with respect to the origin:

$$f(-t) = (-1)^m f(t), \qquad t \in [0, \Gamma]. \tag{2.9}$$

The following result is a version of Lemma 0.2.1 in the Hölder classes.

LEMMA 1.2.3. *Let $A > 0$, $B > 0$, and*

$$C = (r + \alpha) \left(\frac{A}{r + \alpha - m} \right)^{\frac{r+\alpha-m}{r+\alpha}} \left(\frac{B}{m} \right)^{\frac{m}{r+\alpha}}. \tag{2.10}$$

The following statements are equivalent for a function $f \in W^r H^\alpha(I)$ in the case of $I = \mathbb{R}$ or $\mathbb{R}_+$:

$$\|f^{(m)}\|_{L_\infty(I)} \le A h^{-m} \|f\|_{L_\infty(I)} + B h^{r+\alpha-m}, \quad h > 0, \tag{2.11}$$

and

$$\|f^{(m)}\|_{L_\infty(I)} \le C \|f\|_{L_\infty(I)}^{\frac{r+\alpha-m}{r+\alpha}}. \tag{2.12}$$

Indeed, minimizing the right-hand side of the inequality (2.11) with respect to h, we obtain the inequality (2.12). On the other hand, an application of the Young inequality to (2.12) yields the inequality (2.11). As we have already explained, the sharp Kolmogorov inequalities in $W^r H^\omega$ have the additive form. The relation (2.10) between constants A, B in (2.11) and C in (2.12) enables us to make transitions from sharp additive inequalities to the exact multiplicative inequalities in $W^r H^\alpha(I)$ for $I = \mathbb{R} \vee \mathbb{R}_+$.

Let us describe a method of the extension of functions $f \in H^\omega[a, b]$ to larger intervals.

LEMMA 1.2.4. *Let $a < b < c$ and ω be a concave modulus of continuity. Let a function $h \in H^\omega[a, b]$ be extended to the interval $[a, c]$ by the formula*

$$h(t) = h(b) + \int_b^t \chi \omega'(t - a)\, dt, \qquad t \in [b, c], \quad \chi \in \{-1, 1\}. \tag{2.13}$$

Then, $h \in H^\omega[a, c]$.

Proof. Notice that $h\big|_{[a,b]} \in H^\omega[a,b]$, and $h\big|_{[b,c]} \in H^\omega[b,c]$. Thus, we need to verify the inequality $|h(y) - h(x)| \leq \omega(y - x)$ only for $x \in [a,b]$ and $y \in [b,c]$. Using the concavity of ω, for such x and y we have

$$|h(y) - h(x)| \leq |h(b) - h(x)| + |h(y) - h(b)| \leq \omega(b - x) + [\omega(y) - \omega(b)] \leq \omega(y - x).$$

$\square$

The elementary proof of the following property of concave functions is suggested by B. S. Mityagin.

PROPOSITION 1.2.5. *Let* $k_1, k_2 : k_1 > k_2 > 0$, *and* $a, b : 0 < a < b < 1$ *be such that*

$$k_1 \omega(a) = k_2 \omega(b). \qquad (2.14)$$

Then,

$$k_1^2 \int_0^a \omega(t)\, dt \leq k_2^2 \int_0^b \omega(t)\, dt. \qquad (2.15)$$

Proof. It is sufficient to give the proof for the modulii of continuity ω, endowed with the following properties:

$$\begin{aligned}&(i)\ \ \omega'(t) > 0, \qquad t \in (0,1);\\ &(ii)\ \ \lim_{t \to 0+} \omega'(t) < +\infty.\end{aligned} \qquad (2.16)$$

Let us introduce the notations

$$\xi := \omega(t), \quad A := \omega(a), \quad B = \omega(b), \quad k := \frac{B}{A}. \qquad (2.17)$$

In these notations, the inequality

$$k^2 \int_0^a \omega(t)\, dt \leq \int_0^b \omega(y)\, dy \qquad (2.18)$$

is equivalent to the inequality

$$k^2 \int_0^A \frac{\xi\, d\xi}{\omega'(\omega^{-1}(\xi))} \leq \int_0^B \frac{\xi\, d\xi}{\omega'(\omega^{-1}(\xi))} = k^2 \int_0^A \frac{\xi\, d\xi}{\omega'(\omega^{-1}(k\xi))}, \qquad (2.19)$$

where ω^{-1} stands for the function inverse to ω. But $k > 1$, so

$$\omega^{-1}(\xi) < \omega^{-1}(k\xi), \qquad 0 \leq \xi \leq A, \qquad (2.20)$$

and

$$\omega'(\omega^{-1}(\xi)) > \omega'(\omega^{-1}(k\xi)), \qquad (2.21)$$

proving the desired inequality (2.15). $\square$

Chapter 2

Maximization of Functionals in $H^\omega[a, b]$ and Perfect Ω-Splines

Our goal in this chapter is to introduce the reader to the notion of *perfect ω-splines* as extremal functions of *linear integral functionals*. We also give a comprehensive list of various properties of ω-splines used in our arguments.

2.1. Introduction to the theory of functional classes $H^\omega[a, b]$

2.1.1. Simple kernels $\Psi(\cdot)$ and their rearrangements $\Re(\Psi; \cdot)$

The Korneichuk lemma describes extremal functions of the functional

$$h \mapsto \int\limits_a^b h(t)\psi(t)\, dt, \qquad h \in H^\omega[a, b], \tag{1.1}$$

where ψ is the derivative of *a simple kernel* on $[a, b]$.

DEFINITION 2.1.1.   Let the kernel $\psi(\cdot) \in \mathbb{L}_1[a, b]$ be such that $\int_a^b \psi(x)\, dx = 0$ and for some points $a', b' : a < a' \le b' < b$,

$$
\begin{array}{lll}
(i) & \psi(x) < 0, & \text{for a.e. } x \in [a, a']; \\
(ii) & \psi(x) = 0, & \text{for a.e. } x \in [a', b']; \\
(iii) & \psi(x) > 0, & \text{for a.e. } x \in [b', b].
\end{array}
\tag{1.2}
$$

Then $\Psi(x) = \xi \int\limits_a^x \psi(t)\, dt,\ a \le x \le b,\ \xi \in \{1, -1\}$, is called *a simple kernel*.

If Ψ is a simple kernel, the equation $|\Psi(t)| = y$, for $0 < y < \|\Psi\|_{\mathbb{C}[a,b]}$, has precisely two solutions: $\alpha_y \in (a, a')$ and $\beta_y \in (b', b)$ (see Figure 2.1.1). The value of the maximum of the functional (1.1) is expressed in terms of *the rearrangement of the simple kernel Ψ*.

DEFINITION 2.1.2. Let $\Psi(x)$ be a simple kernel on $[a, b]$ introduced in Definition 2.1.1, and $c := \dfrac{1}{2}(a' + b')$. Let the function $\rho : [a, c] \longrightarrow [c, b]$ be derived from

$$
\begin{cases}
\Psi(t) = \Psi(\rho(t)), & t \in [a, a'], \quad \rho(t) \in [b', b]; \\
\rho(t) = a' + b' - t, & t \in [a', c].
\end{cases}
\tag{1.3}
$$

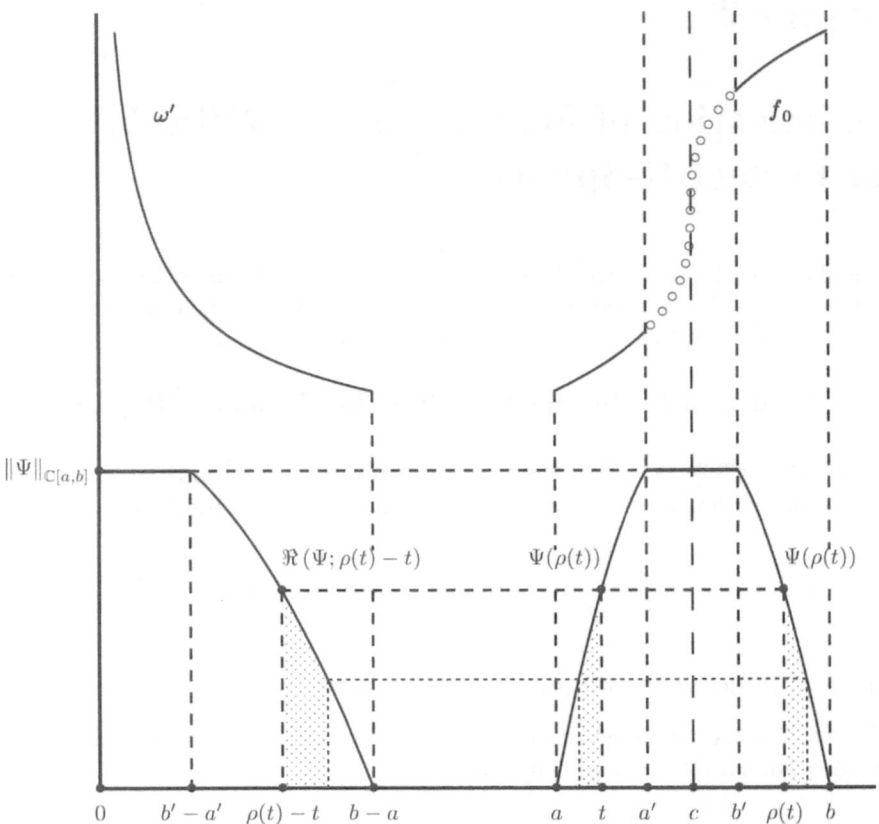

FIGURE 2.1.1. Simple kernel Ψ, its rearrangement $\Re(\Psi;\cdot)$ and the function f_0

The rearrangement $\Re(\Psi;t)$ of the simple kernel $\Psi(t)$ is defined on the interval $[0,b-a]$ as follows:

$$\Re(\Psi;t) := \begin{cases} \|\Psi\|_{\mathbb{C}[a,b]}, & t \in [0,b'-a'], \\ |\Psi(y_t)|, & \begin{cases} t \in (b'-a',b-a], & y_t \in [a,a'] : \\ \rho(y_t) - y_t = t. \end{cases} \end{cases} \tag{1.4}$$

Figure 2.1.1 also illustrates the graph of the rearrangement of a simple kernel Ψ. Notice that the property

$$\Re(\Psi;\rho(t)-t) = \Psi(t) = \Psi(\rho(t)), \qquad 0 \le t \le c,$$

along with the Cavalieri theorem implies that

$$\|\Psi\|_{\mathbb{L}_1[a,b]} = \|\Re(\Psi;\cdot)\|_{\mathbb{L}_1[0,b-a]}. \tag{1.5}$$

Systematic exposition of properties of rearrangements is given in the G. G. Hardy, G. E. Littlewood, G. Polya monograph [32] and in the A. Zygmund's book [93], and in Kong-Ming Shong's article [43].

2.1.2. The Korneichuk lemma

The following result, due to N. P. Korneichuk ([46], [48], [50]), turns out to be very useful in the characterization of extremal functions of the functional (1.1) for kernels ψ with a finite or countable set of points of sign change.

LEMMA 2.1.1. *Let* $\Psi(t) := \int_a^t \psi(y)\, dy$, $a \leq t \leq b$, *be a simple kernel whose derivative* ψ *satisfies (1.2). Let* $\omega(t)$ *be a concave modulus of continuity. Then,*

$$\sup_{f \in H^\omega[a,b]} \int_a^b f(t)\psi(t)\, dt = \int_0^{b-a} \Re(\Psi; t)\, \omega'(t)\, dt, \qquad (1.6)$$

where $\rho(\cdot)$ *and* $\Re(\Psi; \cdot)$ *are introduced by (1.3) and (1.4), respectively. The upper bound in (1.6) is attained on those functions whose derivative is given by the formula*

$$\frac{d}{dx} f_0(x) = \begin{cases} \omega'(\rho(x) - x), & a \leq x \leq c, \\ \omega'(x - \rho^{-1}(x)), & c \leq x \leq b, \quad c = (a' + b')/2. \end{cases} \qquad (1.7)$$

By Definition 2.1.1, the kernel ψ has the zero mean on $[a, b]$, so extremal functions of the functional (1.1) are determined up to an additive constant. Therefore,

$$\sup_{h \in H^\omega[a,b]} \int_a^b h(t)\psi(t)\, dt = \sup_{h \in H^\omega_a[a,b]} \int_a^b h(t)\psi(t)\, dt. \qquad (1.8)$$

Moreover, it can be observed from (1.3) and (1.7) that the derivative $\dfrac{d}{dt} f_0(t)$ of the extremal function of the functional (1.1) is determined *uniquely* by (1.7) only on *the support* $[a, a'] \cup [b', b]$ of the kernel ψ. We illustrated this phenomenon on Figure 2.1.1 by graphing $f_0(t)$ with solid lines on the support of ψ and by putting circles along the graph of f_0 on the *zero-interval* $[a', b']$ of ψ.

We formulate some corollaries from Lemma 2.1.1 frequently used in this monograph.

COROLLARY 2.1.2. *If the kernel* ψ *in Lemma 2.1.1 is symmetric with respect to the midpoint* $\gamma = \dfrac{1}{2}(a + b)$ *of the interval* $[a, b]$, *then* $\rho(x) - x = 2(\gamma - x)$, *and the derivative of the extremal function* $f_0(x)$ *is expressed by the formula*

$$\frac{d}{dx} f_0(x) = \begin{cases} \omega'(2(\gamma - x)), & a \leq x \leq \gamma; \\ \omega'(2(x - \gamma)), & \gamma \leq x \leq b. \end{cases} \qquad (1.9)$$

COROLLARY 2.1.3. *Let the function* $\dfrac{d}{dx} f_0(x)$ *be defined by* (1.7). *Then,* f_0 *has the full modulus of continuity on* $[0, b - a]$: $\omega(f_0; t) = \omega(t)$, $0 \leq t \leq b - a$, *or, more precisely,*

$$f_0(\rho(t)) - f_0(t) = \omega(\rho(t) - t), \qquad 0 \leq t \leq c. \tag{1.10}$$

Proof. By (1.7), for any $x :\ 0 \leq x \leq c := \dfrac{1}{2}(a' + b')$, we have

$$f_0(\rho(x)) - f_0(x) = \int_c^{\rho(x)} \omega'(u - \rho^{-1}(u))\, du - \int_c^{x} \omega'(\rho(u) - u)\, du =$$

$$\int_{c\cdot}^{x} \omega'(\rho(u) - u)\rho'(u)\, du - \int_c^{x} \omega'(\rho(u) - u)\, du = \int_c^{x} \omega'(\rho(u) - u)\, d(\rho(u) - u) =$$

$$= \omega(\rho(x) - x). \quad (1.11)$$

It remains to notice that the function $\rho(t) - t$ increases from 0 to $b - a$, as t decreases from c to 0. $\qquad\qquad\square$

From (1.11) we derive the following property of the extremal function f_0: *for any point* $v \in [a, c)$ *there exists such a point* $w = w(v) \in (c, b]$, $w \neq v$, *that two conditions are satisfied:*

$$\begin{aligned} f_0(w) - f_0(v) &= \omega(w - v); \\ f_0'(w) &= f'(v) = \omega'(w - v). \end{aligned} \tag{1.12}$$

According to (1.11), we can put $w = \rho(v)$ in (1.12) for $v \in [a, c)$.

2.2. Maximization of integral functionals in $H^\omega[a, b]$, $-\infty < a < b \leq +\infty$

Throughout this section we fix an interval $[a, b]$, $-\infty < a < b \leq +\infty$. Our objective in the section is to characterize extremal functions of the problem

$$\int_a^b h(t)\psi(t)\, dt \to \sup, \qquad h \in H_a^\omega[a, b], \tag{2.0}$$

for integrable kernels ψ with a finite number or a countable set of points of sign change on $[a, b]$, accumulating to the endpoint b. In partucular, we give the solution of the problem (2.0) on the half-line $\mathbb{R}_+$.

2.2.1. Notations and definitions

We adopt the following notations.

NOTATION 2.2.1. Let $N_1 \in \mathbb{N}$, $N_2 \in \mathbb{N} \cup \{\infty\}$, and $\xi = \{\xi_i = \xi(i)\}_{i=N_1}^{N_2}$ be a collection of points. Then, $\alpha \triangleright \xi \blacktriangleright \beta$, if and only if

$$\alpha = \xi(N_1) \leq \xi(N_1 + 1) \leq \cdots \leq \xi(N_2) = \beta, \quad \text{for } N_2 \in \mathbb{N};$$
$$\alpha = \xi(N_1); \quad \xi(i) \leq \xi(i+1), \quad i \geq N_1; \quad \lim_{i \to \infty} \xi(i) = \beta; \quad \text{for } N_2 = +\infty.$$

Analogously, $\alpha \blacktriangleleft \xi \triangleleft \beta$, if and only if

$$\beta = \xi(N_1) \geq \xi(N_1 - 1) \geq \cdots \geq \xi_{N_2} = \alpha, \quad \text{for } N_2 \in \mathbb{N};$$
$$\beta = \xi(N_1); \quad \xi(i) \geq \xi(i+1), \quad i \geq N_1; \quad \lim_{i \to \infty} \xi(i) = \alpha; \quad \text{for } N_2 = +\infty.$$

NOTATION 2.2.2. Let $[\alpha, \beta]$ be a finite interval, and $\psi \in \mathbb{L}_1[\alpha, \beta]$. By the definition,

$$\operatorname{sign} \psi = 1 \ \text{ on } [\alpha, \beta] \iff \operatorname{meas} \{t \in [\alpha, \beta] : \psi(t) > 0\} = \beta - \alpha;$$
$$\operatorname{sign} \psi = -1 \ \text{ on } [\alpha, \beta] \iff \operatorname{sign}(-\psi) = 1 \ \text{ on } [\alpha, \beta].$$

NOTATION 2.2.3. The following notation will be used for intervals $I = [\gamma, \gamma]$ with coincident endpoints: $I = \square$.

NOTATION 2.2.4. Let $E \subset \mathbb{R}$. The function

$$\mathcal{X}(E; t) = \begin{cases} 1, & t \in E; \\ 0, & t \notin E; \end{cases} \tag{2.1}$$

is called *the indicator of the set E*.

DEFINITION 2.2.1. Let $j \in \{-1, 0, +1\}$, $n \in \mathbb{N} \cup \{\infty\}$, and $\psi \in \mathbb{L}_1[a, b]$. Then, ψ belongs to the class $\mathcal{M}_n^j[a, b]$ for $n \geq 2$, if and only if $\operatorname{sign} \int_a^b \psi(x)\, dx = j$, and there exist such $\alpha = \{\alpha_i\}_{i=0}^n$ that $\alpha_{i-1} < \alpha_i$, $i = 1, \ldots, n$, $a \triangleright \alpha \blacktriangleright b$, and

$$\operatorname{sign} \psi = (-1)^i \ \text{ on } [\alpha_{i-1}, \alpha_i], \quad i = 1, \ldots, n.$$

By the definition,

$$\mathcal{M}_1^l[a, b] := \varnothing, \ l = 0, 1, \ \mathcal{M}_1^{-1}[a, b] := \{\psi \in \mathbb{L}_1[a, b] \,|\, \operatorname{sign} \psi = -1 \ \text{on } [a, b]\}.$$

We also introduce the class of kernels

$$\mathbb{M}_n[a, b] := \bigcup_{j=-1}^{1} \mathcal{M}_n^j[a, b]. \tag{2.2}$$

DEFINITION 2.2.2. Let $N \in \mathbb{N} \cup \{\infty\}$.

The sets of indices $\{J_i(N)\}_{i\in\mathbb{N}}$, $\{L_i\}_{i\in\mathbb{N}}$ and $\mathcal{P}(N)$ are defined as follows.

(1)   For $N = 1,\ 2,\ 3,\quad J_i(N) = L_i = \varnothing,\qquad i = 1, \ldots, N.$

(2)   For $N \geq 4,$

$$J_i(N) = \varnothing,\quad i = N-2,\ N-1,\ N;\quad L_i = \varnothing,\quad l = 1,\ 2,\ 3;$$
$$J_i(N) = \{j = i+1+2k,\ k \in \mathbb{N} \mid j \leq N\},\quad 1 \leq i \leq N-3;$$
$$L_i = \{l = i-1-2k,\ k \in \mathbb{N} \mid l \geq 1\},\quad 4 \leq i \leq N;$$

(3) $\mathcal{P}(N) = \{(i,j) \in \mathbb{N} \times \mathbb{N} \mid 1 \leq i \leq N-3,\ j \in J_i(N)\ \}.$

2.2.2. V_n^j-partitions of the interval $[a,b]$

The structure of extremal functions of the problem (2.0) will be characterized in terms of special partitions of the interval $[a,b]$.

DEFINITION 2.2.3. Let $n \in \mathbb{N} \cup \{\infty\}$.

A partition $\mathcal{V} = \left(\{A_i,\ B_i,\ C_i,\ D_i\}_{i=1}^n;\ \{B_{ij},\ C_{ji}\}_{(i,j)\in\mathcal{P}(N)}\right)$ of the interval $[a,b]$ into subintervals is called a V_n^j-partition, $j \in \{-1,0,1\}$, if the following conditions are satisfied.

(A) $C_i = [\gamma_{4i-4}, \gamma_{4i-3}];\ D_i = [\gamma_{4i-3}, \gamma_{4i-2}];$
       $B_i = [\gamma_{4i-2}, \gamma_{4i-1}];\quad A_i = [\gamma_{4i-1}, \gamma_{4i}];$
       for $i = 1, \ldots, n$ and such $\gamma = \{\gamma_i\}_{i=0}^{4n}$ that $a \triangleright \gamma \blacktriangleright b;$

(B) $C_i = \square,\quad i = 1,\ 2,\ 3;\quad B_i = \square,\quad i = n-2,\ n-1,\ n;$

(C_1) $j = 0 \implies D_i = \square,\quad i = 1, \ldots, n;$

(C_2) $j = -1 \implies D_{2k} = \square,\quad k = 1, \ldots, [n/2];$

(C_3) $j = +1 \implies D_{2k-1} = \square,\quad k = 1, \ldots, \lceil n/2 \rceil;$

(D) $B_i = \bigcup_{j\in J_i(n)} B_{ij},\quad 1 \leq i \leq n-3,$ where

$$B_{ij} = [\xi_i(\tfrac{j-i+1}{2}), \xi_i(\tfrac{j-i-1}{2})],\qquad j \in J_i(n),$$

for such $\xi_i = \{\xi_i(k)\}_{k=1}^{|J_i(n)|}$ that $\gamma_{4i-2} \blacktriangleleft \xi_i \triangleleft \gamma_{4i-1};$

(E) $C_i = \bigcup_{l\in L_i} C_{il},\quad 4 \leq i \leq n,$ where

$$C_{il} = [\varkappa_i(\tfrac{i-l-1}{2}), \varkappa_i(\tfrac{i-l+1}{2})],\qquad l \in L_i,$$

for such $\varkappa_i = \{\varkappa_i(k)\}_{k=1}^{|L_i|}$ that $\gamma_{4i-4} \triangleright \varkappa_i \blacktriangleright \gamma_{4i-3}.$

REMARK 2.2.1. We list the atoms of V_n^j-partitions of the interval $[a,b]$ into the intervals $\{A_i,\ B_i,\ C_i,\ D_i\}_{i=1}^n$ in their natural order and without the degenerated intervals

$A_N, \{B_i\}_{i=N-2}^N, \{C_i\}_{i=1}^3:$

$N = 2:\ D_1 A_1 D_2;$

$N = 3:\ D_1 A_1 D_2 A_2 D_3;$

$N = 4:\ D_1 B_1 A_1 D_2 A_2 D_3 A_3 C_4 D_4;$

$N = 5:\ D_1 B_1 A_1 D_2 B_2 A_2 D_3 A_3 C_4 D_4 A_4 C_5 D_5;$

$N = 6:\ D_1 B_1 A_1 D_2 B_2 A_2 D_3 B_3 A_3 C_4 D_4 A_4 C_5 D_5 A_5 C_6 D_6,$

$N \geq 7:\ D_1 B_1 A_1 D_2 B_2 A_2 D_3 B_3,\qquad A_k C_{k+1} D_{k+1} B_{k+1},\qquad 3 \leq k \leq N-4,$
       $A_{N-3} C_{N-2} D_{N-2} A_{N-2} C_{N-1} D_{N-1} A_{N-1} C_N D_N.$

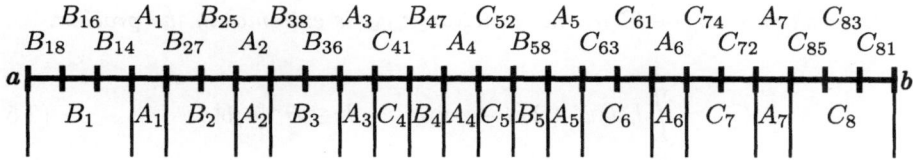

FIGURE 2.2.1. V_8^0-partition

Figure 2.2.1 clarifies the order of atoms in a V_8^0-partition.

2.2.3. Theorem X and perfect ω-splines

Given a $\psi \in \mathcal{M}_n^j[a, b]$, $j \in \{-1,\, 0 + 1\}$, we describe extremal functions of the problem (2.0). In particular, in Theorem X below we show that

if $\psi \in \mathcal{M}_n^0[a, b]$ and $n \ge 2$, then the kernel $\Psi(t) = \int\limits_a^t \psi(x)\, dx$ can be decomposed into the sum of

$$
I_n =
\begin{cases}
\dfrac{n^2}{4}, & n \text{ is even;} \\[2ex]
\dfrac{n^2 - 5}{4}, & n \text{ is odd;}
\end{cases}
\tag{2.3}
$$

simple kernels $\{\Phi_i(\cdot) = \Phi_i(\omega; \cdot)\}_{i=1}^{I_n}$ such that

$$
(i) \quad \Psi(t) = \sum_{i=1}^{I_n} \Phi_i(t), \qquad a \le t \le b;
$$

$$
(ii) \quad \sup_{h \in H^\omega[a,b]} \int\limits_a^b h(t)\psi(t)\, dt = \sum_{i=1}^{I_n} \sup_{h \in H^\omega[a,b]} \int\limits_a^b h(t)\Phi_i'(t)\, dt\,.
$$

$$\tag{2.4}$$

In the following theorem we describe the structure of extremal functions of the problem (2.0) called *perfect ω-splines*.

THEOREM X. *Let $\psi \in \mathcal{M}_n^j[a, b]$, $j \in \{-1, 0, +1\}$, and $\{\alpha_i\}_{i=1}^{n-1}$ be the points of sign change of ψ as in Definition 2.2.1. Then, there exist a solution $x_{\omega, \psi}$ of the problem (2.0) and a V_n^j-partition $\mathcal{V}$ of the interval $[a, b]$ with the following properties:*

(A) $\alpha_i \in A_i,$ $i = 1, \ldots, n - 1$;

(B) $\int\limits_{B_{ij} \cup C_{ji}} \psi(t)\, dt = 0,$ $(i, j) \in \mathcal{P}(n)$;

(C) $\int\limits_{A_i} \psi(t)\, dt = 0,$ $i = 1, \ldots, n - 1$;

(D$_1$) $j = -1 \implies x_{\omega, \psi}(t) = -\omega(t - a),\ t \in D_{2k-1} \ne \square,\ k = 1, \ldots, \lceil n/2 \rceil$;

(D$_2$) $j = 1 \implies x_{\omega, \psi}(t) = \omega(t - a),\ t \in D_{2k} \ne \square,\ k = 1, \ldots, \lceil n/2 \rceil$;

(E) *for each pair $(i,j) \in \mathcal{P}(n)$, the function $x_{\omega,\psi}$ is extremal in the problem*

$$\int_a^b h(t)\psi_{ij}(t)\,dt \to \sup, \qquad h \in H^\omega[a,b], \tag{2.5}$$

$$\psi_{ij}(t) := \psi(t) \cdot \mathcal{X}(B_{ij} \cup C_{ji}; t), \qquad t \in [a,b]; \tag{2.6}$$

(F) *for each $i = 1, \ldots, n-1$, the function $x_{\omega,\psi}$ is extremal in the problem*

$$\int_a^b h(t)\psi_i(t)\,dt \to \sup, \qquad h \in H^\omega[a,b], \tag{2.7}$$

$$\psi_i(t) := \psi(t) \cdot \mathcal{X}(A_i; t), \qquad t \in [a,b]. \tag{2.8}$$

Notice that all kernels

$$\Psi_{ij}(t) = \int_a^t \psi_{ij}(y)\,dy, \quad (i,j) \in \mathcal{P}(n), \qquad \Psi_i(t) = \int_a^t \psi_i(y)\,dy, \quad i = 1, \ldots, n-1, \tag{2.9}$$

are simple on their respective supports in the sense of Definition 2.1.1. Indeed, from the inclusions $\alpha_i \in A_i = [\gamma_{4i-1}, \gamma_{4i}]$, $i = 1, \ldots, n-1$, and the order of atoms in the V_n^j-partition, shown in Remark 2.2.1, it follows that

$$\operatorname{sign} \psi = \begin{cases} (-1)^i & \text{on} & \begin{cases} B_{ij}, & (i,j) \in \mathcal{P}(n); \\ [\gamma_{4i-1}, \alpha_i], & i = 1, \ldots, n-1; \end{cases} \\ (-1)^{i+1} & \text{on} & \begin{cases} C_{ji}, & (i,j) \in \mathcal{P}(n); \\ [\alpha_i, \gamma_{4i}], & i = 1, \ldots, n-1; \end{cases} \end{cases} \tag{2.10}$$

where $[\gamma_{4i-1}, \gamma_{4i}] := A_i$, $i = 1, \ldots, n-1$, and

$$\operatorname{sign} \psi(t) = \begin{cases} -1 \text{ on } D_{2i-1}, & i = 1, \ldots, \lceil n/2 \rceil; \\ 1 \text{ on } D_{2i}, & i = 1, \ldots, \lfloor n/2 \rfloor. \end{cases} \tag{2.11}$$

Therefore, by (2.10) and the statements (B) and (C) of Theorem X, each of the kernels $\{\Psi_{ij}\}_{(i,j) \in \mathcal{P}(n)}$ and $\{\Psi_i\}_{i=1}^{n-1}$ in (2.9) is simple. Then, Korneichuk's Lemma 2.1.1 provides us with the following formulas for the derivative $\frac{d}{dt}x_{\omega,\psi}(t)$:

$$\frac{d}{dt}x_{\omega,\psi}(t) = \begin{cases} (-1)^{i+1}\omega'(\rho_{ij}(t) - t), & t \in B_{ij}; \\ (-1)^{i+1}\omega'(t - \rho_{ij}^{-1}(t)), & t \in C_{ji}; \end{cases} \tag{2.12}$$

for all $(i,j) \in \mathcal{P}(n)$, where $\rho_{ij} : B_{ij} \to C_{ji}$ is determined from the equation

$$\Psi_{ij}(t) = \Psi_{ij}(\rho_{ij}(t)), \quad t \in B_{ij}, \quad \rho_{ij}(t) \in C_{ji}, \tag{2.13}$$

and

$$\frac{d}{dt} x_{\omega,\psi}(t) = \begin{cases} (-1)^{i+1} \omega'(\rho_i(t) - t), & t \in [\gamma_{4i-1}, \alpha_i]; \\ (-1)^{i+1} \omega'(t - \rho_i^{-1}(t)), & t \in [\alpha_i, \gamma_{4i}]; \end{cases} \tag{2.14}$$

where $\rho_i : [\gamma_{4i-1}, \alpha_i] \to [\alpha_i, \gamma_{4i}]$ is determined from the equation

$$\Psi_i(t) = \Psi_i(\rho_i(t)), \quad t \in [\gamma_{4i-1}, \alpha_i], \quad \rho_i(t) \in [\alpha_i, \gamma_{4i}]. \tag{2.15}$$

Schematic graphs of the extremal function $x_{\omega,\psi}$, $\psi \in \mathcal{M}_n^j[a,b]$ for various values of n and j are illustrated on Figures 2.2.2–2.2.6.

2.2.4. Structural properties of extremal functions $x_{\omega,\psi}$

First of all, the extremal function of the problem (2.0) is *unique*.

COROLLARY 2.2.1. *The derivative of the extremal function $x_{\omega,\psi}$ and the V_n^j-partition $\mathcal{V} = \mathcal{V}(\omega, \psi)$ of the problem (2.0) are unique.*

The V_n^j-partition $\mathcal{V}(\omega, \psi)$ from Theorem X is called *the extremal V_n^j-partition of the problem* (2.0).

REMARK 2.2.2. By the formulas (2.12)–(2.15),

$$\operatorname{sign} \frac{d}{dt} x_{\omega,\psi}(t) = \begin{cases} (-1)^{i+1}, & t \in A_i, \quad i = 1, \ldots, n-1; \\ (-1)^{i+1}, & t \in B_{ij}, \ C_{ji}, \quad (i,j) \in P(n); \end{cases} \tag{2.16}$$

while according to statements (D_1), (D_2) of Theorem X,

$$\operatorname{sign} \frac{d}{dt} x_{\omega,\psi}(t) = \begin{cases} -1, & t \in D_{2i-1}, \quad i = 1, \ldots, \lceil n/2 \rceil; \\ 1, & t \in D_{2i}, \quad i = 1, \ldots, [n/2]. \end{cases} \tag{2.17}$$

Therefore, by (2.16) and (2.17), the function $\frac{d}{dt} x_{\omega,\psi}(t)$ can have at most $n-2$ sign changes on the interval $[a,b]$, if ψ belongs to $\mathcal{M}_n^0[a,b]$ or $\mathcal{M}_n^{+1}[a,b]$, and $n-1$ sign changes on $[a,b]$, if $\psi \in \mathcal{M}_n^{-1}[a,b]$. These values are the upper bounds for the number of sign changes of $\frac{d}{dt} x_{\omega,\psi}(t)$, because some of the intervals of the extremal V_n^j-partition may degenerate into points. However, the following result shows that in the case of *a strictly concave* modulus of continuity ω, the function $\frac{d}{dt} x_{\omega,\psi}(t)$ has the maximum possible number of sign changes.

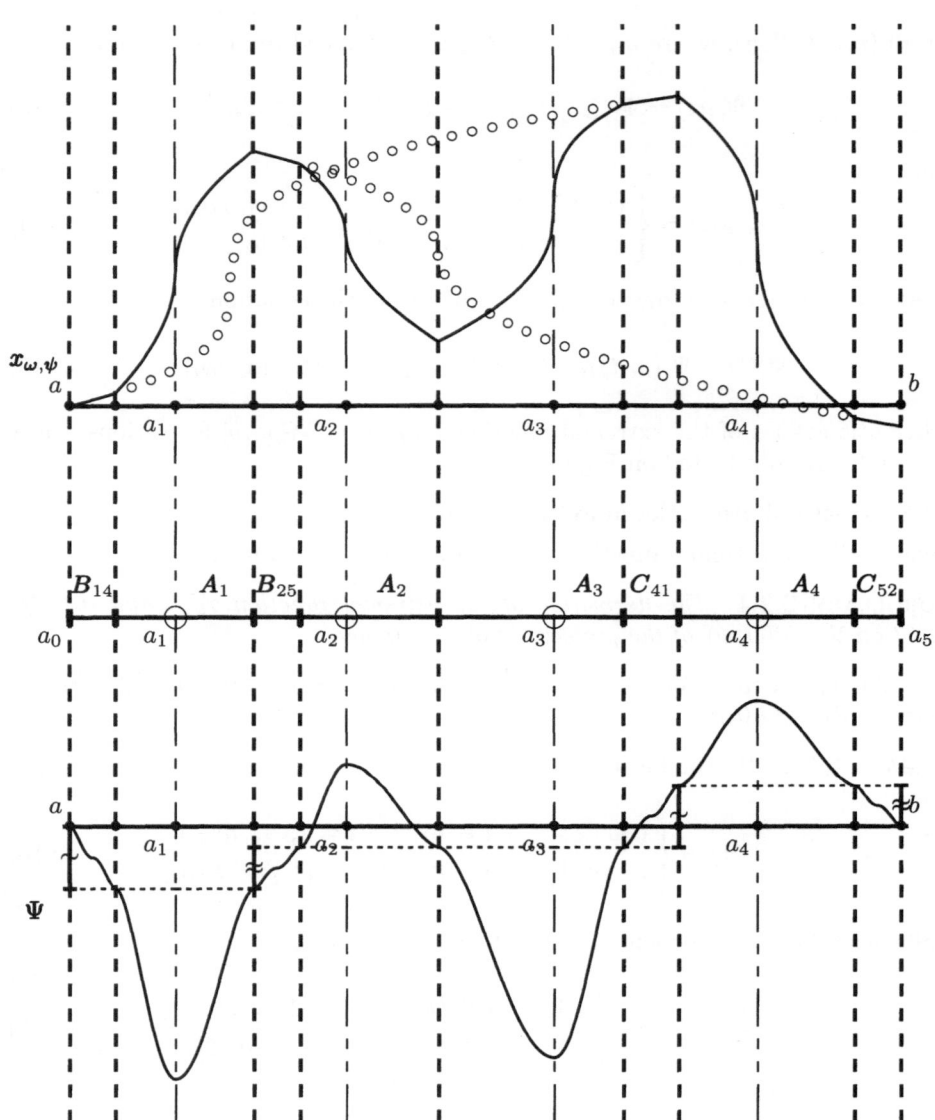

FIGURE 2.2.2. V_5^0-partition and graphs of $x_{\omega,\psi}$ and Ψ for $\psi \in \mathcal{M}_5^0[a, b]$

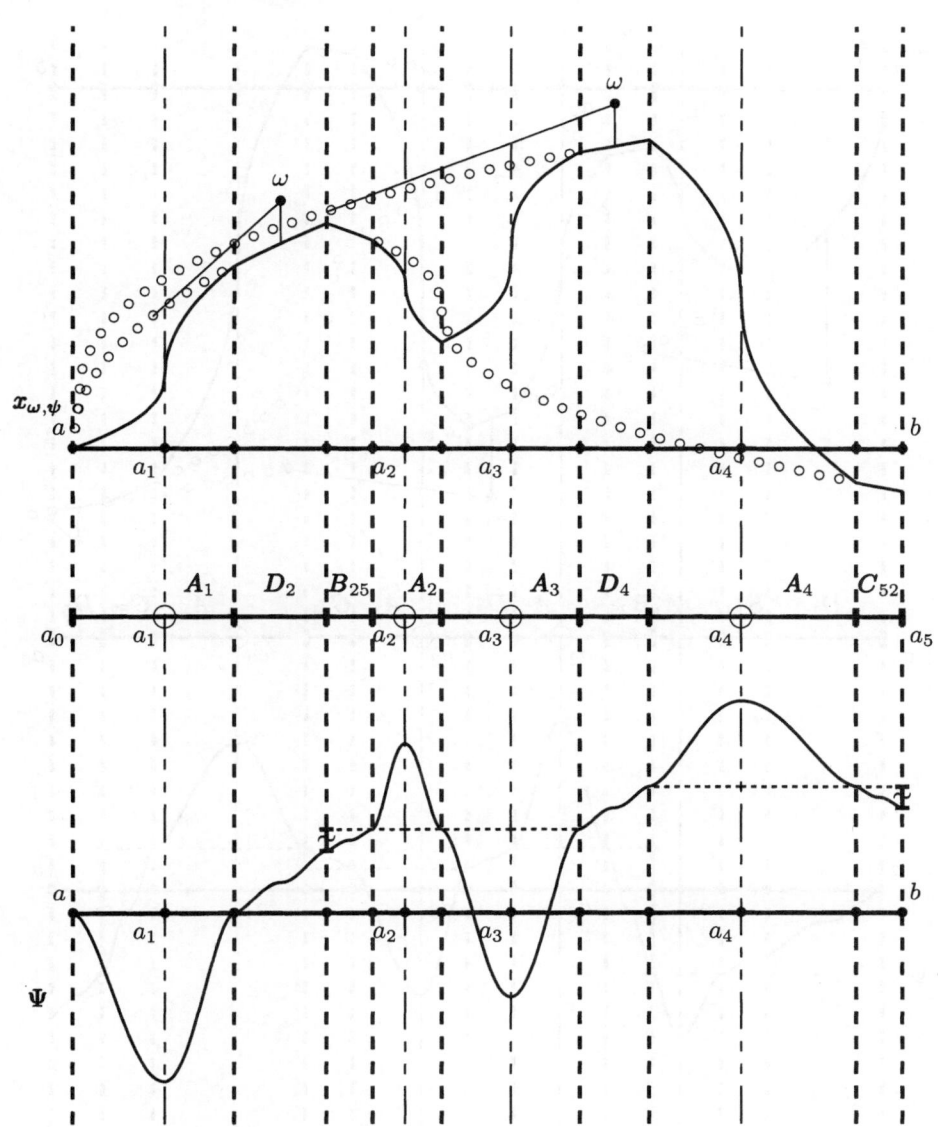

FIGURE 2.2.3. V_5^{+1}-partition and graphs of $x_{\omega,\psi}$ and Ψ for $\psi \in \mathcal{M}_5^{+1}[a,b]$

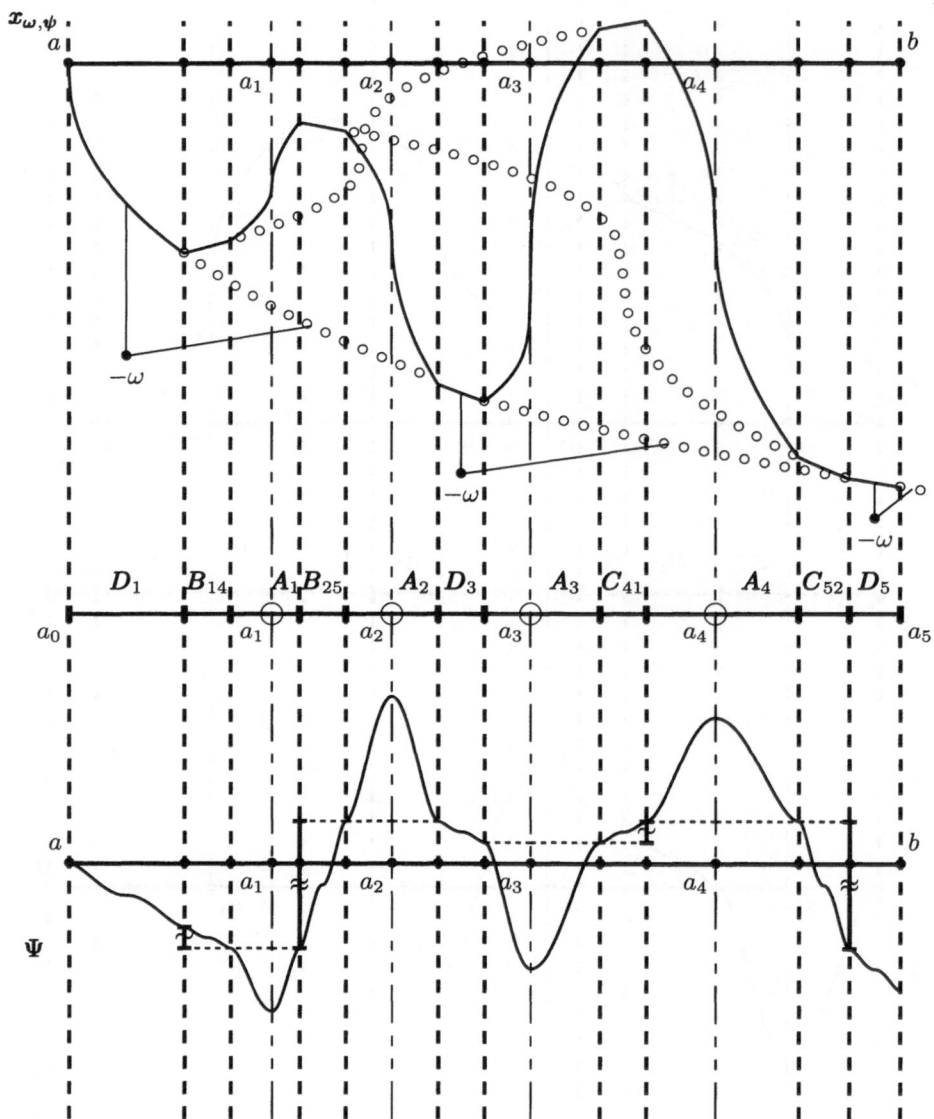

FIGURE 2.2.4. V_5^{-1}-partition and graphs of $x_{\omega,\psi}$ and Ψ for $\psi \in \mathcal{M}_5^{-1}[a,b]$

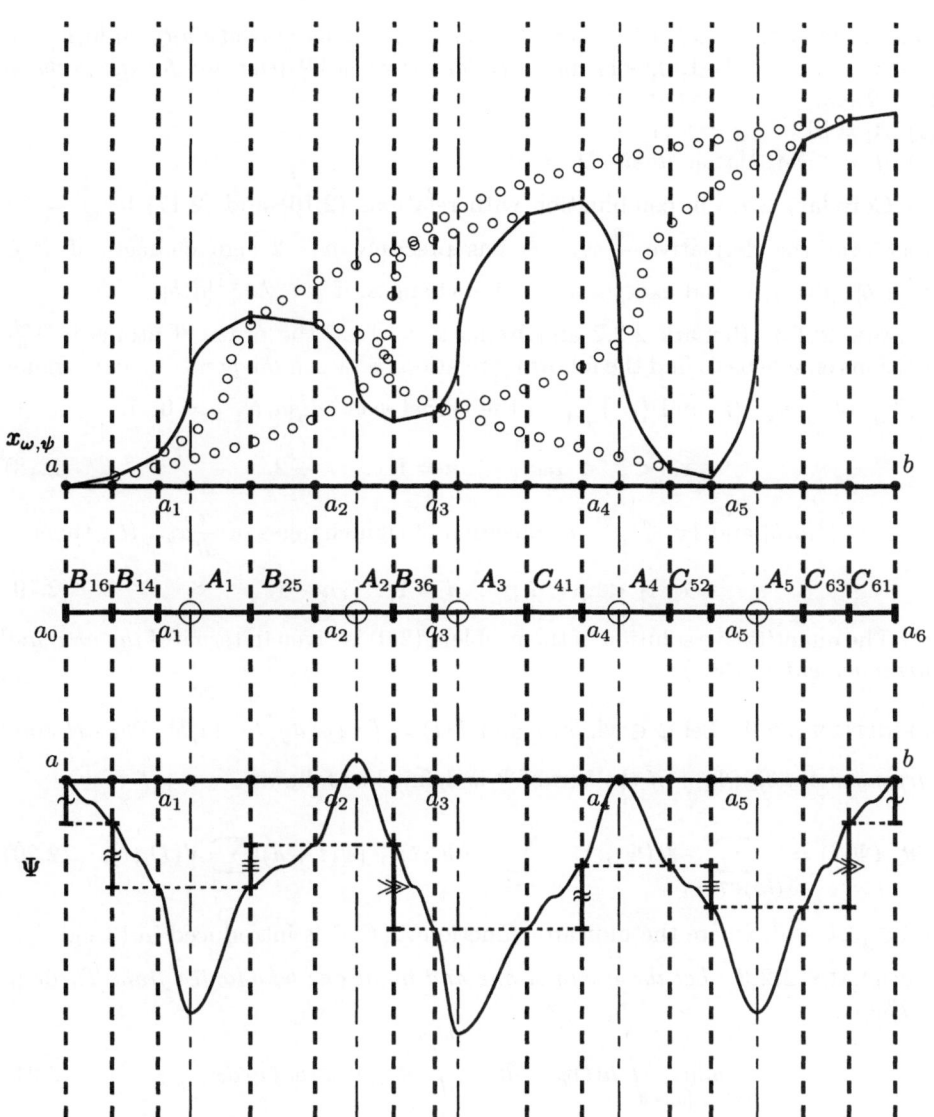

FIGURE 2.2.5. V_6^0-partition and graphs of $x_{\omega,\psi}$ and Ψ for $\psi \in \mathcal{M}_6^0[a, b]$

COROLLARY 2.2.2. *Let ω be a strictly concave modulus of continuity on $[0, b - a]$, $\psi \in \mathcal{M}_n^j[a, b]$, $j \in \{-1, 0, +1\}$ and $\mathcal{V}$ be the extremal V_n^j-partition for the problem (2.0). Then,*

  (I) $A_i \neq \square, \qquad i = 1, \ldots, n - 1;$
  (II) *if* $\psi \in \mathcal{M}_n^{-1}[a, b]$, *then* $D_1 \neq \square$.

Corollary 2.2.2 in combination with relations (2.16) and (2.17) for $j = -1$ imply that the derivative $\dfrac{d}{dt} x_{\omega, \psi}(t)$ has precisely $n - 2$ sign changes, if $\psi \in \mathcal{M}_n^j[a, b]$, $j = 0, 1$, and exactly $n - 1$ sign changes, if $\psi \in \mathcal{M}_n^{-1}[a, b]$.

REMARK 2.2.3. Remark 2.2.2 and Remark 2.2.1 on the order of atoms in V_n^j-partitions enable us to find the following relations between *the points of sign change* $\{z_i\}_{i=1}^{n-2}$ of $\dfrac{d}{dt} x_{\omega, \psi}(t)$ and $\{\alpha_i\}_{i=1}^{n-1}$ of the kernel $\psi \in \mathcal{M}_n^j[a, b]$, $j = 0, 1$:

$$\alpha_i < z_i < \alpha_{i+1}, \qquad i = 1, \ldots, n - 2. \tag{2.18}$$

If $\psi \in \mathcal{M}_n^{-1}[a, b]$ and $\{z_i\}_{i=1}^{n-1}$ is a collection of sign changes of $\dfrac{d}{dt} x_{\omega, \psi}(t)$, then

$$\alpha_{i-1} < z_i < a_i, \qquad i = 1, \ldots, n - 1. \tag{2.19}$$

The quantitative solution of the problem (2.0) is given in terms of *the extremal rearrangement* $\Re_\omega(\Psi; \cdot)$.

DEFINITION 2.2.4. *Let $\psi \in \mathbb{M}_n[a, b]$ and $\Psi(t) = \int_b^t \psi(y)\,dy$, $t \in [a, b]$. The extremal rearrangement $\Re_\omega(\Psi; \cdot)$ of the kernel Ψ is defined as follows:*

$$\Re_\omega(\Psi; t) = \sum_{(i,j) \in \mathcal{P}(n)} \Re(\Psi_{ij}; t) + \sum_{i=1}^{n-1} \Re(\Psi_i; t) + |\Psi(t + a)| \sum_{i=0}^{n} \mathcal{X}(D_i; t), \tag{2.20}$$

for $t \in [0, b - a]$, where the indicator function $\mathcal{X}(E; \cdot)$ is introduced in (2.1).

COROLLARY 2.2.3. *Let the assumptions and notations be adopted from Theorem X. Then,*

$$\sup_{h \in H_a^\omega[a,b]} \int_a^b h(t)\psi(t)\,dt = \int_0^{b-a} \Re_\omega(\Psi; x)\,\omega'(x)\,dx. \tag{2.21}$$

Unlike the $\sum$-rearrangement $\Phi(\Psi; \cdot)$, defined by N. P. Korneichuk in (6.28) of [48], p. 144, the extremal rearrangement $\Re_\omega(\Psi; \cdot)$ does depend on ω. Furthermore, the $\mathbb{L}_1$-norm of $\Re_\omega(\Psi; \cdot)$ *exceeds* the $\mathbb{L}_1$-norm of the kernel Ψ if one of the intervals $\{B_{ij}\}_{(i,j) \in \mathcal{P}(n)}$ is nondegenerate:

$$\|\Re_\omega(\Psi, \cdot)\|_{\mathbb{L}_1[0, b-a]} \geq \|\Psi\|_{\mathbb{L}_1[a, b]}. \tag{2.22}$$

The following result can be derived either directly from the formulas of Theorem X and or from the property (0.4.7) of the Hölder classes.

COROLLARY 2.2.4. *Let $\alpha \in (0,1]$, and $\psi \in \mathcal{M}_n^j[a,b]$. Let $z_{\alpha,\psi}(t) \in H^\alpha[0,1]$ be the solution of the problem*

$$\int_0^1 h(t)\psi(t)\,dt \to \sup, \qquad h \in H_0^\alpha[0,1].$$

Then, for all $\zeta > 0$, the function $z_\zeta(t) := \zeta^\alpha z_{\alpha,\psi}(t/\zeta)$, $0 \le t \le \zeta$, is a solution of the problem

$$\int_0^\zeta h(t)\psi(t/\zeta)\,dt \to \sup, \qquad h \in H_0^\alpha[0,\zeta]. \tag{2.23}$$

COROLLARY 2.2.5. *Let v be any point on $[a,b]$ other than the points of sign change of the kernel ψ and the endpoints of atoms of the extremal partition of the interval $[a,b]$. Then, there exists such a point $w = w(v) \in [a,b]$, $w \ne v$, and $\chi \in \{-1,1\}$ that two conditions are satisfied:*

$$\begin{aligned} x_{\omega,\psi}(w) - x_{\omega,\psi}(v) &= \chi\omega(|w-v|); \\ x'_{\omega,\psi}(w) = x'_{\omega,\psi}(v) &= \chi\omega'(|w-v|). \end{aligned} \tag{2.24}$$

The proof of Corollary 2.2.5 follows immediately from the relations (1.12) and the fact that the function $x_{\omega,\psi}(t)$ is a solution of a finite number of problems

$$\int_\alpha^\beta h(t)\Phi'(t)\,dt \to \sup, \quad h \in H^\omega[\alpha,\beta],$$

where $\Phi(t)$ is one of the simple kernels $\{\Psi_i\}_{i=1}^{n-1}$ or $\{\Psi_{ij}\}_{(i,j)\in\mathcal{P}(n)}$.

REMARK 2.2.4. *The solution of the problem (2.0) for the linear modulus of continuity.*

The identity

$$\int_a^b h(t)\psi(t)\,dt = -\int_a^b h'(t)\Psi(t)\,dt$$

for $h \in H_a^\omega[a,b]$ and $\Psi(x) = \int_b^x \psi(t)\,dt$ implies that we have the following formulae in the case of the linear modulus of continuity $\omega_K(t) = Kt$, $K > 0$:

$$\frac{d}{dt}x_{\omega_K,\psi}(t) = -K\,\operatorname{sign}\Psi(t), \qquad a \le t \le b; \tag{2.25}$$

and

$$\sup_{h\in H_a^{\omega K}[a,b]} \int_a^b h(t)\psi(t)\,dt = K\|\Psi\|_{\mathbb{L}_1[a,b]} = K\|\Re_{\omega_K}(\Psi;\cdot)\|_{\mathbb{L}_1[a,b]}. \tag{2.26}$$

We conclude the subsection with the identification of the extremal function $x_{\omega,\psi}$ in two trivial cases $\psi \in \pm\mathcal{M}_1^i[a,b]$, $j=\pm1$.

REMARK 2.2.5. If $\xi\psi > 0$ on $[a,b]$ for $\xi \in \{-1,1\}$, then $x_{\omega,\psi}(t) = \xi\omega(t-a)$.

2.2.5. Limiting properties of extremal functions $x_{\omega,\psi}$

In this section we touch on some limiting properties of perfect ω-splines. First, we define the class $M_n[a,b]$ of those integral kernels whose number of sign changes does not exceed $n-1$.

DEFINITION 2.2.5. For $n \in \mathbb{N}$, the class $M_n[a,b]$ is defined as follows:

$$M_n[a,b] := \bigcup_{k=1}^{n} (\pm\mathbb{M}_k[a,b]),$$

where classes of kernels $\mathbb{M}_n[a,b]$ are introduced in (2.2), and $-F$ denotes the set $\{-f \mid f \in F\}$ for a functional class F.

The following limiting properties of extremal functions of the problem (2.0) are crucial in various geometric constructions involving perfect ω-splines.

COROLLARY 2.2.6. Let $\mathbb{S}$ be a compact in $\mathbb{R}^d$, and the family of integrable kernels $\psi_s(t)$, defined on $a_s \le t \le b_s$ for $s \in \mathbb{S}$, be endowed with the following properties:

(i) the endpoints a_s and b_s are continuous functions of s on $\mathbb{S}$, and
$$a_s < a < b < b_s, \qquad s \in \mathbb{S}, \qquad \text{for some } a < b.$$

(ii) there exists an $n \in \mathbb{N}$ such that $\psi_s \in M_n[a_s,b_s]$ for all $s \in \mathbb{S}$;

(iii) the family $\{\psi_s\}_{s\in\mathbb{S}}$ depends continuously in s on $\mathbb{S}$ in the integral metrics in the following sense: for all $s \in \mathbb{S}$,
$$\lim_{\mathbb{S}\ni s'\to s} \|\bar{\psi}_{s'} - \bar{\psi}_s\|_{\mathbb{L}_1[a,b]} \to 0,$$

where $\bar{\psi}_s(x) := \psi_s\left(\dfrac{b_s - a_s}{b-a}(x-a)+a_s\right)$, $a \le x \le b$, $s \in \mathbb{S}$.

Let z_s be the solution of the problem

$$\int_{a_s}^{b_s} f(t)\psi_s(t)\,dt \to \sup, \qquad f \in H^\omega[a_s,b_s], \quad f(a_s) = 0.$$

Then, functions z_s depend continuously on s on $\mathbb{S}$ in the uniform metrics, i.e., for all $s \in \mathbb{S}$,

$$\|z_{s'} - z_s\|_{\mathbb{C}[\max\{a_s,a_{s'}\},\min\{b_s,b_{s'}\}]} \to 0, \qquad as \quad \mathbb{S}\ni s' \to s.$$

COROLLARY 2.2.7. *Let $\{\omega_l\}_{l\in\mathbb{N}}$ be a sequence of concave modulii of continuity on $[0, b-a]$ convergent in the metrics $\mathbb{C}[0, b-a]$ to a concave modulus of continuity ω, and let the sequence $\{\psi_l\}_{l\in\mathbb{N}}$ of kernels from $M_n[a,b]$ converge in $\mathbb{L}_1[a,b]$ to the kernel $\psi \in M_n[a,c]$ which vanishes outside the interval $[a,c] \subset [a,b]$. For each $l \in \mathbb{N}$, let x_l be the solution of the problem*

$$\int_a^b h(t)\psi_l(t)\, dt \rightarrow \sup, \qquad h \in H_a^{\omega_l}[a,b].$$

Then, there exists a subsequence $\{x_{l_k}\}_{k\in\mathbb{N}}$ convergent in $\mathbb{C}[a,c]$ to the solution of the problem

$$\int_a^c h(t)\psi(t)\, dt \rightarrow \sup, \qquad h \in H_a^\omega[a,c].$$

2.2.6. Criterion for triviality of the extremal V_n^0-partition

In this subsection we offer a criterion for extremal V_n^0-partitions $\mathcal{V} = \mathcal{V}_n(\omega, \psi)$ to have atoms $\{A_i\}_{i=1}^{n-1}$ as their only nondegenerating intervals.

Let the kernel $\psi \in \mathcal{M}_n^0[a,b]$ be such that the kernel $\Psi(x) = \int_a^x \psi(t)\, dt$ on $[a,b]$ has $n-2$ interior zeroes $\{b_i\}_{i=1}^{n-2}$,

$$\Psi(b_i) = 0, \quad i = 1, \ldots, n-2, \quad b_i \in (a_i, a_{i+1}), \quad b_0 := a, \ b_{n-1} := b. \qquad (2.27)$$

Then, the extremal V_n^0-partition $\mathcal{V}$ contains the atoms $\{A_i\}_{i=1}^{n-1}$ as the only non-degenerating atoms, if and only if

$$A_i = [b_{i-1}, b_i], \quad i = 1, \ldots, n-1, \qquad B_i = C_i = \{b_{i-1}\}, \quad i = 1, \ldots, n. \qquad (2.28)$$

DEFINITION 2.2.6. *Extremal partitions of the form (2.28) are called the trivial partitions of the interval $[a,b]$.*

COROLLARY 2.2.8. *Let $\psi \in \mathcal{M}_n^0[a,b]$, and $\{b_i\}_{i=0}^{n-1}$ be the zeroes of the kernel Ψ as in (2.27). Then, the extremal partition $\mathcal{G}_n = \{\{A_i, B_i, C_i\}_{i=1}^n\}$ is of the form (2.28), if and only if*

$$\sum_{l=i}^{j-1} (-1)^{i+l}\omega(b_{l+1} - b_l) \le \omega(b_j - b_i), \quad \forall(i,j) \in \mathcal{P}(n). \qquad (2.29)$$

N. P. Korneichuk earlier encountered conditions similar to (2.29) in estimating approximations of the classes H^ω by piecewise constant functions with the knots $\{b_i\}_{i=1}^{n-2}$ (see inequalities (27) on p. 163 in [49]).

We mention two cases of the triviality of extremal partitions.

1. If ω is a linear modulus of continuity, i.e. $\omega(t) = Kt$, $K > 0$, then the condition (2.29) is satisfied for any choice of the points $\{b_i\}_{i=0}^{n-1}$ as in (2.27).

2. If the points $\{b_i\}_{i=0}^{n-1}$ from in (2.27) are such that

$$b_{i+1} - b_i \geq b_i - b_{i-1}, \qquad i = 1, \ldots, n-2, \tag{2.30}$$

then the condition (2.29) is satisfied for any concave modulus of continuity ω.

It turns out (cf. [12]) that for any fixed concave modulus of continuity ω on $\mathbb{R}_+$ and $B > 0$, the knots of the extremal function R_B in the problem

$$\|f\|_{\mathbb{L}_\infty(\mathbb{R}_+)} \to \sup, \qquad f \in W^1 H^\omega(\mathbb{R}_+), \qquad \|f\|_{\mathbb{L}_1(\mathbb{R}_+)} \leq B, \tag{2.31}$$

satisfies inequalities $b_{i+1} - b_i > \dfrac{1}{2}(b_i - b_{i-1})$, $i \in \mathbb{N}$ (with $b_0 := 0$). Consequently, by Corollary 2.2.8, $\dfrac{d}{dt} R_B(t) \in H^\omega(\mathbb{R}_+)$ and the support of R_B is compact. We also remark that in the case of Hölder classes $W^1 H^\alpha(\mathbb{R}_+)$, the knots $\{b_i\}_{i \in \mathbb{N}}$ of R_B constitute a *geometric mesh*.

Corollary 2.2.8 is used in Chapter 15 to estimate the N-widths of functional classes $W^1 H^\omega[-1, 1]$ from above.

2.2.7. Properties of extremal rearrangements $\Re_\omega(\Psi; \cdot)$

COROLLARY 2.2.9. *Let $\Psi(t) = \int\limits_b^t \psi(y)\, dy$ on $[a, b]$ be such that $\psi \in \mathcal{M}_n^0[a, b]$, and $\sigma(t)$, $\mu(t)$ be two modulii of continuity on $[0, b-a]$ such that $\sigma(t) \leq \mu(t)$ for all $t \in [0, b-a]$. Then,*

$$(1) \quad \sup_{h \in H^\sigma[a, b]} \int\limits_a^b h(t)\psi(t)\, dt = \int\limits_0^{b-a} \Re_\sigma(\Psi; t)\sigma'(t)\, dt \leq \int\limits_0^{b-a} \Re_\sigma(\Psi; t)\, \mu'(t)\, dt;$$

$$(2) \quad \int\limits_0^{b-a} \Re_\mu(\Psi; t)\sigma'(t)\, dt \leq \int\limits_0^{b-a} \Re_\sigma(\Psi; t)\mu'(t)\, dt. \tag{2.32}$$

Corollary 2.2.9 comes in handy when estimating the value of change in the quantitative solution of the problem (2.0) when perturbating the modulus ω.

Let us introduce classes of functions on $[a, b]$ whose lower and upper bounds are majorized by given constants.

DEFINITION 2.2.7. *Let A, $B > 0$. The class $\Theta_{A,B}[a, b]$ is defined as follows:*

$$\Theta_{A,B}[a, b] := \{\psi \in \mathbb{L}_1[a, b] \mid \int\limits_a^b \psi(t)\, dt = 0, \ \text{ess sup } \psi \leq B, \ \text{ess inf } \psi \geq -A \,\}.$$

$$\tag{2.33}$$

DEFINITION 2.2.8. Let $m \in (0, 1)$, and $a < b$, and $M = M(a, b, m) = a + (b-a)m$. The function $f_{m,a,b}$ is defined as follows:

$$f_{m,a,b}(t) = \begin{cases} -m\omega\left(m^{-1}(M - t)\right), & t \in [a, M]; \\ (1 - m)\omega\left((1 - m)^{-1}(t - M)\right), & t \in [M, b]. \end{cases} \tag{2.34}$$

Let

$$\psi_{A,B}(t) = \begin{cases} -A, & t \in [a, a + \dfrac{B}{A + B}(b - a)]; \\ B, & t \in [a + \dfrac{B}{A + B}(b - a), b]. \end{cases} \tag{2.35}$$

By the Korneichuk Lemma 2.1.1, the function $h_{A,B}(t)$, extremal in the problem $(*)$ for $\psi(t) = \psi_{A,B}(t)$, is (uniquely up to an additive constant) given by the formula

$$h_{A,B}(t) = f_{v,a,b}(t), \qquad \text{where} \quad v = \frac{B}{A + B}, \tag{2.36}$$

(see Figure 2.2.6).

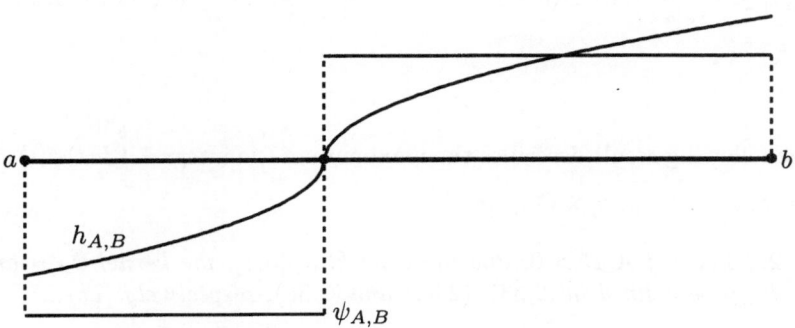

FIGURE 2.2.6. Graphs of functions $h_{A,B}$ and $\psi_{A,B}$

LEMMA 2.2.10. *Let the class* $\Theta_{A,B}[a, b]$ *be defined by* (2.33). *Then,*

$$\sup_{\psi \in \Theta_{A,B}[a,b]} \|\psi\|_{\mathbb{L}_1[a,b]} = \frac{2AB(b - a)}{A + B}. \tag{2.37}$$

Proof. We can take the supremum in (2.37) only over continuous functions from $\Theta_{A,B}[a, b]$. Then, for $\psi \in \mathbb{C}[a, b] \cap \Theta_{A,B}[a, b]$, let

$$I_+ = \{\psi(t) > 0, \ t \in [a, b]\} \quad \text{and} \quad I_- = \{\psi(t) < 0, \ t \in [a, b]\}. \tag{2.38}$$

From the equality $\int_a^b \psi(t)\, dt = 0$ it follows that

$$\int_{I_+} \psi(t)\, dt = -\int_{I_-} \psi(t)\, dt = \frac{1}{2}\|\psi\|_{\mathbb{L}_1[a,b]}. \tag{2.39}$$

On the other hand, by the definition of $\Theta_{A,B[a,b]}$,

$$\int_{I_+} \psi(t)\, dt \le B|I_+|, \quad -\int_{I_-} \psi(t)\, dt \le A|I_-|, \quad |I_+| + |I_-| = (b-a). \tag{2.40}$$

From (2.39), (2.40) it follows that

$$\|\psi\|_{\mathbb{L}_1[a,b]} \le 2 \min_{x \in [a,b]} \{Ax, B(b-a-x)\} = \frac{2AB(b-a)}{A+B}, \qquad \psi \in \Theta_{A,B}[a, b]. \tag{2.41}$$

$\square$

The following result describes the maximum of the integral $\int_a^b h(t)\psi(t)\, dt$ over all pairs $(h, \psi) \in H^\omega[a, b] \times \Theta_{A,B}[a, b]$.

LEMMA 2.2.11. *Let $A, B > 0$, and the class $\Theta_{A,B}[a, b]$, the kernel $\psi_{A,B}$ and the function $h_{A,B}$ be defined in (2.33), (2.34) and (2.36), respectively. Then,*

$$\sup_{\psi \in \Theta_{A,B}[a,b]} \sup_{h \in H^\omega[a,b]} \int_a^b h(t)\psi(t)\, dt = \frac{A^2 + B^2}{(A+B)^2}(b-a) \int_a^b \omega(t)\, dt. \tag{2.42}$$

The supremums in (2.42) are attained only on the pairs

$$(\pm\psi_{A,B}(t), \pm h_{A,B}(t) + C), \quad (\pm\psi_{A,B}(b+a-t), \pm[h_{A,B}(b+a-t)+C]), \qquad C \in \mathbb{R}.$$

Proof. For $c < d$ and $E > 0$, let us introduce the function

$$R_{c,d,E} = \min\{A(x-c), B(d-x), E\}, \qquad c \le x \le d. \tag{2.43}$$

By the definition, $R_{c,d,E}(x)$ is a simple kernel whose rearrangement is given by the formula

$$\mathcal{R}(R_{c,d,E}; x) = \min\{E, \frac{AB}{A+B}(d-c-x)\}, \qquad 0 \le x \le d-c. \tag{2.44}$$

The set $U := \left(\bigcup_{i=2}^{\infty} M_n[a,b] \right) \cap \mathbb{C}[a,b]$ is dense in $\mathbb{L}_1[a,b]$. Thus, it suffices to find the supremum in (2.42) only over continuous functions from $\Theta_{A,B}[a,b]$ with a finite number of points of sign change, i.e.

$$\sup_{n \ge 2} \quad \sup_{\psi \in \mathcal{M}_n^0[a,b] \cap \mathbb{C}[a,b]} \quad \sup_{h \in H^\omega[a,b]} \int_a^b h(t)\psi(t)\, dt.$$

Fix $\psi \in \mathcal{M}_n^0[a,b] \cap \mathbb{C}[a,b]$, $n \in \mathbb{N}$, and let $\Psi(x) = \int_a^x \psi(t)\, dt$, $a \le x \le b$. Let $\Phi(t)$ be one of the simple kernels $\{\Psi_i\}_{i=1}^{n-1}$ or $\{\Psi_{ij}\}_{(i,j)\in\mathcal{P}(n)}$ from Theorem X with nondegenerated supports. Then, supp $\Phi = [\alpha, \beta]$, for some $a \le \alpha < \beta \le b$. By the definition of simple functions, there exist such α', β', $\alpha < \alpha' \le \beta' < \beta$, that

$$\frac{d}{dx}\Phi(x) = \begin{cases} \psi(x), & x \in [\alpha, \alpha'] \cap [\beta', \beta]; \\ 0, & \text{otherwise.} \end{cases} \tag{2.45}$$

The function $\psi \in \Theta_{A,B}[a,b]$ has opposite signs on $[\alpha, \alpha']$ and $[\beta', \beta]$. Thus, one of the following two cases occur:

$$\begin{cases} 0 < \psi(x) < B, & x \in (\alpha, \alpha'); \\ -A < \psi(x) < 0, & x \in (\beta', \beta); \end{cases} \quad \begin{cases} 0 < \psi(x) < B, & x \in (\beta', \beta); \\ -A < \psi(x) < 0, & x \in (\alpha, \alpha'). \end{cases} \tag{2.46}$$

Let $D = \|\Phi\|_{\mathbb{C}[\alpha,\beta]}$. Then,

$$D = |\Phi(\alpha')| = \|\psi\|_{\mathbb{L}_1[\alpha,\alpha']} = \Phi(\beta') = \|\psi\|_{\mathbb{L}_1[\beta',\beta]}. \tag{2.47}$$

Each of the inequalities (2.46) implies that

$$|\Phi(t)| \le R_{\alpha,\beta,D}(t) \quad \text{or} \quad |\Phi(t)| \le R_{\alpha,\beta,D}(\beta + \alpha - t), \qquad t \in [\alpha, \beta]. \tag{2.48}$$

Consequently, both cases in (2.46) lead us to the same estimate for rearrangements:

$$\Re(\Phi; t) \le \Re(R_{\alpha,\beta,D}; t), \qquad 0 \le t \le \beta - \alpha. \tag{2.49}$$

For $(i,j) \in \mathcal{P}(n)$ put

$$[\alpha_{ij}, \beta_{ij}] := \operatorname{supp} \Psi_{ij}, \quad D_{ij} := \|\Psi_{ij}\|_{C[\alpha_{ij}, \beta_{ij}]}, \tag{2.50}$$

and for $1 \le k \le n-1$, let

$$[\alpha_i, \beta_i] := \operatorname{supp} \Psi_i, \quad D_i := \|\Psi_i\|_{C[\alpha_i, \beta_i]}. \tag{2.51}$$

As before, let $I_+ = \{\psi(t) > 0,\ t \in [a,b]\}$ and $I_- = \{\psi(t) < 0,\ t \in [a,b]\}$. From the property (2.47) and Lemma 2.2.10 it follows that

$$\sum_{i=1}^{n-1} D_i + \sum_{(i,j)\in\mathcal{P}(n)} D_{ij} = \int_{I_+} \psi(t)\,dt = -\int_{I_-} \psi(t)\,dt = \frac{1}{2}\|\psi\|_{L_1[a,b]} \le \frac{AB(b-a)}{A+B}. \tag{2.52}$$

Consecutively using (2.49), (2.50), (2.52), we obtain the chain of inequalities

$$\sup_{h\in H^\omega[a,b]} \int_a^b h(t)\psi(t)\,dt =$$

$$= \sum_{i=1}^{n-1} \int_0^{b-a} \Re\left(\Psi_i; t\right) \omega'(t)\,dt + \sum_{(i,j)\in\mathcal{P}(n)} \int_0^{b-a} \Re\left(\Psi_{i,j}; t\right) \omega'(t)\,dt$$

$$\le \sum_{i=1}^{n-1} \int_0^{\beta_i-\alpha_i} \Re\left(R_{\alpha_i,\beta_i,D_i}; t\right) \omega'(t)\,dt + \sum_{(i,j)\in\mathcal{P}(n)} \int_0^{\beta_{ij}-\alpha_{ij}} \Re\left(R_{\alpha_{ij},\beta_{ij},D_{ij}}; t\right) \omega'(t)\,dt =$$

$$= \sum_{i=1}^{n-1} \int_0^{\beta_i-\alpha_i} \min\{D_i, \frac{AB}{A+B}(\beta_i - \alpha_i - t)\}\omega'(t)\,dt+$$

$$+ \sum_{(i,j)\in\mathcal{P}(n)} \int_0^{\beta_{ij}-\alpha_{ij}} \min\{D_{ij}, \frac{AB}{A+B}(\beta_{ij} - \alpha_{ij} - t)\}\omega'(t)\,dt \le$$

$$\le \int_0^{b-a} \min\left\{\sum_{i=1}^{n-1} D_i + \sum_{(i,j)\in\mathcal{P}(n)} D_{ij}, \frac{AB}{A+B}(b-a-t)\right\} \omega'(t)\,dt =$$

$$= \int_0^{b-a} \min\left\{\frac{1}{2}\|\psi\|_{L_1[0,1]}, \frac{AB}{A+B}(b-a-t)\right\} \omega'(t)\,dt \le$$

$$\le \int_0^{b-a} \min\left\{\frac{AB(b-a)}{A+B}, \frac{AB}{A+B}(b-a-t)\right\} \omega'(t)\,dt = \int_0^{b-a} \Re(\Psi_{A,B}; t)\omega'(t)\,dt. \tag{2.53}$$

Moreover, one can observe that we have the equalities in (2.53), if and only if Ψ is a simple function ($\Psi'(t) \in \mathcal{M}_2^0[a,b]$) coinciding with $\chi\Psi_{A,B}(t)$ or $\chi\Psi_{A,B}(1-t)$ for a fixed $\chi \in \{1, -1\}$. $\qquad\square$

The functions $h_{A,B}$ also feature in the following proposition used in our estimates of the n-widths of the class $W^r H^\omega[-1, 1]$ from above.

PROPOSITION 2.2.12. *Let* $0 < a < 1$, $k > 0$. *Let* $f \in H^\omega[0, 1]$, *be such that*

$$(I)\quad f(t) \leq 0, \quad 0 \leq t \leq a, \qquad f(t) \geq 0, \quad a \leq t \leq 1;$$

$$(II)\quad \int_a^1 f(t)\, dt = k \int_a^0 f(t)\, dt.$$

Then

$$\int_a^0 f(t)\, dt \leq \frac{1}{(1 + \sqrt{k})^2} \int_0^1 \omega(t)\, dt. \tag{2.54}$$

Proof. The inclusion $f \in H^\omega[0, 1]$ along with the assumption (II) enables us to obtain the following (in-)equalities

$$\int_a^0 f(t)\, dt = \frac{a}{a-1} \int_a^1 f\left(\frac{a}{a-1}(t-1)\right) dt =$$

$$= \frac{a}{a-1} \int_a^1 \left[f\left(\frac{a}{a-1}(t-1)\right) - f(t)\right] dt + \frac{a}{a-1} \int_a^1 f(t)\, dt \leq$$

$$\leq \frac{a}{1-a} \int_a^1 \omega(\frac{a-t}{a-1})\, dt + \frac{k \cdot a}{a-1} \int_a^0 f(t)\, dt =$$

$$= a \int_0^1 \omega(t)\, dt + \frac{k \cdot a}{a-1} \int_a^0 f(t)\, dt. \tag{2.55}$$

Comparing the leftmost and the rightmost expressions in (2.55), we obtain the estimate

$$\int_a^0 f(t)\, dt \leq \frac{a(1-a)}{1 + (k-1)a} \int_0^1 \omega(t)\, dt. \tag{2.56}$$

The maximum of the rational function $\dfrac{x(1-x)}{1 + (k-1)x}$ equals $(1 + \sqrt{k})^{-2}$ and is attained for $x = \dfrac{1}{1 + \sqrt{k}}$. This observation concludes the proof of the result.   □

REMARK 2.2.6. The proof of Proposition 2.2.12 shows that the equality sign in (2.55) for a fixed $k > 0$ is possible, if and only if

$$a = a_k = \frac{1}{1 + \sqrt{k}}, \quad f(t) = h_{1-a_k, a_k}(t), \quad \text{where } l_k := \frac{\sqrt{k}}{1 + \sqrt{k}}. \tag{2.57}$$

Chapter 3

Fredholm Kernels

Due to the exceptional role of polynomial spline kernels in generating extremal functions of various extremal functions in $W^r H^\omega$, we reserved the entire chapter for the presentation of properties of different kinds of *Fredholm kernels*.

3.1. Kernels of type I

Fix $r, m \in \mathbb{N} : 0 < m < r$. Let $\{\nu_i\}_{i=0}^n$ and $\{\vartheta_i\}_{i=0}^{n-r+1}$ satisfy

$$
\begin{aligned}
& 0 =: \nu_0 < \nu_1 < \cdots < \nu_n; \quad 0 =: \vartheta_0 < \vartheta_1 < \cdots < \vartheta_{n-r+1} < \nu_n; \\
& \nu_{i-1} < \vartheta_i < \nu_{i+r-1}, \qquad i = 1, \ldots, n - r + 1.
\end{aligned}
\tag{1.1}
$$

We derive the coefficients $\{\alpha_i\}_{i=0}^n$ from the system of linear equations

$$
\begin{cases}
\displaystyle\sum_{i=0}^n \alpha_i \nu_i^j = m! \cdot \delta_{m,j}, & j = 0, \ldots, r-1; \\
\displaystyle\sum_{i=0}^n \alpha_i (\nu_i - \vartheta_l)_+^{r-1} = 0, & l = 1, \ldots, n-r+1.
\end{cases}
\tag{1.2}
$$

The inequalities between the points $\{\nu_i\}_{i=0}^n$ and $\{\vartheta_i\}_{i=1}^{n-r+1}$ in (1.1) are precisely the conditions which guarantee that the system of equations (1.2) has a unique solution (cf. [37], [38]). The kernel $K(t)$ is defined as follows:

$$
K(t) := -\frac{1}{(r-1)!} \sum_{i=0}^n \alpha_i (\nu_i - t)_+^{r-1}, \qquad 0 \le t \le \nu_n.
\tag{1.3}
$$

From (1.2) we observe that the kernel K enjoys the properties

$$
\begin{aligned}
K^{(i)}(\nu_n) &= 0, & i = 0, \ldots, r-2. \\
K^{(i)}(0) &= \frac{(-1)^{r-m}}{m!} \delta_{r-i-1,m}, & i = 0, \ldots, r-2. \\
K(\vartheta_l) &= 0, & l = 1, \ldots, n-r+1.
\end{aligned}
\tag{1.4}
$$

Put

$$
k_m(j) := \begin{cases}
j, & 0 \le j < m; \\
j - 1, & m \le j \le r - 1.
\end{cases}
\tag{1.5}
$$

Recall Definition 1.1.1 of *points of sign change* of an integrable function f without zero intervals and *the number of sign changes* $\mu(f; [a, b])$ on the interval $[a, b]$.

PROPOSITION 3.1.1. *Let the collections of points*

$$\bar{\nu} = (\nu_0, \ldots, \nu_n), \qquad \bar{\vartheta} = (\vartheta_0, \ldots, \vartheta_{n-r+1})$$

satisfy inequalities (1.1), *and the kernel* $K = K_{\bar{\nu}, \bar{\vartheta}}$ *be defined in* (1.2), (1.3). *Then,*

(i) $\operatorname{supp} K = [0, \nu_j]$,   *for some* $j : r - 1 \le j \le n$;

(ii) $\mu\left(K^{(r-l)}; [0, \nu_j]\right) = j - 1 - k_m(l-1)$, $l = 1, \ldots, r$,

(iii) $\operatorname{sign} \alpha_i = (-1)^{i+m}$, $i = 1, \ldots, j$;   (1.6)

(iv) $\operatorname{sign} K(t) = (-1)^{i+r+m}$, $\hat{\vartheta}_i < t < \hat{\vartheta}_{i+1}$, $i = 0, \ldots, j - r + 1$,

where $\hat{\vartheta}_i := \vartheta_i$, $i = 0, \ldots, j - r + 1$; $\hat{\vartheta}_{j-r+2} := \nu_j$.

Proof. The proof is subdivided into consideration of two possible cases.

Case 1. supp K(t) $= [0, \nu_j]$, $j = 1, \ldots, n$.

Notice that

$$K^{(r-1)}(t) = (-1)^r \sum_{i=0}^{n} \alpha_i \chi_{[0,\nu_i]}(t). \tag{1.7}$$

Thus, the step function $K^{(r-1)}$ can have at most $j - 1$ sign changes on the interval $[0, \nu_j]$ at the points $\{\nu_i\}_{i=1}^{j-1}$. Using the boundary conditions (1.4) and employing the Rolle theorem (Proposition 1.1.3), we obtain the chain of inequalities between the number of sign changes of the consecutive derivatives of the kernel K:

$$j - 1 \ge \mu\left(K^{(r-1)}; [0, \nu_j]\right) \ge \mu\left(K^{(r-2)}; [0, \nu_j]\right) + k_m(1) \ge \cdots$$

$$\cdots \ge \mu\left(K^{(r-l)}; [0, \nu_j]\right) + k_m(l-1) \ge \cdots \ge \mu\left(K'; [0, \nu_j]\right) + k_m(r-2). \tag{1.8}$$

We distinguish two subcases: $0 < m < r$ and $m = r$.

Subcase 1.1. $0 < m < r - 1$.

In this case $k_m(r - 2) = r - 3$, and the inequalities (1.8) imply that

$$\mu\left(K'; [0, \nu_j]\right) \le j - r + 2. \tag{1.9}$$

On the other hand, by (1.4) and inequalities (1.1), K has at least $j - r + 3$ distinct zeroes on the interval $[0, \nu_j]$ at the points $\{\vartheta_i\}_{i=0}^{j-r+1}$ and ν_j. Thus, by the Rolle theorem, the derivative K' has at least $j - r + 2$ sign changes on the interval $[0, \nu_j]$, i.e.,

$$\mu\left(K'; [0, \nu_j]\right) \ge j - r + 2, \qquad \text{for } 0 < m < r - 1. \tag{1.10}$$

From (1.9), (1.10) we infer that

$$\mu\left(K'; [0, \nu_j]\right) = j - r + 2, \tag{1.11}$$

and we have the equalities everywhere in (1.8) for $0 < m < r - 1$.

Subcase 1.2. $m = r - 1$.

In this case, by the definition of $k_m(\cdot)$ in (1.5), $k_{r-1}(r-2) = r-2$. Therefore, from the inequalities (1.8) it follows that

$$\mu\left(K'; [0, \nu_j]\right) \le j - r + 1. \tag{1.12}$$

However, by (1.1), K has at least $j - r + 2$ distinct zeroes on the interval $[0, \nu_j]$ at the points $\{\vartheta_i\}_{i=1}^{j-r+1}$ and ν_j. Thus, by the Rolle theorem, the derivative K' exhibits at least $j - r + 1$ sign changes on the interval $[0, \nu_j]$, i.e.,

$$\mu\left(K'; [0, \nu_j]\right) \ge j - r + 1, \qquad \text{for } m = r - 1. \tag{1.13}$$

By (1.12), (1.13),

$$\mu\left(K'; [0, \nu_j]\right) = j - r + 1, \tag{1.14}$$

and we have the equalities everywhere in (1.8) in the case of $m = r - 1$, as well.

Therefore, we proved in both Subcases 1.1 and 1.2 of Case 1 that

$$\mu\left(K^{(r-l)}; [0, \nu_j]\right) = j - 1 - k_m(l - 1), \qquad l = 0, \dots, r - 1. \tag{1.15}$$

It follows immediately from the equalities (1.15) that

$$\mu\left(K; [0, \nu_j]\right) = j - r + 1, \tag{1.16}$$

i.e. all the interior zeroes $\{\vartheta_i\}_{i=1}^{j-r+1}$ are of multiplicity 1.

Notice that in both cases $\mu\left(K; [0, \nu_j]\right) = j - r + 1 \ge 0$, so

$$j \ge r - 1. \tag{1.17}$$

Also by (1.15), $\mu\left(K^{(r-1)}; [0, \nu_j]\right) = j - 1$, implying that

$$\left(\sum_{l=1}^{i} a_l\right)\left(\sum_{l=1}^{i+1} a_l\right) < 0, \qquad i = 1, \dots, j - 1. \tag{1.18}$$

Therefore,

$$a_l \cdot a_{l+1} < 0, \qquad l = 1, \dots, j - 1. \tag{1.19}$$

Using the fact that $\operatorname{sign} K^{(r-m-1)}(0) = (-1)^{r-m}$, we arrive at the conclusion that

$$\operatorname{sign} \alpha_i = (-1)^{m+i}, \qquad i = 1, \dots, j. \tag{1.20}$$

By (1.2), $\alpha_0 = -\sum_{l=1}^{n} \alpha_i$. The relations (1.18) and (1.20) then imply that $\operatorname{sign} \alpha_0 = (-1)^m$.

In addition, we have demonstrated that the kernel K changes the sign only at the points $\{\vartheta_i\}_{i=1}^{j-r+1}$ on the interval $(0, \nu_j)$ and

$$\text{sign}\, K(t) = (-1)^{i+r}, \qquad \hat{\vartheta}_i \leq t \leq \hat{\vartheta}_{i+1}, \qquad i = 0, \ldots, j - r + 1, \qquad (1.21)$$

where $\hat{\vartheta}_i := \vartheta_i$, $i = 0, \ldots, j - r + 1$; $\hat{\vartheta}_{j-r+2} := \nu_j$.

Case 2. There exists an interval $[a, b] \subset \text{supp}\, K$:

$$K^{(i)}(a) = K^{(i)}(b) = 0, \qquad i = 0, \ldots, r - 2. \qquad (1.22)$$

The Rolle theorem then implies that $K^{(r-1)}$ has at least $r - 1$ sign changes on the interval $[a, b]$, i.e., K displays at least $r - 1$ knots $\{\nu_i\}_{i_1+1}^{i_1+r-1}$ on the interval $[a, b]$. Since the kernel K is a polynomial of degree $\leq r - 1$ on each of the intervals $[\nu_i, \nu_{i+1}]$, $i = 0, \ldots, n - 1$, there exists an interval $[\nu_{j_1}, \nu_{j_2}] \subset \text{supp}\, K$, $j_1 \geq 1$, such that

$$K^{(i)}(\nu_{j_1}) = K^{(i)}(\nu_{j_2}) = 0, \qquad i = 0, \ldots, r - 2; \quad j_2 - j_1 \geq r. \qquad (1.23)$$

The step function $K^{(r-1)}$ can have at most $j_2 - j_1 - 1$ sign changes on the interval (ν_{j_1}, ν_{j_2}) at the points $\{\nu_i\}_{i=j_1+1}^{j_2-1}$. Taking into account the relations (1.23) and repeatedly applying the Rolle theorem as in Case 1, we can infer then that K can have *at most* $j_2 - j_1 - r - 1$ zeroes on the open interval (ν_{j_1}, ν_{j_2}). However, the inequalities $\nu_{i-1} < \vartheta_i < \nu_{i+r-1}$, $i = 1, \ldots, n - r$, imply that K exhibits at least $j_2 - j_1 - r + 1$ distinct zeroes $\{\vartheta_i\}_{i=j_1+1}^{j_2+1-r}$ on the interval (ν_{j_1}, ν_{j_2}). This contradiction precludes Case 2 from arising. $\qquad \square$

COROLLARY 3.1.2. *Let the kernel K, defined in (1.2), (1.3) have the zero mean on the interval $[0, \nu_n]$, i.e.*

$$\int\limits_0^{\nu_n} K(t)\, dt = -\frac{1}{r!} \sum_{i=0}^n a_i \nu_i{}^r = 0. \qquad (1.24)$$

Then, the support of the kernel K contains the interval $[0, \nu_r]$, i.e.

$$\text{supp}\, K(t) = [0, \nu_j], \qquad \text{for some } j : r \leq j \leq n.$$

Proof. By Proposition 1.1, $\text{supp}\, K(t) = [0, \nu_j]$, for some $j : r \leq j \leq n$. Because $\int_0^{\nu_j} K(t)\, dt = \int_0^{\nu_n} K(t)\, dt = 0$, the kernel K has at least one sign change on the interval $[0, \nu_j]$. Thus, by (1.30), (ii) for $l = r$,

$$\mu(K, [0, \nu_j]) = j - r + 1 \geq 1, \qquad (1.25)$$

and the result follows. $\qquad \square$

COROLLARY 3.1.3. *Let $[0, \nu_j]$ be the support of the kernel K defined in* (1.2), (1.3). *Then,*

$$\vartheta_i < \nu_{i+r-2}, \qquad i = 1, \ldots, j - r. \tag{1.26}$$

Proof. The inequalities (1.26) follow from the Rolle theorem and the property $K^{(i)}(\nu_j) = 0$, $j = 0, \ldots, r - 2$.

Indeed, for $i = 0, \ldots, r-1$, let $\xi_0^i := \nu_j$, $i = 0, \ldots, r-1$, and $\{\xi_l^i\}_{l=1}^{j-r}$ be the $j - r + i$ rightmost points of sign change of $K^{(i)}$ on the interval $[0, \nu_j)$ enumerated in *the decreasing order:* $\xi_{j-r}^i < \xi_{j-r-1}^i < \cdots < \xi_0^i = \nu_j$, $i = 0, \ldots, r - 1$.

By the Rolle theorem, for $i = 0, \ldots, r - 2$, we have:

$$\xi_{l-1}^i > \xi_l^{i+1} > \xi_l^i, \qquad l = 1, \ldots, j - r + i. \tag{1.27}$$

In particular,

$$\nu_{j-l} =: \xi_l^{r-1} > \xi_l^{r-2} > \cdots > \xi_l^0 := \vartheta_{j-r+2-l}, \qquad l = 1, \ldots, j - r + 1, \tag{1.28}$$

so we obtain the inequalities (1.26). ☐

In Chapters 5 and 6 we show that kernels $K(t)$ of the form (1.2), (1.3) generate the rth derivative of extremal functions in the problem

$$f^{(m)}(0) \to \sup, \qquad f \in W^r H^\omega [0, \Gamma], \quad \|f\|_{C[0,\Gamma]} \le B, \tag{1.29}$$

for $0 < m < r$.

3.2. Kernels of type II

Let the collections of points $\{\nu_i\}_{i=0}^{n+1}$ and $\{\vartheta_i\}_{i=0}^{n-r+1}$ be such that

$$0 =: \nu_0 < \nu_1 < \cdots < \nu_n; \quad 0 =: \vartheta_0 < \vartheta_1 < \cdots < \vartheta_{n-r} < \vartheta_{n-r+1} := \nu_n;$$
$$\nu_{i-1} < \vartheta_i < \nu_{i+r-1}, \qquad i = 1, \ldots, n - r. \tag{2.1}$$

Derive the coefficients $\{\alpha_i\}_{i=0}^n$ from the system of linear equations

$$
\begin{cases}
\displaystyle\sum_{i=0}^n \alpha_i \nu_i^j = 0, & j = 0, \ldots, r - 1; \\[2mm]
\displaystyle\sum_{i=0}^n \alpha_i (\nu_i - \vartheta_l)_+^{r-1} = 0, & l = 1, \ldots, n - r + 1; \\[2mm]
\displaystyle\sum_{i=0}^n \alpha_i \nu_i^r = r!
\end{cases}
\tag{2.2}
$$

The last equation in (2.2) is added for the normalization of coefficients $\{\alpha_i\}_{i=0}^n$. Let

$$K(t) = -\frac{1}{(r-1)!} \sum_{i=0}^n \alpha_i (\nu_i - t)_+^{r-1}, \qquad 0 \le t \le \nu_n. \tag{2.3}$$

The verification of the following properties of the kernel K proceeds along the lines of the proof of Proposition 3.1.1.

PROPOSITION 3.2.1. *If the kernel $K(t)$ is defined by* (2.2), (2.3), *then*

(i) $\operatorname{supp} K = [0, \nu_j]$,   *for some* $j : r \leq j \leq n$;

(ii) $\mu\left(K^{(r-l)}; [0, \nu_j]\right) = j - l$, $l = 1, \ldots, r$,

(iii) $\operatorname{sign} \alpha_i = (-1)^{i+r}$, $i = 1, \ldots, j$;   (2.4)

(iv) $\operatorname{sign} K(t) = (-1)^i$, $\hat{\vartheta}_i < t < \hat{\vartheta}_{i+1}$, $i = 0, \ldots, j - r + 1$,

where $\hat{\vartheta}_i := \vartheta_i$, $i = 0, \ldots, j - r + 1$; $\hat{\vartheta}_{j-r+2} := \nu_j$.

The kernels $K(t)$ of the type (2.2), (2.3) generate the rth derivative of extremal functions of the problem

$$f^{(r)}(0) \to \sup, \qquad f \in W^r H^\omega[0, \Gamma], \quad \|f\|_{C[0,\Gamma]} \leq B, \tag{2.5}$$

for $0 < m < r$ (see Chapters 5 and 6).

3.3. Kernels of type III

Let the collections of points $\{\nu_i\}_{i=0}^n$ and $\{\vartheta_i\}_{i=0}^{n-r+1}$ lie on the interval $[0, 1)$ and satisfy the inequalities (1.1). Let also $\nu_{n+1} := 1$.

We determine the coefficients $\{\alpha_i\}_{i=0}^{n+1}$ of the kernel

$$K(t) = -\frac{1}{(r-1)!} \sum_{i=1}^{n+1} \alpha_i (\nu_i - t)_+^{r-1}, \tag{3.1}$$

from the equations

$$\begin{cases} \displaystyle\sum_{i=0}^{n+1} (-1)^i \alpha_i = 1; \\[2mm] \displaystyle\sum_{i=0}^{n+1} \alpha_i \nu_i{}^j = 0, \qquad j = 0, \ldots, r; \\[2mm] \displaystyle\sum_{i=0}^{n+1} \alpha_i (\nu_i - \vartheta_l)_+^r = 0, \qquad l = 1, \ldots, n - r, \end{cases} \tag{3.2}$$

PROPOSITION 3.3.1. *The kernels $K(t)$ from* (3.1), (3.2) *enjoy the properties*

(i) $\operatorname{supp} K = [0, 1]$;

(ii) $\mu\left(K^{(r-l)}; [0, 1]\right) = n - l$, $l = 1, \ldots, r$,

(iii) $\operatorname{sign} \alpha_i = (-1)^i$, $i = 0, \ldots, n + 1$;   (3.3)

(iv) $\operatorname{sign} K(t) = (-1)^{i+r}$, $\vartheta_i < t < \vartheta_{i+1}$, $i = 0, \ldots, n - r + 1$,

where $\vartheta_{n-r+2} := 1$.

The kernels $K(t)$ of the form (3.1) with coefficients derived from (3.2) generate the rth derivative of the function related to *the problem of n-widths* of nonperiodic classes $W^r H^\omega[0, 1]$ (consult Section 13.2 and Chapter 16).

Chapter 4

Review of Classical Chebyshev Polynomial Splines

In this chapter we review the notion and an elementary construction of the Chebyshev perfect spline T_n of degree $r + 1$ with $n - r \geq 0$ knots and $n + 1$ points of alternance on the interval $[0, 1]$. Relying on the Rolle theorem or an application of Fredholm kernels, we give two proofs of extremality of Chebyshev perfect splines of the problem

$$f^{(m)}(0) \to \sup, \qquad f \in W_\infty^{r+1}[0,1], \qquad \|f\|_{C[0,1]} \leq \rho_n := \|T_n\|_{C[0,1]},$$

for all $0 < m \leq r$. Then, we discuss the possibility of application of these two methods to the solution of the corresponding problem in $W^r H^\omega$. In Section 4.5 we discuss the elementary proof of the original exact Kolmogorov inequalities for intermediate derivatives due to A. S. Cavaretta [18]. Finally, we derive some special technical results of the general theory of perfect splines employed in the proof of the main results of the paper.

4.1. Construction of Chebyshev perfect splines

DEFINITION 4.1.1. Let $l, n \in \mathbb{Z}_+$. The function $T(x)$ is called *a perfect spline of degree l with n knots* $\{t_i\}_{i=1}^n$, $-\infty =: t_0 < t_1 < t_1. < \cdots < t_n < t_{n+1} := +\infty$, if for some $\chi \in \{\pm 1\}$,

$$T(x) = \frac{\chi x^l}{l!} + \sum_{i=0}^{l-1} a_i x^i + \frac{2\chi}{l!} \sum_{i=1}^n (-1)^i (x - t_i)_+^l, \qquad x \in \mathbb{R}.$$

The perfect spline is distinguished among other functions from $\chi W_\infty^r(\mathbb{R})$ by the fact that its rth derivative alternates between χ and $-\chi$ on the consecutive intervals (t_i, t_{i+1}), $i = 0, \ldots, n$.

Various interpolating properties, including *the fundamental theorem of algebra for perfect splines*, and applications of polynomial perfect splines in *the optimal recovery* of functions from Sobolev classes and other extremal problems are highlighted in the papers of A. Pinkus [74], S. Karlin [37], C. A. Michelli, T. J. Rivlin, S. Winograd [67], I. J. Schoenberg, A. Whitney [79], and in the books by V. N. Malozemov, A. B. Pevny [61] and V. M. Tihomirov [88].

The following proof of the existence of polynomial perfect splines with the given number of alternance points proceeds along the lines of the construction of Chebyshev polynomial splines in [61].

LEMMA 4.1.1. , *Let* r, m, $n \in \mathbb{N}$, $m \leq r$, $n \geq r$.

Then, there exist a collection of points $\{\nu_i\}_{i=0}^{n+1}$, $0 = \nu_0 < \nu_1 < \cdots < \nu_{n+1} = 1$, *and a perfect spline* $T_n(x) = T_{n,r}(x)$ *of degree* $r + 1$ *with* $n - r$ *knots on the interval* $[0,1]$, *such that*

$$T_n(\nu_i) = (-1)^{m+i} \|T_n\|_{C[0,1]}, \qquad i = 0, \ldots, n+1. \tag{1.1}$$

Proof. The proof of Lemma 4.1.1 is based on Borsuk's Theorem 1.1.1.

Let $\mathbb{S}^{n-r} = \{s = (s_1, \ldots, s_{n-r+1}) \in \mathbb{R}^{n-r+1} \mid \sum\limits_{i=1}^{n-r+1} |s_i| = 1\}$.

Given an $s \in \mathbb{S}^{n-r}$, generate the partition $\{t_i = t_i(s)\}_{i=0}^{n-r+1}$ of the interval $[0,1]$ according to the rules:

$$t_0(s) = 0, \quad t_j(s) = \sum_{i=1}^{j} |s_i|, \quad j = 1, \ldots, n-r+1. \tag{1.2}$$

Put

$$g_s(x) = \operatorname{sign} s_j, \qquad x \in (t_{j-1}, t_j), \quad j = 1, \ldots, n-r+1. \tag{1.3}$$

Let

$$f_s(x) = \frac{1}{r!} \int_0^1 (x - t)_+^r g_s(t)\, dt, \qquad 0 \leq x \leq 1. \tag{1.4}$$

Define the polynomial $P_s(x) = \sum\limits_{i=0}^{n} a_i(s) x^i$ as the polynomial of the best approximation for the function $f_s(x)$ on the interval $[0,1]$. Set

$$T_s(x) = f_s(x) - P_s(x), \qquad 0 \leq x \leq 1. \tag{1.5}$$

Consider the mapping $\varkappa : \mathbb{S}^{n-r} \mapsto \mathbb{R}^{n-r}$ defined as follows:

$$\varkappa(s) = (a_{r+1}(s), a_{r+1}(s), \ldots, a_n(s)), \qquad s \in \mathbb{S}^{n-r}. \tag{1.6}$$

Clearly, the mapping $\varkappa(s)$ is *continuous and odd* on $\mathbb{S}^{n-r}$. Therefore, by the Borsuk theorem, there exists an $s^* \in \mathbb{S}^{n-r}$ such that $\varkappa(s^*) = 0$. This means that $P_{s^*}(x)$ is a polynomial of degree not greater that r. By Chebyshev's Theorem 1.1.2, $T_{s^*}(x)$ exhibits $n + 2$ points of alternance on the interval $[0,1]$. Therefore, the derivative T'_{s^*} has at least n distinct zeroes at the interior points of alternance on the open interval $(0,1)$. By the Rolle theorem, $T_{s^*}^{(r+1)}$ has *at least* $n - r$ distinct sign changes on $(0,1)$:

$$\mu(T_{s^*}^{(r+1)}; [0,1]) \geq n - r. \tag{1.7}$$

.

On the other hand, by (1.3), the step function $T_{s^*}^{(r+1)}$ can have *at most* $n - r$ sign changes at the points $\{t_i(s^*)\}_{i=1}^{n-r}$:

$$\mu\left(T_{s^*}^{(r+1)}; [0,1]\right) \leq n - r. \tag{1.8}$$

The juxtaposition of properties (1.7) and (1.8) leads us to the conclusion that the function $g_{s^*} = T_{s^*}^{(r+1)}$ does change its sign precisely $n - r$ times on the interval $[0,1]$ at $\{t_i(s^*)\}_{i=1}^{n-r}$. Thus, by (1.3), all the entries $s_1^*, \ldots, s_{n-r+1}^*$ of the vector s^* are different from zero and have alternating signs:

$$\text{sign } s_j^* = (-1)^j \chi, \qquad j = 1, \ldots, n - r + 1, \ \chi = 1 \text{ or } -1, \text{ fixed}. \tag{1.9}$$

Since $\varkappa$ is an odd mapping on $\mathbb{S}^{n-r}$, $\varkappa(-s^*) = -\varkappa(s^*) = 0$, so we can assume without loss of generality that $\chi = (-1)^{m+r+1}$, i.e.

$$\text{sign } T_{s^*}^{(r+1)}(x) = (-1)^{j+m+r+1}, \qquad t_{j-1}(s^*) < x < t_j(s^*), \qquad j = 1, \ldots, n-r+1. \tag{1.10}$$

By the Chebyshev theorem, T_{s^*} has $n + 2$ alternance points $\{\nu_i\}_{i=0}^{n+1} : 0 \leq \nu_0 < \nu_1 < \cdots < \nu_n < \nu_{n+1} \leq 1$:

$$T_{s^*}(\nu_i) = (-1)^i \eta \|T_{s^*}\|_{C[0,1]}, \qquad i = 0, \ldots, n+1; \quad \eta = 1 \text{ or } -1, \text{ fixed}. \tag{1.11}$$

Since the derivative vanishes at the interior extremal points, $T_{s^*}'(\nu_i) = 0$ for all $i = 1, \ldots, n$. By (1.8), T_{s^*}' can have at most n zeroes on the interval $[0,1]$. Thus, the points $\{\nu_i\}_{i=1}^n$ exhaust the set of zeroes of T_{s^*}'. Therefore, $\nu_0 = 0$ and $\nu_{n+1} = 1$. Also, the juxtaposition of (1.10) and (1.11) enables us to conclude that $\eta = (-1)^m$ in (1.11). It remains to rename the function T_{s^*} and the knots $\{t_i(s^*)\}_{i=1}^{n-r}$:

$$T_n := T_{s^*}, \qquad \vartheta_i := t_i(s^*), \qquad i = 0, \ldots, n - r + 1. \tag{1.12}$$

$\square$

Let $\rho_n := \|T_n\|_{C[0,1]}$, $n \geq r$. We show two ways of proving that T_n is extremal in the problem

$$f^{(m)}(0) \to \sup, \qquad f \in W_\infty^{r+1}[0,1], \qquad \|f\|_{C[0,1]} \leq \rho_n. \tag{1.13}$$

4.2. Zero count argument

The first method employs the standard *zero* or *sign change counting argument* based on the Rolle theorem. Variations of this technique play an essential role in the proof of many results in the theory of spline functions and, more generally, in the theory of extremal problems in Sobolev classes ([23], [37], [61], [64]).

Let us assume that the function T_n from Lemma 4.1.1 is not extremal in the problem (1.13). In this case, there exists a function $f \in W_\infty^{r+1}[0,1]$ with the following properties:

$$
\begin{aligned}
&(a) \quad f^{(m)}(0) > T_n^{(m)}(0) > 0; \\
&(b) \quad \|f\|_{C[0,1]} < \rho_n; \\
&(c) \quad \|f^{(r+1)}\|_{L_\infty[0,1]} < 1.
\end{aligned}
\tag{2.1}
$$

Put

$$
H(t) = T_n(t) - f(t), \qquad 0 \leq t \leq 1.
\tag{2.2}
$$

Let, as before, $\{\vartheta_i\}_i^{n-r}$ be the points of sign change of $T_{s*}^{(r+1)}$ such that $0 =: \vartheta_0 < \vartheta_1 < \cdots < \vartheta_{n-r+1} := 1$. By the property (1.10) and the notation (1.12),

$$
T_n^{(r+1)}(t) = (-1)^{i+m+r+1}, \qquad t \in (\vartheta_{i-1}, \vartheta_i), \qquad i = 1, \ldots, n-r+1.
\tag{2.3}
$$

Then, from the strict inequality (2.1), (c) and the property (2.3) we infer that

$$
\operatorname{sign} H^{(r+1)}(t) = (-1)^{i+m+r+1}, \qquad t \in (\vartheta_{i-1}, \vartheta_i), \qquad i = 1, \ldots, n-r+1.
\tag{2.4}
$$

Therefore, $H^{(r+1)}$ has precisely $n - r$ sign changes on $[0,1]$ at the points $\{\vartheta_i\}_{i=1}^{n-r}$. On the other hand, the strict inequality (2.1), (b) and the property (1.1) of the function T_n imply that

$$
(-1)^{i+m} H(\nu_i) = \rho_n - f(\nu_i) \geq \rho_n - \|f\|_{C[0,1]} > 0, \qquad i = 0, \ldots, n+1.
\tag{2.5}
$$

Therefore, H alternates its sign at the points $\{\nu_i\}_{i=0}^{n+1}$. Thus, H has *at least* $n + 1$ distinct zeroes $\{\xi_i^0\}_{i=1}^n$ on the interval $[0,1]$, $0 =: \xi_0^0 < \xi_1^0 < \cdots < \xi_{n+1}^0 < \xi_{n+1}^0 := 1$. By (2.4) $H^{(r+1)}$ has precisely $n - r$ sign changes. Consequently, H can have *at most* $n+1$ zeroes on $[0,1]$. Analogically, from the Rolle theorem we infer that each of the derivatives $H^{(k)}$, $k = 0, \ldots, r$, has precisely $n + 1 - k$ simple (of multiplicity one) zeroes $\{\xi_i^k\}_{i=1}^{n+1-k} : 0 =: \xi_0^k < \xi_1^k < \cdots < \xi_{n+1-k}^k < \xi_{n=1.k}^k := 1$. From the sign distribution in (2.4) or (2.5) we find that

$$
\operatorname{sign} H^{(k)}(t) = (-1)^{k+m+i+1}, \qquad i = 1, \ldots, n+1.k.
\tag{2.6}
$$

In particular, by (2.6) for $k = m$ and $i = 1$, we have

$$
T^{(m)}(0) - f^{(m)}(0) > 0.
\tag{2.7}
$$

This contradiction with the property (2.7), (a) proves that our assumption was wrong, and T_n is extremal in the problem (1.13).

However, this method, so popular and widespread in Approximation Theory, is not applicable in the theory of extremal problems in $W^r H^\omega$. Let us mention some examples. In $W_\infty^{r+1}[0,1]$, the Chebyshev perfect splines T_n and $-T_n$ are the only two functions that have the minimal norm among all perfect splines with at most $n - r$ knots on the interval $[0, 1]$. In particular, the norms $\rho_n := \|T_n\|_{C[0,1]}$ strictly decrease to zero as n increases to infinity. It is even possible to prove that the ith knot of the Chebyshev spline T_n is a decreasing function of n. A standard method of proving all these properties (cf. [37] and [61]) involves the examination of the difference of two perfect splines and accurate estimates of the number of sign changes (zeroes) of this difference and its $(r + 1)^{\text{st}}$ derivative.

A more complicated structure of the extremal perfect ω-splines, described in Theorem X of Chapter 2, makes it impossible to compare not only the extremal functions and arbitrary functions from H^ω, but even the perfect ω-splines between themselves. Also notice that the same function T_n solves the problem (1.13) for all m's. However, as Theorem 6.0.1 demonstrates, the corresponding Chebyshev ω-splines in $W^r H^\omega[0, 1]$ *do differ for different* m's.

Moreover, the structure of Chebyshev ω-splines depends on the type of the extremal problem under consideration. As we show in Chapters 6, 9 and 12, *the problem of n-widths of* $W^r H^\omega[0, 1]$ and each of *the Kolmogorov–Landau problem* has its own family of Chebyshev ω-splines.

4.3. Application of the Fredholm kernels

For $k = 0, \ldots, r + 1$, the derivatives $T_n^{(k)}$ have precisely $n + 1 - k$ points $\{\eta_i^k\}_{i=1}^{n+1-k}$ of sign change on $[0, 1]$ such that $0 =: \eta_0^k < \eta_1^k < \cdots < \eta_{n+1-k}^k < \eta_{n+1.k}^k := 1$, $k = 0, \ldots, r + 1$. In these notations, $\nu_i = \eta_i^1$, $i = 0, \ldots, n + 1$, and $\vartheta_i = \nu_i^{r+1}$, $i = 0, \ldots, n - r + 1$, where $\{\nu_i\}_{i=1}^{n-r}$ are the points of alternance of T_n, and $\{\vartheta_i\}_{i=1}^{n-r}$ are the knots of T_n. By the Rolle theorem, we have the following relations between the points of sign change of consecutive derivatives:

$$\eta_i^k < \eta_i^{k+1} < \eta_{i+1}^k, \qquad i = 1, \ldots, n - k, \qquad k = 0, \ldots, r. \tag{3.1}$$

In particular,

$$\nu_i = \eta_i^1 < \vartheta_i = \eta_i^{r+1} < \eta_{i+r}^1 = \nu_{i+r}, \qquad i = 1, \ldots, n - r. \tag{3.2}$$

Let $\{\alpha_i\}_{i=0}^n$ be derived from the system of linear equations

$$\begin{cases} \displaystyle\sum_{i=0}^n \alpha_i \nu_i^j = m! \, \delta_{m,j}, & j = 0, \ldots, r; \\[2mm] \displaystyle\sum_{i=0}^n \alpha_i (\nu_i - \vartheta_l)_+^r = 0, & l = 1, \ldots, n - r. \end{cases} \tag{3.3}$$

The inequalities (3.2) between the points $\{\nu_i\}_{i=0}^n$ and $\{\vartheta_i\}_{i=1}^{n-r}$ guarantee that the system of equations (3.3) has a unique solution. In fact, the inequalities $\nu_{i-1} < \vartheta_i < \nu_{i+r}$, $i = 1, \ldots, n - r$, are precisely the conditions that assure the nonsingularity of the matrix in (3.3)) (cf. [39]).

According to S. Karlin [38], we define the kernel $F(t)$:

$$F(t) := -\frac{1}{r!} \sum_{i=0}^{n} \alpha_i(\nu_i - t)_+^r, \quad 0 \le t \le \nu_n. \tag{3.4}$$

From the system of linear equations (3.3) it follows that

$$F^{(i)}(\nu_n) = 0, \qquad i = 0, \dots, r-1.$$

$$F^{(i)}(0) = \frac{m!}{(r-i)!} \sum_{j=1}^{n} \alpha_j \nu_j^{r-i} = (-1)^{r-m} \delta_{r-i,m}, \qquad i = 0, \dots, r-1. \tag{3.5}$$

$$F(\vartheta_l) = 0, \qquad l = 1, \dots, n-r.$$

In Proposition 3.1.1 we mentioned the following properties of the kernel F:

(A) $(-1)^{i+m} \operatorname{sign} \alpha_i \ge 0, \qquad i = 0, \dots, n+1;$

(B) $(-1)^{i+r+m+1} \operatorname{sign} F(t) \ge 0, \qquad \vartheta_{i-1} \le t \le \vartheta_i, \qquad i = 1, \dots, n-r+1.$
$$\tag{3.6}$$

By Taylor's formula,

$$f(t) = \sum_{i=0}^{r} \frac{f^{(i)}(0)}{i!} t^i + \frac{1}{r!} \int_0^1 f^{(r+1)}(y)(t-y)_+^r \, dy, \qquad 0 \le t \le 1, \tag{3.7}$$

and from equations (3.3) we obtain the formula for the mth derivative of the function $f \in W_\infty^{r+1}[0,1]$ at the origin:

$$f^{(m)}(0) = \sum_{i=0}^{n} \alpha_i f(\nu_i) + \int_0^1 f^{(r+1)}(y)F(y) \, dy. \tag{3.8}$$

Therefore, by (3.8), any function $f \in W_\infty^{r+1}[0,1]$ with $\|f\|_{C[0,1]} \le \rho_n$, satisfies the inequality

$$|f^{(m)}(0)| \le \sum_{i=0}^{n} |\alpha_i|\rho_n + \int_0^1 |F(y)| \, dy. \tag{3.9}$$

Properties (1.1) and (2.3) of the function T_n and (3.6) of the kernel F guarantee that the function T_n is extremal in the inequality (3.9), i.e.

$$T_n(0) = \sum_{i=0}^{n} |\alpha_i|\rho_n + \int_0^1 |F(y)| \, dy. \tag{3.10}$$

However, we have to make some adjustments to applications of this method in the theory of extremal problems in $W^r H^\omega$ for nonlinear ω. Notice that the $(r+1)^{\text{st}}$ derivative of the function T_n is extremal in the problem

$$\int_0^1 h(t) F(t)\, dt \to \sup, \qquad h \in \mathbb{L}_\infty[0,1] : \|h\|_{\mathbb{L}_\infty[0,1]} \le 1. \qquad (3.11)$$

Thus, $T_n^{(r+1)}(t) = \operatorname{sign} F(t)$, $t \in \operatorname{supp} F(t)$, and the points of sign change of the step function $T_n^{(r+1)}$ coincide with the simple zeroes of the kernel F. Let

$$K(t) = -F'(t), \qquad t \in (0,1). \qquad (3.12)$$

By (3.5), $F(0) = F(1) = 0$, for $0 < m < r$, so the problem (3.11) is equivalent to the problem

$$\int_0^1 g(t) K(t)\, dt \to \sup, \qquad g \in H^1[0,1], \qquad (3.13)$$

where the Hölder classes $H^\alpha[a,b]$, $\alpha \in (0,1]$, are introduced in (0.3.4) of Definition 0.3.2. The derivative of the extremal function of the problem (3.13) is extremal in the problem (3.11).

The following examples explain why in Chapter 3 we will rather use kernels $K(t)$ and solutions of the problem

$$\int_0^1 g(t) K(t)\, dt \to \sup, \qquad g \in H^\omega[0,1], \qquad (3.14)$$

in generating the corresponding numerical differentiation formulae for nonlinear ω.

EXAMPLE 4.3.1. Let us show that for a strictly concave modulus of continuity ω, the points of sign change of the derivative of the extremal function in the problem (3.11) no longer coincide with the zeroes of the kernel $F(t) = -\int_0^1 K(y)\, dy$.

Let the kernel K be an odd function on the interval $[-10, 10]$, given by the following formula on the interval $[0, 10]$:

$$K(t) = \begin{cases} 1, & t \in [0,5]; \\ -1, & t \in (5,10]. \end{cases} \qquad (3.15)$$

The kernel $F(t) = -\int_0^t K(y)\, dy$, $t \in [-10,10]$, is even and given by the following formula on $[0,10]$:

$$F(t) = -\min\{t; 10 - t\}, \qquad t \in [0,10]. \qquad (3.16)$$

Thus, by (3.15), the kernel K has sign changes at the points $0, \pm 5$, while the kernel F is nonpositive on $[-10, 10]$. The extremal function h^* in the problem

$$\int_{-10}^{10} h(t)K(t)\,dt \rightarrow \sup, \qquad h \in H_0^{1/2}[-10, 10], \tag{3.17}$$

is odd and admits the following representation on $[0, 10]$ (see Figure 4.3.1):

$$\frac{d}{dt}h^*(t) = \begin{cases} 2^{-\frac{1}{2}}t^{\frac{1}{2}}, & 0 < t \leq 1; \\ 2^{-\frac{1}{2}}(5-t)^{\frac{1}{2}} - 1, & 1 \leq t < 5; \\ -2^{-\frac{1}{2}}(t-5)^{\frac{1}{2}} - 1, & 5 \leq t < 9; \\ -2^{-\frac{1}{2}}t^{\frac{1}{2}}, & 9 < t \leq 10. \end{cases} \tag{3.18}$$

As can be seen from (3.18), the function $\dfrac{d}{dt}h^*(t)$ does have two points of sign change ± 1 located between the neighboring points -5 and 0, 0 and 5 of sign change of the kernel K. The general relations between the points of sign change of $\dfrac{d}{dt}x_{\omega,\psi}$ and ψ have already been described in (2.2.18).

REMARK 4.3.1. The extremal V_4^0-partition of the interval $[-10, 10]$ (as in Theorem X) is as follows:

$$\begin{aligned} B_{14} &= [-10, -9], & A_1 &= [-9, -1], & A_2 &= [-1, 1], \\ A_3 &= [1, 9], & C_{41} &= [9, 10]. \end{aligned} \tag{3.19}$$

All other intervals of the V_4^0-partition degenerate into points. Figure 4.3.1 also illustrates the graph of the function $h^*(t)$ and the extremal rearrangement $\Re_{\omega_{1/2}}(F; t)$.

The following remark indicates how to derive the solution of the Kolmogorov–Landau problem on $\mathbb{R}$ and $\mathbb{R}_+$ by using the Chebyshev splines.

REMARK 4.3.2. Let us introduce the rescaled Chebyshev splines $\{\hat{T}_n\}_{n \geq r}$ of the norm 1:

$$\hat{T}_n(t) = \rho_n^{-1} T_n\left(\rho_n^{\frac{1}{r+1}} t\right), \qquad 0 \leq t \leq \rho_n^{-\frac{1}{r+1}}, \tag{3.20}$$

where $\rho_n := \|T_n\|_{C[0,1]}$. By taking the limit as $n \rightarrow \infty$, we obtain the extremal Chebyshev spline in the Kolmogorov inequalities on the entire half-line $\mathbb{R}_+$.

Let $\{Z_n\}_{n \geq r}$ be the Chebyshev splines on the interval $[-1, 1]$, and $\{\hat{Z}_n\}_{n \geq r}$ be the corresponding rescaled Chebyshev splines of the norm 1:

$$Z_n(t) = 2^{r+1} T_n(1/2 \cdot t + 1/2), \qquad -1 \leq t \leq 1; \tag{3.21}$$

$$\hat{Z}_n(t) = \rho_n^{-1} Z_n(\rho_n^{\frac{1}{r+1}} t), \qquad -\rho_n^{-\frac{1}{r+1}} \leq t \leq \rho_n^{-\frac{1}{r+1}}. \tag{3.22}$$

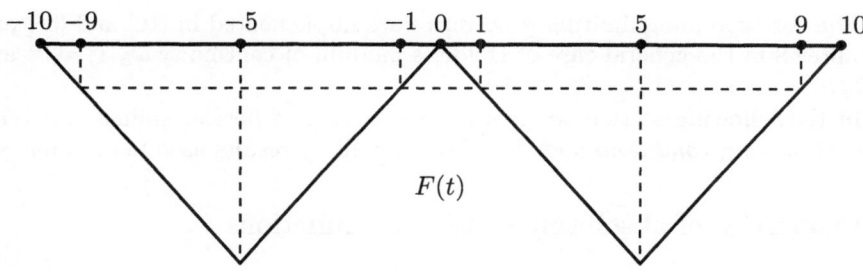

$F(t)$

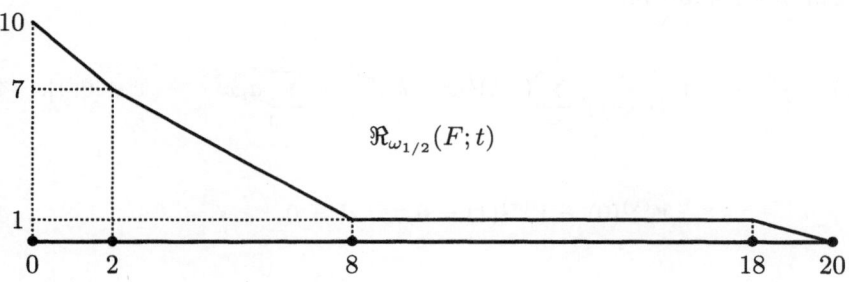

$\Re_{\omega_{1/2}}(F;t)$

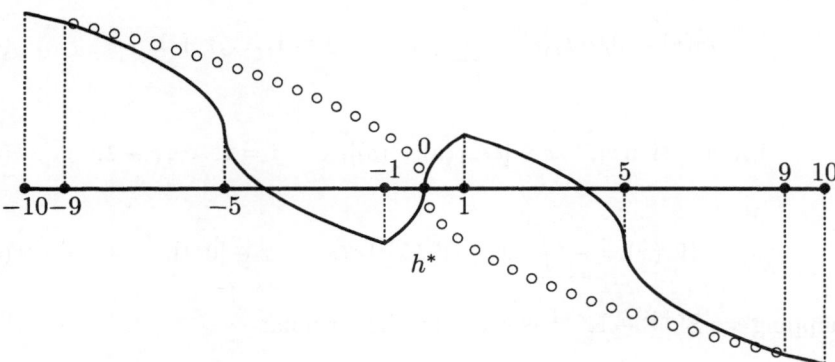

h^*

FIGURE 4.3.1. Kernel F, its rearrangement $\Re_{\omega_{1/2}}(F;\cdot)$ and the function h^*

The extremal Euler splines in the original Kolmogorov problem

$$\|f^{(m)}\|_{L_\infty(\mathbb{R})} \to \sup, \qquad f \in W^r_\infty(\mathbb{R}), \quad \|f\|_{L_\infty(\mathbb{R})} \le 1$$

can be obtained as limits of the sequences $\{\hat{Z}_{2n}\}_{n\in\mathbb{N}}$ for odd m and $\{\hat{Z}_{2n+1}\}_{n\in\mathbb{N}}$ for even m.

The corresponding limiting procedures are implemented in [19] and [38] and in Chapter 8 in the general case of Hölder's modulii of continuity $\omega_\alpha(t) = t^\alpha$ and $I = \mathbb{R}_+$.

In the following section we give a construction of perfect splines satisfying *the zero boundary conditions* and also obtain auxiliary results used in our analysis.

4.4. Properties of absolutely continuous functions

LEMMA 4.4.1. *For any integer $r \geq 0$, there exists such a perfect spline of degree $r + 1$ with $r + 1$ knots on $[0, 1]$,*

$$Y_r(x) = \frac{x^{r+1}}{(r+1)!} + \frac{1}{(r+1)!} \sum_{j=1}^{r+1} (-1)^j (x - t_j)_+^{r+1} + \sum_{i=0}^{r} a_i x^i, \qquad x \in [0,1], \quad (4.1)$$

that

$$Y_r^{(k)}(0) = Y_r^{(k)}(1) = 0, \qquad k = 0, \ldots, r. \qquad (4.2)$$

Proof. Let $\mathbb{S}^{r+1} := \{ s = (s_1, \ldots, s_{r+2}) \in \mathbb{R}^{r+2} \mid \sum_{i=1}^{r+2} |s_i| = 1 \}$.

Given an $s = (s_1, \ldots, s_{r+2}) \in \mathbb{S}^{r+1}$, we generate a subpartition of the interval $[0,1]$:

$$t_0(s) = 0, \quad t_j(s) = \sum_{i=1}^{j} |s_j|, \qquad j = 1, \ldots, r + 1. \qquad (4.3)$$

Let

$$V_s(x) = \operatorname{sign} s_i, \quad x \in [t_{j-1}(s), t_j(s)], \qquad j = 1, \ldots, r + 1. \qquad (4.4)$$

and

$$W_s(x) = \frac{1}{r!} \int_0^1 (x - t)_+^r V_s(t)\, dt, \qquad x \in [0, 1]. \qquad (4.5)$$

The mapping $\varkappa : \mathbb{S}^{r+1} \to \mathbb{R}^{r+1}$ is given by the formula:

$$\varkappa(s) := \left(W_s(1), W_s^{(2)}(1), \ldots, W_s^{(r)}(1) \right). \qquad (4.6)$$

Clearly, $\varkappa(s)$ is a continuous and odd function of s on the $\mathbb{S}^{r+1}$.

By the Borsuk Theorem 1.1.1, there exists an $s^* \in \mathbb{S}^{r+1}$ such that $\varkappa(s^*) = 0$. Now we can put

$$Y_r(t) = W_{s^*} \cdot \operatorname{sign} s_1^*. \qquad (4.7)$$

□

REMARK 4.4.1. If Y_r is a perfect spline from Lemma 4.4.1, then

$$\mu\left(Y_r^{(k)}; (0,1)\right) = k, \tag{4.8}$$

where, as before, $\mu\left(Y_r^{(k)}; [0,1]\right)$ stands for the number of sign changes of the function $Y_r^{(k)}(x)$ on the open interval $(0,1)$. The property (4.8) can be easily verified by the standard application of Rolle's Proposition 1.1.3:

$$r \geq \mu\left(Y_r^{(r)}; (0,1)\right) \geq \cdots \geq \mu\left(Y_r^{(k)}; (0,1)\right) + r - k \geq \cdots \geq \mu\left(Y_r; (0,1)\right) + r \geq r. \tag{4.9}$$

The proof of Corollary 4.4.2 below employs the following fact.

Let $L, M \in \mathbb{N}$, $A > 0$, and

$$\Gamma_L := \left\{ (d_1, \ldots, d_L) \in \mathbb{R}^L \;\middle|\; d_1 \geq 0, \ldots, d_L \geq 0, \;\; \sum_{i=1}^{L} d_i = A \right\}. \tag{4.10}$$

Then

$$\min_{(d_1,\ldots,d_L) \in \Gamma_L} \sum_{l=1}^{L} d_l^M = \frac{A^M}{L^{M-1}}, \tag{4.11}$$

and the minimum is attained on the vector $\hat{d} = (\frac{A}{L}, \ldots, \frac{A}{L})$.

COROLLARY 4.4.2. *Let* $f \in \mathbb{AC}^r[a,b]$ *and the points* $\{\sigma_i\}_{i=0}^N$ *be such that* $a = \sigma_0 < \sigma_1 < \cdots < \sigma_N$ *and*

$$(-1)^i f^{(r)}(x) \geq P > 0, \quad \text{for a.e. } x \in [\sigma_{i-1}, \sigma_i], \quad i = 1, \ldots, N. \tag{4.12}$$

Then,

$$\|f\|_{\mathbb{L}_1[a,b]} \geq \frac{d_r(b-a)^{r+1}}{N^r}, \tag{4.13}$$

where $u_r = \|Y_r\|_{\mathbb{L}_1[0,1]}$.

Proof. Put

$$\delta_l = \sigma_l - \sigma_{l-1}, \quad l = 1, \ldots, N. \tag{4.14}$$

Notice that $\sum_{l=1}^{N} \delta_l = b - a$.

For each $l = 1, \ldots, N$, we define perfect splines Λ_l as follows:

$$\Lambda_l(t) = \delta_l^r \, Y_r\left(\frac{t - \sigma_{l-1}}{\delta_l}\right), \quad \sigma_{l-1} \leq t \leq \sigma_l, \quad l = 1, \ldots, N, \tag{4.15}$$

where $Y_r(t)$ is a perfect spline from Lemma 4.4.1. By (4.9), $\Lambda_l(t) > 0$, $t \in (\sigma_{l-1}, \sigma_l)$ for $l = 1, \ldots, N$. Using these inequalities, the boundary conditions (4.2), the property (4.12), and the fact (4.11) for $L = N$, $M = r + 1$, $A = b - a$, we obtain the chain of inequalities:

$$\|f\|_{\mathbb{L}_1[a,b]} \geq (-1)^r \sum_{l=1}^{N} (-1)^l \int_{\sigma_{l-1}}^{\sigma_l} f(t) \Lambda_l^{(r)}(t)\, dt = \sum_{l=1}^{N} \int_{\sigma_{l-1}}^{\sigma_l} (-1)^l f^{(r)}(t) \Lambda_l(t)\, dt \geq$$

$$\geq P \sum_{l=1}^{N} \int_{\sigma_{l-1}}^{\sigma_l} \Lambda_l(t)\, dt = P \sum_{l=1}^{N} \delta_l^{r+1} u_r \geq \frac{P u_r (b-a)^{r+1}}{N^r}, \quad (4.16)$$

$\square$

Corollary 4.4.2 and the following result are used in the proof of the construction of Chebyshev ω-splines with a given number of alternance points.

COROLLARY 4.4.3. *Let a function* $f \in \mathbb{AC}^{r-1}[a,b]$ *have precisely n distinct zeroes* $\{\tau_i\}_{i=1}^{n}$ *and* $f^{(r)}$ *have precisely $n-r$ points of sign change* $\{\sigma_i\}_{i=1}^{n-r}$*, arranged in the increasing order. If*

$$(A) \quad \|f'\|_{\mathbb{C}[a,b]} \leq M, \quad and \quad |f^{(r)}(t)| \geq P, \quad for\ a.e.\ t \in [a,b],$$
$$(B) \quad \tau_{i+1} - \tau_i > \delta > 0, \quad i = 1, \ldots, n-1, \qquad (4.17)$$

then there exists such a constant $\hat{\delta} = \hat{\delta}(n, r, M, P, \delta)$ *that*

$$\sigma_i > \tau_i + \hat{\delta}, \quad i = 1, \ldots, n - r. \qquad (4.18)$$

Proof. By the Rolle theorem, each of the derivatives $f^{(l)}$, $l = 0, \ldots, r$, has precisely $n - l$ points of sign change at some monotonely arranged points $\{\eta_i^l\}_{i=1}^{n-l}$. Notice that $\eta_i^0 = \tau_i$, $i = 1, \ldots, n$, and $\eta_i^r = \sigma_i$, $i = 1, \ldots, n - r$.

The Rolle theorem also gives us inequalities between the points of sign change of the consecutive derivatives:

$$\eta_i^l < \eta_i^{l+1} < \eta_{i+1}^l, \quad i = 1, \ldots, n - l. \qquad (4.19)$$

In particular, $f'(t)$ does not change its sign on each of the intervals (τ_i, η_i^1) and (η_i^1, τ_{i+1}), $i = 1, \ldots, n - 1$. Because $f(\tau_i) = 0$, $i = 1, \ldots, n - 1$, we have the following equalities:

$$\int_{\tau_i}^{\eta_i^1} |f'(t)|\, dt = |f(\eta_i^1)| = \int_{\eta_i^1}^{\tau_{i+1}} |f'(t)|\, dt. \qquad (4.20)$$

Therefore,

$$\int_{\tau_i}^{\eta_i^1} |f'(t)|\, dt = \frac{1}{2}|f'|_{L_1[\tau_i, \tau_{i+1}]}, \qquad i = 1, \dots, n-1. \tag{4.21}$$

Put $L := \dfrac{Pu_r}{(n-r+1)^r}$, where u_r is defined in Corollary 4.4.2. From Corollary 4.4.2 and our assumption (4.17), (B) it follows that for all $i = 1, \dots, n-1$,

$$\|f'\|_{L_1[\tau_i, \tau_{i+1}]} \geq L(\tau_{i+1} - \tau_i)^r \geq L\delta^r. \tag{4.22}$$

On the other hand, by our assumption (4.17), (A), for $i = 1, \dots, n-1$,

$$\int_{\tau_i}^{\eta_i^1} |f'(t)|\, dt \leq \|f'\|_{\mathbb{C}[a,b]}(\eta_i^1 - \tau_i) \leq M(\eta_i^1 - \tau_i). \tag{4.23}$$

Combining the identity (4.21) and the estimates (4.22) from below and (4.23) from above, we come to the conclusion that

$$\eta_i^1 - \tau_i \geq \frac{L\delta^r}{M} =: \hat{\delta}, \qquad i = 1, \dots, n-1. \tag{4.24}$$

Finally, the Rolle inequalities (4.19) and (4.24) imply that

$$\eta_i^r > \eta_i^1 \geq \tau_i + \hat{\delta}, \tag{4.25}$$

$\square$

Properly modifying notations in [61], we introduce the functional classes $\{H_m[a,b]\}_{m \in \mathbb{N}}$.

DEFINITION 4.4.1. *Let $m \in \mathbb{N}$. The function h belongs to the class $H_m[a,b]$ provided that there exists a collection of $m+1$ points $\{\tau_i\}_{i=0}^{m}$, $a = \tau_0 < \tau_1 < \cdots < \tau_m = b$ such that*

(i) $h \in \mathbb{C}[a,b]$;

(ii) $h\big|_{[\tau_{j-1}, \tau_j]} \in \mathbb{AC}^1[\tau_{j-1}, \tau_j], \qquad j = 1, \dots, m;$ \hfill (4.26)

(iii) $\operatorname{sign} h''(x) = (-1)^j, \qquad$ *for a.e.* $x \in (\tau_{j-1}, \tau_j), \quad j = 1, \dots, m.$

The derivation of the estimate from above for the n-widths of Sobolev classes will be based on the following property of the class $H_m[a,b]$ (see [61]).

LEMMA 4.4.4. *The following inequality holds for the number $Z(h; [\tau_0, \tau_j])$ of zeroes of any function $h \in H_m[a,b]$ on the interval $[\tau_0, \tau_j]$:*

$$Z(h; [\tau_0, \tau_j]) \leq j + 1, \qquad j = 1, \dots, m. \tag{4.27}$$

If (4.27) holds as the equality, then $(-1)^j h(\tau_j) \geq 0$.

Proof. The proof is carried out by the induction in j. The assertion (4.27) holds for $j = 1$, since h is strictly concave on $[\tau_0, \tau_1]$.

Let the property (4.27) hold for some $j \in \{1, \ldots, m+1\}$. For definiteness, let j be odd. Two possibilities arise.

I. $Z(h; [\tau_0, \tau_j]) \leq j$. If $h(\tau_j) > 0$, then $Z(h; [\tau_0, \tau_{j+1}]) \leq j+2$, and the equality $Z(h; [\tau_0, \tau_{j+1}]) = j+2$ implies that $h(\tau_{j+1}) \geq 0$. This observation is a reflection of the fact that h is strictly convex on $[\tau_j, \tau_{j+1}]$, so h can have one or two simple zeroes or one double zero. If $h(\tau_j) \leq 0$, then $Z(h; [\tau_0, \tau_{j+1}]) \leq j+1$.

II. $Z(h; [\tau_0, \tau_j]) = j+1$. By the inductive assumption, $h(\tau_j) \leq 0$. Therefore, $Z(h; [\tau_0, \tau_{j+1}]) \leq j+2$, and the equality holds only if $h(\tau_{j+1}) \geq 0$.   □

4.5. Sharp Kolmogorov inequalities in $W_\infty^r(\mathbb{R})$ and Cavaretta's proof

Let $m, r \in \mathbb{N} : 1 \leq m < r - 1$. In this section we give the proof of Kolmogorov's inequalities (0.1.5) due to A. S. Cavaretta [18].

Let w_r be the 2π-periodic function on $\mathbb{R}$ such that

$$w_r^{(r)}(t) = \operatorname{sign}\sin t, \qquad t \in \mathbb{R}, \qquad \int\limits_{-\pi}^{\pi} w_r(t)\, dt.$$

REMARK 4.5.1. The function $w_r(t)$ is the Euler spline $f_{1,r-1}(\widehat{\omega}; t)$ for $\widehat{\omega}(t) = t$, where the Euler ω-splines $\{f_{n,r}(\omega, \cdot)\}_{n \in \mathbb{N}}$ are introduced in Definition 0.3.2.

We mention two properties of the function w_n:

Property A. $\|w_r^{(k)}\|_{\mathbb{L}_\infty(\mathbb{R})} = K_{r-k}, \qquad 0 \leq k \leq r,$

where the Favard constants $\{K_l\}_{l \in \mathbb{N}}$ are introduced in (0.1.5);

Property B. $w^{(r)}$ has $2l$-alternance on the interval $[-l\pi, l\pi]$, i.e.

it alternates $2l$ times between the values K_r and $-K_r$.

The Cavaretta's proof divides into two parts.

4.5.1. A property of periodic functions

First, we give the solution of the Kolmogorov problem in the periodic Sobolev classes.

LEMMA 4.5.1. *Let $l \in \mathbb{N}$, $g \in W_\infty^r(\mathbb{R})$ be a $2\pi l$-periodic function with all its derivatives up to the order $r - 1$, and $\|g\|_{\mathbb{C}[-l\pi, l\pi]} \leq K_r$. Then,*

$$\|g^{(m)}\|_{\mathbb{C}[-l\pi, l\pi]} \leq K_{r-m}. \tag{5.1}$$

Proof. We shall argue by contradiction. Let us assume that there exists a function F, satisfying the conditions of the lemma and such that

$$F^{(m)}(0) = \|F^{(m)}\|_{L_\infty(\mathbb{R})} > K_{r-m}. \tag{5.2}$$

If necessary, we can choose a $\beta \in \mathbb{R}$ so that the translated function $\bar{w}_r(t) := w_r(t + \beta)$ enjoys the property $w_r^{(m)}(0) = K_{r-m}$. Put

$$Y(t) := F(t) - \alpha \bar{w}_r(t), \quad t \in \mathbb{R}, \qquad \text{where } \alpha := F^{(m)}(0)/K_{r-m}. \tag{5.3}$$

Clearly, $\alpha > 1$, and

$$Y^{(m)}(0) = Y^{(m+1)}(0) = 0. \tag{5.4}$$

By Property B of the Euler function w_r, the number $\mu\left(Y; [-l\pi, l\pi]\right)$ of sign changes of the function Y on the interval $[-l\pi, l\pi]$ is no less than $2l$. Taking into account the periodicity of Y and its derivatives, and the Rolle theorem, we conclude that

$$\mu\left(Y^{(m)}; [-l\pi, l\pi]\right) \geq 2l. \tag{5.5}$$

Therefore, by (5.4) and (5.5), the number of zeroes of $Y^{(m+1)}$ and (by the Rolle theorem) $Y^{(r-1)}$ exceeds $2l$.

On the other hand, $|\alpha \bar{w}_r^{(r)}(t)| \equiv \alpha > 1$, and $|F^{(r)}(t)| \leq 1$ on each of the intervals $[k\pi + \beta, (k+1)\pi + \beta]$ for all $|k| \leq l$. Consequently, by (5.3), the function $Y^{(r-1)}$ is monotone on each of the intervals $[k\pi + \beta, (k+1)\pi + \beta]$ for $-l \leq k \leq l$, so

$$\mu\left(Y^{(r-1)}; [-l\pi, l\pi]\right) \leq 2l. \tag{5.6}$$

This contradiction with the previous finding shows that our assumption on the existence of the $2\pi l$-periodic function F with the property $\|F^{(m)}\|_{L_\infty(\mathbb{R})} > K_{r-m}$ was wrong. □

4.5.2. Reduction of the Kolmogorov problem to the periodic case

Suppose that there exists such a function $\hat{f} \in W_\infty^r(\mathbb{R})$ that $\|\hat{f}\|_{L_\infty(\mathbb{R})} \leq K_r$. By a translation and multiplication by -1, if necessary, we can arrange that $\hat{f}^{(m)}(0) > K_{r-m}$. Let $\zeta(t)$ be an infinitely differentiable function on $\mathbb{R}$ with the properties

$$
\begin{aligned}
&(i) \quad \text{supp}\,\zeta = [-1, 1]; \\
&(ii) \quad \zeta(t) \equiv 1, \quad t \in [-1/2, 1/2]; \\
&(iii) \quad 0 \leq \zeta(t) \leq 1, \quad t \in [-1, 1].
\end{aligned}
\tag{5.7}
$$

For a natural N, put $f_N(t) = \hat{f}(t)\zeta(t/N)$, $t \in \mathbb{R}$. The Leibnitz formula gives explicit formulas for the rth derivatives of f_N:

$$f_N^{(r)}(t) = \sum_{j=0}^r \binom{n}{j} \zeta^{r-j}(t/N) N^{-(r-j)}. \tag{5.8}$$

From (5.8) we derive the existence of numbers $j \in \mathbb{N}$ and $\kappa_m > 0$, $\kappa_r > 0$, such that

$$\|f_j^{(r)}\|_{L_\infty(\mathbb{R})} \le 1 + \kappa_r, \qquad f_j^{(m)}(0) > \gamma(1 - \kappa_m), \quad \text{where} \quad \frac{\gamma(1 - \kappa_m)}{1 + \kappa_r} > K_{r-m}. \tag{5.9}$$

By (5.9), $\|f_j\|_{L_\infty(\mathbb{R})} \le K_r$. Let $\rho(\cdot) := f_j(\cdot)(1 + \kappa_r)^{-1}$. Then, the function ρ belongs to the Sobolev class $W_\infty^r(\mathbb{R})$, and it can be extended to the entire line $\mathbb{R}$ with the period $2\pi j$ (since $\operatorname{supp} \rho = [-\pi j, \pi j]$). In addition, $\|\rho\|_{L_\infty(\mathbb{R})} \le K_r$ and $\rho^{(m)}(0) > K_{r-m}$. This contradiction with the assertion of Lemma 4.5.1 concludes the Cavaretta's proof of the Kolmogorov inequalities (0.1.5).

Chapter 5

Additive Kolmogorov–Landau Inequalities

In this chapter we first derive the numerical differentiation formulae of the form

$$f^{(m)}(0) = \sum_{i=0}^{n} \alpha_i f(\nu_i) + \int_0^1 f^{(r)}(t) K(t) \, dt.$$

Then we give sufficient conditions of extremality of a function $f \in W^r H^\omega[0,1]$ in the Kolmogorov–Landau inequalities.

5.1. Numerical differentiation formulae

Let $m, r : 0 < m \le r$, be integers, and $n \in \mathbb{N} : n \ge r$. H. Kallioniemi [34] discusses numerical differentiation formulae of the form

$$f^{(m)}(0) = \sum_{i=0}^{n} \alpha_i f(\nu_i) + \int_0^1 f^{(r+1)}(t) F(t) \, dt$$

for functions from the Sobolev class $W_\infty^{r+1}[0,1] = W^r H^1[0,1]$. In obtaining the formulas for the mth derivative of a function from $W^r H^\omega[0,1]$, we consider two different cases.

Case 1. $0 < m < r$.
Let collections of points $\bar{\nu} = \{\nu_i\}_{i=0}^n$ and $\bar{\vartheta} = \{\vartheta_i\}_{i=0}^{n-r+1}$ on $[0,1]$ be such that

$$0 =: \nu_0 < \nu_1 < \cdots < \nu_n;$$

$$0 =: \vartheta_0 < \vartheta_1 < \cdots < \vartheta_{n-r+1} < \vartheta_{n-r+2} := \nu_n; \tag{1.1}$$

$$\nu_{i-1} < \vartheta_i < \nu_{i+r-1}, \qquad i = 1, \ldots, n-r+1.$$

Let the coefficients $\{\alpha_i\}_{i=0}^n$ satisfy the system of linear equations

$$\begin{cases} \displaystyle\sum_{i=0}^{n} \alpha_i \nu_i{}^j = m! \delta_{m,j}, & j = 0, \ldots, r-1; \\[4mm] \displaystyle\sum_{i=0}^{n} \alpha_i (\nu_i - \vartheta_l)_+^{r-1} = 0, & l = 1, \ldots, n-r+1. \end{cases} \tag{1.2}$$

The inequalities between the points $\{\nu_i\}_{i=0}^n$ and $\{\vartheta_i\}_{i=1}^{n-r+1}$ in (1.1) guarantee that the system of equations (3.2) has a unique solution.

Then we introduce kernels $K(t)$ and $F(t)$:

$$K(t) = -\frac{1}{(r-1)!} \sum_{i=0}^n \alpha_i (\nu_i - t)_+^{r-1}, \quad 0 \le t \le \nu_n.$$

$$F(t) = \int_{\nu_n}^t K(y)\, dy = \frac{1}{r!} \sum_{i=0}^n \alpha_i (\nu_i - t)_+^r, \quad 0 \le t \le \nu_n. \tag{1.3}$$

From the system of linear equations (1.2) it follows that

$$K^{(i)}(\nu_n) = 0, \qquad i = 0, \ldots, r-2.$$

$$K^{(i)}(0) = \frac{(-1)^{i+1}}{(r-i)!} \sum_{j=0}^n \alpha_j \nu_j^{\,r-i} = (-1)^{r-m} \delta_{r-1-i,m}, \qquad i = 0, \ldots, r-2. \tag{1.4}$$

Next, the Taylor formula reads:

$$f(\tau) = \sum_{j=0}^{r-1} \frac{f^{(j)}(0)}{j!} \tau^j + \frac{1}{(r-1)!} \int_0^{\nu_n} f^{(r)}(x)(\tau - x)_+^{r-1}\, dx, \tag{1.5}$$

for any function $f \in \mathbb{C}^r[0,1]$.

Therefore,

$$\sum_{i=1}^n \alpha_i f(\nu_i) = \sum_{j=0}^{r-1} \left[\sum_{i=1}^n \alpha_i \nu_i^j \right] \frac{f^{(j)}(0)}{j!} + \frac{1}{(r-1)!} \int_0^{\nu_n} f^{(r)}(x) \sum_{i=1}^n \alpha_i (\nu_i - x)_+^{r-1}\, dx. \tag{1.6}$$

From the identity (1.6) and equations (1.2) we obtain the formula for the mth derivative of the function f at the origin:

$$f^{(m)}(0) = \sum_{i=0}^j \alpha_i f(\nu_i) + \int_0^{\nu_n} f^{(r)}(x) K(x)\, dx, \tag{1.7}$$

where the kernel K is defined in (1.2) and (1.3).

Thus, for any function $f \in W^r H^\omega[0,1]$, the following estimate holds:

$$|f^{(m)}(0)| \le \sum_{i=0}^n |\alpha_i| \cdot \|f\|_{\mathbb{C}[0,1]} + \sup_{h \in H^\omega[0,\nu_n]} \int_0^{\nu_n} h(x) K(x)\, dx. \tag{1.8}$$

In Proposition 3.1.1 of we listed the properties of the kernel $K(t)$:

(i) supp $K = [0, \nu_j]$,   for some $j : r - 1 \le j \le n$;

(ii) the kernel K has precisely $j - r + 1$ simple zeroes

$$\{\vartheta_i\}_{i=1}^{j-r+1} \text{ on the interval } (0, \nu_j); \tag{1.9}$$

(iii) sign $\alpha_i = (-1)^{i+m}$, $i = 0, \ldots, j$;

(iv) $(-1)^{i+r+m}$ sign $K(t) \ge 0$, $\vartheta_i \le t \le \vartheta_{i+1}$, $i = 0, \ldots, n - r + 1$.

In addition, let us assume that the kernel K has the zero mean on $[0, \nu_n]$:

$$\int_0^{\nu_n} K(t)\, dt = -\frac{1}{r!} \sum_{i=0}^{n} \alpha_i \nu_i{}^r = 0. \tag{1.10}$$

Then, by Corollary 3.1.2, supp $K = [0, \nu_j]$ for some $j \ge r$.

By (1.9) and (1.10), $(-1)^{r+m+1} K(x) \in \mathcal{M}_{j-r+2}^0 [0, \nu_j]$, where classes $\mathcal{M}_l^K[a, b]$ for $l \in \mathbb{N}$ and $K \in \{-1, 0, +1\}$, are introduced in Definition 2.2.1. Since the kernel $K(t)$ has the zero mean on $[0, 1]$,

$$\sup_{h \in H^\omega [0, \nu_j]} \int_0^{\nu_j} h(t) K(t)\, dt =$$

$$= \sup_{h \in H^\omega [0, \nu_j]} \int_0^{\nu_j} [h(t) - h(0)] K(t)\, dt = \sup_{h \in H_0^\omega [0, \nu_j]} \int_0^{\nu_j} h(t) K(t)\, dt, \tag{1.11}$$

Theorem X describes the structure of the function realizing the supremum in (1.11). Finally, from (1.8) and Corollary 2.2.3 of Theorem X we derive the following estimate for the value of the mth derivative of the function $f \in W^r H^\omega[0, 1]$ at the origin:

$$|f^{(m)}(0)| \le \sum_{i=0}^{j} |\alpha_i| \cdot \|f\|_{C[0,1]} + \int_0^1 \Re_\omega (F; t)\, \omega'(t)\, dt, \tag{1.12}$$

where $\Re_\omega (F; t)$ is the rearrangement of the kernel $F(t)$ as defined in (2.20) of Definition 2.2.4.

Case 2. m=r.
Let collections of points $\bar\nu = \{\nu_i\}_{i=0}^n$ and $\bar\vartheta = \{\vartheta_i\}_{i=0}^{n-r}$ on $[0, 1]$ satisfy inequalities

$$0 =: \nu_0 < \nu_1 < \cdots < \nu_n < \nu_n;$$
$$0 =: \vartheta_0 < \vartheta_1 < \cdots < \vartheta_{n-r} < \vartheta_{n-r+1} := \nu_n; \tag{1.13}$$
$$\nu_{i-1} < \vartheta_i < \nu_{i+r-1}, \qquad i = 1, \ldots, n - r.$$

Let coefficients $\{\alpha_i\}_{i=0}^n$ solve the system of linear equations

$$\begin{cases} \displaystyle\sum_{i=0}^n \alpha_i \nu_i{}^j = 0, & j = 0, \ldots, r-1; \\[2ex] \displaystyle\sum_{i=0}^n \alpha_i (\nu_i - \vartheta_l)_+^{r-1} = 0, & l = 1, \ldots, n-r; \\[2ex] \displaystyle\sum_{i=0}^n \alpha_i \nu_i{}^r = r! \end{cases} \tag{1.14}$$

The last equation in (1.14) is added for the normalization of coefficients $\{\alpha_i\}_{i=0}^n$. Kernels $K(t)$ and $F(t)$ are introduced as follows:

$$K(t) = -\frac{1}{(r-1)!} \sum_{i=0}^n \alpha_i (\nu_i - t)_+^{r-1},$$

$$F(t) = \int_{\nu_n}^t K(y)\,dy = \frac{1}{r!} \sum_{i=0}^n \alpha_i (\nu_i - t)_+^r. \tag{1.15}$$

As in the case $0 < m < r$, Taylor's formula enables us to derive the numerical differentiation formula for $f^{(r)}(0)$:

$$f^{(r)}(0) = \sum_{i=0}^j \alpha_i f(\nu_i) + \int_0^1 [f^{(r)}(t) - f^{(r)}(0)] K(t)\,dt. \tag{1.16}$$

In Proposition 3.2.1 we mentioned the following properties of the kernel K:

   (i) $\operatorname{supp} K = [0, \nu_j]$,   for some $j : r \le j \le n$;

   (ii) the kernel K has precisely $j - r$ simple zeroes

   $\{\vartheta_i\}_{i=1}^{j-r}$ on the interval $(0, \nu_j)$; $\qquad\qquad\qquad\qquad\qquad$ (1.17)

   (iii) $\operatorname{sign} \alpha_i = (-1)^{i+r}$, $i = 0, \ldots, j$;

   (iv) $(-1)^i \operatorname{sign} K(t) \ge 0$, $\vartheta_i \le t \le \vartheta_{i+1}$, $i = 0, \ldots, n-r$.

Thus, by (1.17), $K(t) \in -\mathcal{M}_{j-r+1}^{-1}[0, \nu_j]$, where classes $\mathcal{M}_l^{-1}[a, b]$, $l \in \mathbb{N}$, are introduced in Definition 2.2.1. Theorem X provides the formula for the extremal function of the problem

$$\int_0^{\nu_j} h(t) K(t)\,dt \to \sup, \qquad h \in H_0^\omega[0, \nu_j]. \tag{1.18}$$

Corollary 2.2.3 and (1.16) lead us to the estimate for the value of the rth derivative of the function $f \in W^r H^\omega[0,1]$ at the origin:

$$|f^{(r)}(0)| \le \sum_{i=0}^j |\alpha_i| \cdot \|f\|_{C[0,1]} + \int_0^1 \Re_\omega(F;t)\omega'(t)\,dt. \tag{1.19}$$

5.2. Sufficient conditions of extremality

Now we describe conditions for the function $\mathcal{Z} \in W^r H^\omega[0,1]$ and the kernel K to be *extremal* in the inequalities (1.12) for $0 < m < r$ and (1.19) for $m = r$, i.e., to *transform those inequalities into equalities.* The analysis of formulae (1.7) and (1.16) for $f^{(m)}(0)$ shows that the function $\mathcal{Z}$ is extremal in the inequalities (1.12) for $0 < m < r$ and (1.19) for $m = r$, if the following two conditions are satisfied for $\chi = 1$ or -1:

$$\text{(i)} \quad \mathcal{Z}(\nu_i) = (-1)^{i+m} \chi \|\mathcal{Z}\|_{\mathbb{C}[0,1]}, \qquad i = 0, \ldots, n;$$

$$\text{(ii)} \quad \sup_{h \in H_0^\omega[0,1]} \int_0^1 h(x) K(x)\, dx = \chi \int_0^1 [\mathcal{Z}^{(r)}(x) - \mathcal{Z}^{(r)}(0)] K(x)\, dx; \tag{2.1}$$

where the kernel $K(t)$ is defined by (1.2), (1.3) for $0 < m < r$, and by (1.14), (1.15) for $m = r$. In Theorem 6.0.1 in the following chapter we prove the existence of such collections of points $\bar{\nu}$ and $\bar{\vartheta}$ that the function $\mathcal{Z} = \mathcal{Z}_{\omega,n,r,m}$ has the complete set of $n+2$ alternance points $\{\nu_i\}_{i=0}^n$ and $\nu_{n+1} := 1$, and enjoys properties (2.1), (ii).

In [15], for each positive B we construct functions Z_B with $\|Z_B\|_{\mathbb{C}[0,1]} = B$ and endowed with the properties (2.1) for some $n = n(B,r,m,\omega)$ and collections of points $\bar{\nu} = \bar{\nu}(B,r,m,\omega)$ and $\bar{\vartheta} = \bar{\vartheta}(B,r,m,\omega)$. These functions, called *Zolotarev ω-splines,* solve the problem of sharp Kolmogorov–Landau inequalities in classes $W^r H^\omega[0,1]$ (see also Karlin's constructiob in [37] case for $\omega(t) = t$).

In Chapter 8 we use the following coarse estimates of the norms $\|f^{(m)}\|_{\mathbb{L}_\infty(\mathbb{R}_+)}$ of functions $f \in W^r H^\omega(\mathbb{R}_+)$ in terms of $\|f\|_{\mathbb{L}_\infty(\mathbb{R}_+)}$.

LEMMA 5.2.1. *Let $f \in W^r H^\omega(\mathbb{R}_+)$. Then for all $m = 1, \ldots, r$ there exist such constants $D_{r,m}$ and $E_{r,m,\omega}$ that*

$$\|f^{(m)}\|_{\mathbb{L}_\infty(\mathbb{R}_+)} \le D_{r,m} \|f\|_{\mathbb{L}_\infty(\mathbb{R}_+)} + E_{r,m,\omega}. \tag{2.2}$$

Proof. By Proposition 1.2.2, in order to obtain an inequality for the norm $\|f^{(m)}\|_{\mathbb{L}_\infty(\mathbb{R}_+)}$, it is sufficient to estimate $f^{(m)}(0)$.

Fix any of the kernels $K(t)$ defined by (1.2), (1.3) for $0 < m < r$, or defined by (1.14), (1.15), if $m = r$. Then estimates (2.2) follow from inequalities (1.12) for $0 < m < r$ and (1.19) for $m = r$ with

$$D_{r,m} := \sum_{i=0}^n |\alpha_i|, \qquad E_{r,m,\omega} := \int_0^1 \Re_\omega(F; t)\, \omega'(t)\, dt. \tag{2.3}$$

$\square$

Chapter 6

Proof of the Main Result

6.0. Formulation of the main result

THEOREM 6.0.1. *Let* m, $r \in \mathbb{N}$, $0 < m \le r$; $n \in \mathbb{N}$, $n \ge r$, *and* ω *be a concave modulus of continuity. There exist collections of points* $\bar{\nu} = \bar{\nu}(B, n, r, m, \omega)$ *and* $\bar{\vartheta} = \bar{\vartheta}(B, n, r, m, \omega)$ *as in* (5.1.1) *for* $0 < m < r$ *and as in* (5.1.13) *for* $m = r$, *and the function* $\mathcal{Z}_n = \mathcal{Z}_{n,r,m,\omega}$, *endowed with the properties*

$$(i) \quad \mathcal{Z}_n(\nu_i) = (-1)^{i+m} \|\mathcal{Z}_n\|_{C[0,1]}, \qquad i = 0, \ldots, n+1;$$

$$(ii) \quad \sup_{h \in H_0^\omega[0,1]} \int_0^1 h(x) K(x) \, dx = \int_0^1 [\mathcal{Z}_n^{(r)}(x) - \mathcal{Z}_n^{(r)}(0)] K(x) \, dx; \qquad (0.1)$$

where the coefficients $\{\alpha_i\}_{i=0}^n$ *of the kernel* $K(t) = -\dfrac{1}{(r-1)!} \sum\limits_{i=0}^n \alpha_i (\nu_i - t)_+^{r-1}$ *satisfy equations* (5.1.2), (5.1.10) *for* $0 < m < r$, *and* (5.1.14) *for* $m = r$.

We give a complete proof of Theorem 6.0.1 and then point out the only distinction between the proofs of Theorems 6.0.1 for $0 < m < r$ and $m = r$.

Proof. We first prove Theorem 6.0.1 for *strictly concave* modulii of continuity ω on $[0, 1]$ endowed with the additional property

$$\omega_0 := \inf_{t \in [0,1]} \omega'(t) > 0. \qquad (0.2)$$

Then, we explain how to construct the extremal function $\mathcal{Z}_n(t)$ in the case of *arbitrary* modulii of continuity ω. We divide the proof into several steps.

6.1. Construction of the Borsuk mapping $\varkappa : \mathbb{S}^{2n-r+1} \longrightarrow \mathbb{R}^{2n-r+1}$

Let $\mathbb{S} = \mathbb{S}^{2n-r+1}$ be the sphere of radius 2:

$$\mathbb{S} := \left\{ s = (s_1, \ldots, s_{2n-r+2}) \in \mathbb{R}^{2n-r+2} \ \middle| \ \sum_{i=1}^{2n-r+2} |s_i| = 2 \right\}. \qquad (1.1)$$

For a given $s = (s_1, \ldots, s_{2n-r+2}) \in \mathbb{S}$, we generate four collections of points $\{t_i = t_i(s)\}_{i=0}^{2n-r+2}$, $\{\bar{t}_i = \bar{t}_i(s)\}_{i=0}^{2n-r+2}$, $\{\gamma_i = \gamma_i(s)\}_{i=0}^{2n-r+2}$ and $\{\bar{\gamma}_i = \bar{\gamma}_i(s)\}_{i=0}^{2n-r+2}$:

$$t_0(s) := 0; \qquad t_i(s) = \sum_{j=1}^{i} |s_j|, \qquad i = 1, \ldots, 2n - r + 2.$$

$$\bar{t}_i(s) = \min\{t_i(s), 1\}, \qquad i = 0, \ldots, 2n - r + 2. \qquad (1.2)$$

$$\gamma_i(s) = 2 - t_{2n-r+2-i}(s), \qquad i = 0, \ldots, 2n - r + 2.$$

$$\bar{\gamma}_i(s) = \min\{\gamma_i(s), 1\}, \qquad i = 0, \ldots, 2n - r + 2.$$

Next, we introduce the collection of points $\{\bar{\beta}_i = \bar{\beta}_i(s)\}_{i=0}^{n-r+1} : \bar{\beta}_0(s) := 0$,

$$\bar{\beta}_i(s) = \begin{cases} \bar{t}_{i-1}(s), & \text{if } \gamma_i(s) \leq \bar{t}_{i-1}(s), \\ \gamma_i(s), & \text{if } \bar{t}_{i-1}(s) < \gamma_i(s) < \bar{t}_{i+r-1}(s), \\ \bar{t}_{i+r-1}(s), & \text{if } \gamma_i(s) \geq \bar{t}_{i+r-1}(s), \end{cases} \quad i = 1, \ldots, n - r + 1. \quad (1.3)$$

The points $\{\bar{\beta}_i(s)\}_{i=1}^{n-r+1}$ are defined in such a way that for all $s \in \mathbb{S}$,

$$\bar{t}_{i-1}(s) \leq \bar{\beta}_i(s) \leq \bar{t}_{i+r-1}(s), \qquad i = 1, \ldots, n - r + 1. \qquad (1.4)$$

Notice that by (1.2) and (1.3),

$$\bar{t}_i(s) \in [0, 1], \quad i = 0, \ldots, 2n-r+2; \qquad \bar{\beta}_i(s) \in [0, 1], \quad i = 0, \ldots, n-r+1. \quad (1.5)$$

The set of points $\{\zeta_i = \zeta_i(s)\}_{i=1}^{2n-r+2}$ is the union of two collections of points $\{\bar{t}_i(s)\}_{i=0}^{n}$ and $\{\bar{\beta}_i(s)\}_{i=1}^{n-r+1}$ with the following *fixed order of enumeration*:

$$\begin{cases} \zeta_{2i-1}(s) = \bar{t}_{i-1}(s), & i = 1, \ldots, n-r+1; \\ \zeta_{2i}(s) = \bar{\beta}_i(s), & i = 1, \ldots, n-r+1; \\ \zeta_{i+n+2-r}(s) = \bar{t}_i(s), & i = n-r+1, \ldots, n; \end{cases} \qquad (1.6)$$

i.e. $(\zeta_1, \ldots, \zeta_{2n-r+2}) = (\bar{t}_0, \bar{\beta}_1, \bar{t}_1, \bar{\beta}_2, \bar{t}_2, \bar{\beta}_2, \ldots, \bar{\beta}_{n-r+1}, \bar{t}_{n-r+1}, \bar{t}_{n-r+2}, \ldots, \bar{t}_n)$.

Fix $\varepsilon > 0$. Let

$$\xi_i(s) = \frac{\zeta_i(s) + \varepsilon i}{1 + \varepsilon(2n - r + 2)}, \qquad i = 1, \ldots, 2n - r + 2. \qquad (1.7)$$

Then, we introduce the sets of points $\{\tau_i = \tau_i(s)\}_{i=0}^{n}$,

$$\begin{cases} \tau_{i-1}(s) = \xi_{2i-1}(s), & i = 1, \ldots, n-r+1; \\ \tau_i(s) = \xi_{i+n+2-r}(s), & i = n-r+1, \ldots, n. \end{cases} \qquad (1.8)$$

and $\{\beta_i = \beta_i(s)\}_{i=0}^{n-r+2}$:

$$\beta_i(s) = \xi_{2i}(s), \quad i = 1, \ldots, n-r+1; \quad \beta_0(s) := 0; \quad \beta_{n-r+2}(s) := \tau_n(s). \quad (1.9)$$

Notice that by (1.4) and definitions (1.6)–(1.9), we have the following inequalities between $\{\tau_i\}_{i=0}^{n}$ and $\{\beta_i\}_{i=1}^{n-r+1}$:

$$\tau_{i-1} = \frac{\bar{t}_{i-1} + \varepsilon(2i-1)}{1 + \varepsilon(2n-r+2)} \le \frac{\bar{\beta}_i + \varepsilon(2i-1)}{1 + \varepsilon(2n-r+2)} < \frac{\bar{\beta}_i + 2i\varepsilon}{1 + \varepsilon(2n-r+2)} = \beta_i, \quad (1.10)$$

and

$$\beta_i = \frac{\bar{\beta}_i + 2i\varepsilon}{1 + \varepsilon(2n-r+2)} \le \frac{\bar{t}_{i+r-1} + 2i\varepsilon}{1 + \varepsilon(2n-r+2)} < \tau_{i+r-1}, \quad (1.11)$$

for $i = 1, \ldots, n-r+1$. Therefore, for all $s \in \mathbb{S}$,

$$\tau_{i-1}(s) < \beta_i(s) < \tau_{i+r-1}(s), \quad i = 1, \ldots, n-r+1. \quad (1.12)$$

By definitions (1.6)–(1.9), all points $\{\tau_i\}_{i=1}^{n+1}$ lie on the interval $(0,1)$. Besides, by (1.7)–(1.9), both sets $\{\tau_i(s)\}_{i=1}^{n}$ and $\{\beta_i(s)\}_{i=1}^{n-r+2}$ are uniformly separated: for all $s \in \mathbb{S}$,

$$|\tau_i(s) - \tau_{i-1}(s)| \ge \frac{\varepsilon}{1 + \varepsilon(2n-r+2)}, \quad i = 1, \ldots, n.$$

$$|\beta_i(s) - \beta_{i-1}(s)| \ge \frac{\varepsilon}{1 + \varepsilon(2n-r+2)}, \quad i = 1, \ldots, n-r+2. \quad (1.13)$$

Let the coefficients $\{\alpha_i\}_{i=0}^{n}$ be derived from the system of linear equations

$$\begin{cases} \displaystyle\sum_{i=0}^{n} \alpha_i(s)[\tau_i(s)]^j = m! \delta_{m,j}, & j = 0, \ldots, r-1; \\ \displaystyle\sum_{i=0}^{n} \alpha_i(s)(\tau_i(s) - \beta_l(s))_+^{r-1} = 0, & l = 1, \ldots, n-r+1. \end{cases} \quad (1.14)$$

The conditions (1.12) guarantee that the matrix of the system (1.14) is nonsingular, so the system of linear equations (1.14) is uniquely solvable.

Then, we introduce kernels $K_s(t)$,

$$K_s(t) = -\frac{1}{(r-1)!} \sum_{i=1}^{n} \alpha_i(s)(\tau_i(s) - t)_+^{r-1}. \quad (1.15)$$

and $\widehat{K}_s(t)$:

$$\widehat{K}_s(t) = \operatorname{sign} s_{2n-r+3-i} \left(|K_s(t)| + \varepsilon \right), \quad \bar{\gamma}_{i-1}(s) < t < \bar{\gamma}_i(s), \quad (1.16)$$

$i = 1, \ldots, 2n - r + 2$. By (1.16), $\widehat{K}_s$ can have at most $2n - r + 1$ sign changes on the interval $[0, 1]$. Therefore, the kernel $\widehat{K}_s$ belongs to the class $M_{2n-r+2}[0, \tau_n(s)]$ in notations of Definition 2.2.5. Let

$$I_s := \int\limits_0^{\tau_n(s)} \widehat{K}_s(t)\, dt. \tag{1.17}$$

Next, we define $G_s(t)$ on $[0, \tau_n(s)]$ as the extremal function of the problem

$$\int\limits_0^{\tau_n(s)} h(t)\widehat{K}_s(t)\, dt \to \sup, \qquad h \in H_0^\omega[0, \tau_n(s)]. \tag{1.18}$$

The formulas for the function $G_s(t)$ are given in Theorem X in Chapter 2. We extend the function $G_s(t)$ to the interval $[\tau_n(s), 1]$ by continuity and the following formulas for the derivative:

$$\frac{d}{dt}G_s(t) = \operatorname{sign} s_{2n-r+3-i}\,\omega'(t), \qquad t \in [\gamma_{i-1}(s), \gamma_i(s)] \cap [\tau_n(s), 1], \tag{1.19}$$

where $i = 0, \ldots, 2n - r + 2$. Consecutive applications of Lemma 1.2.4 enable us to conclude that $G_s \in H^\omega[0, 1]$.

Let $W_s(x)$ be the $(r - 1)^{\text{st}}$ integral of $G_s(x)$,

$$W_s(x) = \frac{1}{(r-2)!} \int\limits_0^1 G_s(t)(x - t)_+^{r-2}\, dt, \qquad x \in [0, 1], \tag{1.20}$$

which vanishes with all its $r - 2$ derivatives $\{W_s^{(k)}\}_{k=1}^{r-2}$ at the origin.

Then we introduce the polynomial $q_s(t) = \sum\limits_{i=0}^{n-1} a_i(s)t^i$ of degree $n - 1$ interpolating the function $W_s(t)$ at the points $\{\tau_i = \tau_i(s)\}_{i=1}^n$,

$$q_s(\tau_i(s)) = W_s(\tau_i(s)), \qquad i = 1, \ldots, n. \tag{1.21}$$

Put

$$U_s(t) = W_s(t) - q_s(t), \qquad 0 \le t \le 1. \tag{1.22}$$

Let a piecewise continuous function $Z_s(t)$ on the interval $[0, 1]$ be given by the formula

$$Z_s(t) = \operatorname{sign} s_i\, |U_s(t)|, \qquad \bar{t}_{i-1}(s) \le t \le \bar{t}_i(s), \qquad i = 1, \ldots, 2n - r + 2. \tag{1.23}$$

We also define the function

$$\widetilde{H}_s(t) = \int\limits_0^t Z_s(\tau)\, d\tau, \qquad 0 \le t \le 1. \tag{1.24}$$

Let the polynomial $p_s(t) = \sum\limits_{i=0}^{n} b_i(s)t^i$ of degree n be the polynomial of the best approximation of the function $\widetilde{H}_s$ on the interval $[0,1]$. Put

$$H_s(t) = \widetilde{H}_s(t) - p_s(t), \qquad 0 \le t \le 1. \tag{1.25}$$

Finally, we define the mapping $\varkappa : \mathbb{S}^{2n-r+1} \mapsto \mathbb{R}^{2n-r+1}$:

$$\varkappa(s) := \{a_i(s)\}_{i=r}^{n-1} \times \{b_i(s)\}_{i=1}^{n} \times I_s, \tag{1.26}$$

where $\{a_i(s)\}_{i=0}^{n-1}$ are the coefficients of the polynomial q_s interpolating W_s at the points $\{\tau_i(s)\}_{i=1}^{n}$, $\{b_i(s)\}_{i=0}^{n}$ are the coefficients of the polynomial p_s of the best approximation for $\widetilde{H}_s$, and I_s is the integral of the kernel $\widehat{K}_s$ over the interval $[0, \tau_n(s)]$.

6.2. Continuity of the Borsuk mapping $\varkappa$

First of all, let us show that the mapping $\varkappa$ is continuous on the sphere $\mathbb{S}$.

LEMMA 6.2.1. *Let $\varkappa$ be defined by (1.26). Then, the mapping $s \mapsto \varkappa(s)$, $s \in \mathbb{S}$, is continuous.*

Proof. First, let us establish the continuity of the mapping $s \mapsto G_s$ in the metrics $\mathbb{C}[0,1]$.

MICROLEMMA 6.2.2. *Let the restrictions $G_s\big|_{[0,\tau_n(s)]}$ and $G_s\big|_{[\tau_n(s),1]}$ of the function $G_s(t)$ be defined as the extremal function of the problem (1.18) and by (1.19), respectively. Then, the mapping $s \mapsto G_s$ is continuous on the sphere $\mathbb{S}$.*

Proof. From inequalities (1.12) it follows that the Fredholm determinant of the system of linear equations (1.14) never vanishes on $\mathbb{S}$. Therefore, the Kramer's formula for the solution of the system (1.14) coupled with the continuity of mappings $s \mapsto \{\tau_i(s)\}_{i=0}^{n}$ and $s \mapsto \{\beta_l\}_{l=1}^{n-r+1}$ implies the continuous dependence of coefficients $\{\alpha_i(s)\}_{i=0}^{n}$ on s.

As we have remarked, the kernel $\widehat{K}_s$ belongs to the class $M_{2n-r+2}[0, \tau_n(s)]$ for all $s \in \mathbb{S}$ in the sense of Definition 2.2.5. By (1.13), $\tau_n(s) \ge \dfrac{\varepsilon}{1 + \varepsilon(2n - r + 3)} =: c$ for all $s \in \mathbb{S}$. Let us define the dylation $\widetilde{K}_s$ of the kernel $\widehat{K}_s$ on $[0, c]$:

$$\widetilde{K}_s(t) = \widehat{K}_s(t\tau_n(s)/c), \qquad t \in [0, c], \quad s \in \mathbb{S}.$$

From the continuous dependence of coefficients $\{\alpha_i(s)\}_{i=0}^{n}$ on s we infer the continuity of the family of kernels $\{\widehat{K}_s\big|_{[0,\tau_n(s)]}\}_{s\in\mathbb{S}}$ in *the integral metrics*: for all $s \in \mathbb{S}$,

$$\|\widetilde{K}_{s'} - \widetilde{K}_s\|_{\mathbb{L}_1[0,c]} \to 0, \quad \mathbb{S} \ni s' \to s.$$

Put

$$\begin{cases} a_{s_1,s_2} := \min\{\tau_n(s_1), \tau_n(s_2)\}, \\ b_{s_1,s_2} := \max\{\tau_n(s_1), \tau_n(s_2)\}, \end{cases} \quad (s_1, s_2) \in \mathbb{S} \times \mathbb{S}. \tag{2.1}$$

An application of Corollary 2.2.6 to the family of kernels $\{\widehat{K}_s\big|_{[0,\tau_n(s)]}\}_{s\in\mathbb{S}}$ implies that the extremal function $G_s\big|_{[0,\tau_n(s)]}$ of the problem (1.18) depends continuously on the parameter s: for all $s_1 \in \mathbb{S}$,

$$\|G_{s_1} - G_{s_2}\|_{\mathbb{C}[0,a_{s_1,s_2}]} \to 0, \quad \text{as} \quad \mathbb{S} \ni s_2 \to s_1. \tag{2.2}$$

The continuity of the restriction $\dfrac{d}{dt}G_s\big|_{[\tau_n(s),1]}$ follows immediately from the definition (1.19) and continuity of the mapping $s \mapsto \{t_i(s)\}_{i=n}^{2n-r+2}$: for all $s_1 \in \mathbb{S}$,

$$\|G_{s_1} - G_{s_2}\|_{\mathbb{L}_1[b_{s_1,s_2},1]} \to 0, \quad \text{as} \quad \mathbb{S} \ni s_2 \to s_1. \tag{2.3}$$

Pick any pair of vectors $(s_1, s_2) \in \mathbb{S} \times \mathbb{S}$, and $x \in [a_{s_1,s_2}, 1]$. Recalling the inclusion $G_s \in H^\omega[0,1]$, we derive the chain of inequalities

$$|G_{s_1}(x) - G_{s_2}(x)| \le |G_{s_1}(a_{s_1,s_2}) - G_{s_2}(a_{s_1,s_2})| + |G_{s_1}(b_{s_1,s_2}) - G_{s_1}(a_{s_1,s_2})| +$$

$$+ |G_{s_2}(b_{s_1,s_2}) - G_{s_2}(a_{s_1,s_2})| + \int\limits_{b_{s_1,s_2}}^{1} |G'_{s_1}(t) - G'_{s_2}(t)|\, dt \le$$

$$\le \|G_{s_1} - G_{s_2}\|_{\mathbb{C}[0,a_{s_1,s_2}]} + 2\omega(|\tau_n(s_1) - \tau_n(s_2)|) + \|G_{s_1} - G_{s_2}\|_{\mathbb{L}_1[b_{s_1,s_2},1]}. \tag{2.4}$$

Then, the combination of limiting properties (2.2), (2.3) and inequalities (2.4) completes the proof of Microlemma 6.2.2. $\square$

By Microlemma 6.2.2 and the definition (1.20), the mapping $s \mapsto W_s$ is continuous in $\mathbb{C}[0,1]$.

The coefficients $\{a_i(s)\}_{i=0}^{n-1}$ of the polynomial $q_s(t)$ are derived from the system of linear equations (1.21). By the Lagrange interpolation formula, we have

$$q_s(t) = \sum_{i=1}^{n} W_s(\tau_i) \prod_{\substack{j=1 \\ j\ne i}}^{n} \frac{t - \tau_j}{\tau_i - \tau_j}. \tag{2.5}$$

The formula (2.5) provides explicit expressions for the coefficients $\{a_i(s)\}_{i=0}^{n-1}$ in terms of interpolation points $\{\tau_i = \tau_i(s)\}_{i=1}^{n}$. Now, the continuity of coefficients $\{a_k(s)\}_{k=0}^{n-1}$ follows from the continuity of mappings $s \mapsto W_s$ in $\mathbb{C}[0,1]$, $s \mapsto \{\tau_i(s)\}_{i=1}^{2n-r}$ in l_1^{2n-r} and the uniform separation (1.13) of points $\{\tau_i(s)\}_{i=0}^{2n-r+1}$.

The continuity of the mapping $s \mapsto U_s$ in $\mathbb{C}[0,1]$ implies the existence of such a constant μ that

$$\|Z_s\|_{\mathbb{L}_\infty[0,1]} = \|U_s\|_{\mathbb{C}[0,1]} \leq \mu, \qquad \text{for all } s \in \mathbb{S}. \tag{2.6}$$

Next, using the definitions (1.22) of the function U_s and (1.23), (1.24) of the function $\widetilde{H}_s(t)$, we derive the chain of inequaities

$$\|\widetilde{H}_{s_1} - \widetilde{H}_{s_2}\|_{\mathbb{C}[0,1]} \leq \|Z_{s_1} - Z_{s_2}\|_{\mathbb{C}[0,1]} \leq 2 \left(\mu \|s_1 - s_2\|_{l_1^{2n-r+2}} + \|U_{s_1} - U_{s_2}\|_{\mathbb{L}_1[0,1]} \right), \tag{2.7}$$

which proves the continuity of the mapping $s \mapsto \widetilde{H}_s$ in $\mathbb{C}[0,1]$. The continuous dependence on s of coefficients $\{b_i(s)\}_{i=0}^{n}$ of the polynomial p_s of the best approximation for $\widetilde{H}_s$ on $[0,1]$ follows from the uniqueness of p_s. Finally, the continuous dependence of the integral I_s of the kernel $\widehat{K}_s$ over $[0, \tau_n(s)]$ follows from the continuity of the mapping $s \mapsto \widehat{K}_s$ in the integral metrics established in Microlemma 6.2.2.

The proof of the continuity of the mapping $\varkappa$ is complete.

6.3. Properties of solutions of the equation $\varkappa(s) = 0$

The mappings $s \mapsto \{a_i(s)\}_{i=0}^{n-1}$, $s \mapsto \{b_i(s)\}_{i=0}^{n}$ and $s \mapsto I_s$ are odd on the sphere $\mathbb{S}$, since the mappings $s \mapsto W_s$, $s \mapsto \widetilde{H}_s$, and $s \mapsto \widehat{K}_s$ are odd, respectively.

Summarizing, we showed that the mapping $\varkappa$ is both *continuous* and *odd* on the sphere $\mathbb{S}$. The Borsuk Theorem 1.1.1 assures the existence of such a vector $s^* \in \mathbb{S}$ that $\varkappa(s^*) = 0$, or, equivalently,

$$\begin{aligned} &(i) && a_i(s^*) = 0, && i = r, \ldots, n-1; \\ &(ii) && b_i(s^*) = 0, && i = 1, \ldots, n; \\ &(iii) && I_{s^*} = 0. \end{aligned} \tag{3.1}$$

We introduce a more convenient notation for collections of points at $s = s^*$.

NOTATIONS. All collections $(\{t_i(s)\}_{i=0}^{2n-r+2}, \{\tau_i(s)\}_{i=0}^{n}, \ldots)$ at the point $s = s^*$ are marked with the asterisk, that is

$$t_i^* := t_i(s^*), \qquad i = 0, \ldots, 2n - r + 2; \quad \tau_i^* := \tau_i(s^*), \qquad i = 0, \ldots, n; \ldots$$

We mention some immediate properties of s^*. By (3.1), (i) and the definition (1.22),

$$U_{s^*}^{(r)}(t) = W_{s^*}^{(r)}(t) - \sum_{i=r}^{n-1} \frac{i!}{(i-r)!} a_i(s^*) t^{i-r} = W_{s^*}^{(r)}(t) = G'_{s^*}(t), \tag{3.2}$$

for $t \in [0, 1]$. By (3.1), (ii) and the definition (1.25),

$$H_{s^*}(t) = \widetilde{H}_{s^*}(t) - b_0(s^*), \qquad 0 \le t \le 1. \tag{3.3}$$

Therefore, by the Chebyshev Theorem 1.1.2, the function $H_{s^*}(t) = \widetilde{H}_{s^*}(t) - b_0(s^*)$ has $n + 2$ points of alternance $\{z_i\}_{i=0}^{n+1}$ on the interval $[0, 1]$:

$$H_{s^*}(z_k) = (-1)^k \iota \|H_{s^*}\|_{\mathbb{C}\,[0,1]}, \qquad \iota = 1 \text{ or } \iota = -1, \qquad k = 0, \ldots, n+1, \tag{3.4}$$

where $0 \le z_0 < z_1 < \cdots < z_{n+1} \le 1$. Since $s \mapsto H_s$ is an odd mapping, we can assume without loss of generality that $\iota = (-1)^m$ in (3.4).

In the following lemma we identify the points of alternance of the function H_{s^*}, find the signs of the entries of $s^* = (s_1^*, \ldots, s_{2n-r+2}^*)$, and specify the location of the points $\{t_i^*\}_{i=0}^{2n-r+2}$ on the interval $[0, 2]$.

LEMMA 6.3.1. *Let $\{z_i\}_{i=0}^{n+1}$ be the points of alternance of the function H_{s^*} on the interval $[0, 1]$. Then*

$$
\begin{aligned}
&I. \quad z_0 = 0, \quad z_{n+1} = 1, \quad z_i = \bar{t}_i^* = t_i^*, \quad i = 1, \ldots, n. \\
&II. \quad 1 < t_{n+1}^* < \cdots < t_{2n-r+1}^* < 2. \\
&III. \quad \operatorname{sign} s_i^* = (-1)^{i+m}, \qquad i = 1, \ldots, 2n - r + 2.
\end{aligned}
\tag{3.5}
$$

Proof. Let the index $\Bbbk$ be defined by

$$\Bbbk := \max\{\, 0 \le i \le 2n - r + 1 \ \big| \ t_i^* < 1\}. \tag{3.6}$$

By the property (3.3) and (1.24), $\dfrac{d}{dt} H_{s^*}(t) = \dfrac{d}{dt} \widetilde{H}_{s^*}(t) = Z_{s^*}(t)$, $t \in [0, 1]$. Thus, by the definition (1.23) of $Z_s(t)$,

$$\operatorname{sign} \frac{d}{dt} H_{s^*}(t) = \operatorname{sign} Z_{s^*}(t) = \operatorname{sign} s_i^*, \quad t \in [\bar{t}_{i-1}^*, \bar{t}_i^*]. \tag{3.7}$$

Therefore, the function $\dfrac{d}{dt} H_{s^*}(t)$ can have *at most* $\Bbbk$ sign changes on the interval $[0, 1]$ at the points $\bar{t}_i^* := \min\{t_i^*, 1\}$, $i = 1, \ldots, \Bbbk$. By (3.4), the function H_{s^*} has $n + 2$ points of alternance $\{z_i\}_{i=0}^{n+1}$ on the interval $[0, 1]$, so the derivative $\dfrac{d}{dt} H_{s^*}(t)$ has *at least* n sign changes on the interval $[0, 1]$. This argument shows that

$$n \le \Bbbk, \tag{3.8}$$

where $\Bbbk$ is defined in (3.6).

On the other hand, by the definition (1.16) of the kernel $\widehat{K}_s$,

$$\operatorname{sign} \widehat{K}_{s^*}(t) = \operatorname{sign} s^*_{2n-r+3-i}, \quad t \in [\bar{\gamma}^*_{i-1}, \bar{\gamma}^*_i], \qquad i = 1, \dots, 2n - r + 2. \quad (3.9)$$

By (1.2), $\gamma_i = \min\{2 - t_{2n-r+2-i}, 1\}$, $i = 0, \dots, 2n - r + 2$, so by the definition (3.6) of the index $\Bbbk$ and (3.9), the number $\widehat{\Bbbk}$ of sign changes of the kernel $\widehat{K}_{s^*}$ on $[0, \tau^*_n]$ does not exceed $2n - r - \Bbbk + 1$:

$$\widehat{\Bbbk} \le 2n - r - \Bbbk + 1. \quad (3.10)$$

By (3.1), (iii), $\int_0^{\tau^*_n} \widehat{K}_{s^*}(t)\, dt = 0$. Therefore, $\widehat{K}_{s^*}(t) \in \mathcal{M}^0_{\widehat{\Bbbk}+1}[0, \tau^*_n]$, where classes $\mathcal{M}^j_l[a, b]$ for $l \in \mathbb{N}$ and $j \in \{-1, 0, +1\}$ are introduced in Definition 2.2.1. The function $G_{s^*}\big|_{[0,\tau^*_n]} = U^{(r-1)}_{s^*}\big|_{[0,\tau^*_n]}$ is defined as the function extremal in the problem

$$\int_0^{\tau^*_n} h(t) \widehat{K}_{s^*}(t)\, dt \to \sup, \qquad h \in H^\omega_0[0, \tau^*_n].$$

Remark 2.2.2 enables us to conclude that $G'_{s^*}\big|_{[0,\tau^*_n]} = U^{(r)}_{s^*}\big|_{[0,\tau^*_n]}$ has one less sign change on the interval $[0, \tau^*_n]$ than the generating kernel $\widehat{K}_{s^*}$, i.e. $\widehat{\Bbbk} - 1$ sign changes. Thus, by (3.10), the total number of sign changes of the function $U^{(r)}_{s^*}$ on the interval $[0, \tau^*_n]$ *does not exceed* $\widehat{\Bbbk} - 1 \le 2n - r - \Bbbk$.

On the other hand, by definitions (1.21), (1.22), U_{s^*} vanishes in n distinct points $\{\tau^*_i\}^n_{i=1}$. The Rolle theorem implies that the rth derivative $\dfrac{d^r}{dt^r} U_{s^*}(t) = G'_{s^*}(t)$ has *at least* $n - r$ sign changes on the interval $[0, \tau^*_n]$. Consequently, $2n - r - \Bbbk \ge n - r$, or, equivalently,

$$n \ge \Bbbk. \quad (3.11)$$

The inequalities (3.8) and (3.11) imply that $\Bbbk = n$, i.e. we have equalities everywhere in (3.8), (3.10) and (3.11). Summarizing, it follows from our analysis that these equalities are possible only if the following properties hold.

Property A.   The derivative $\dfrac{d}{dt} H_{s^*}(t) = Z_{s^*}(t)$ has precisely n sign changes on $[0, 1]$ at the points $\bar{t}^*_i = \min\{t^*_i, 1\} = t^*_i$, $i = 1, \dots, n$. Thus, the function H_{s^*} has the $(n + 2)$-alternance (3.4) at the points

$$z_0 = 0, \quad \{z_i = \bar{t}^*_i = t^*_i\}^n_{i=1}, \quad z_{n+1} = 1. \quad (3.12)$$

Consequently, by (3.7),

$$\operatorname{sign} s^*_i = \operatorname{sign} \frac{d}{dt} H_{s^*}(t) = (-1)^{i+m}, \qquad t \in [\bar{t}^*_{i-1}, \bar{t}^*_i], \quad i = 1, \dots, n. \quad (3.13)$$

Also by (3.12),

$$0 < t_1^* < t_2^* < \cdots < t_n^* < 1. \qquad (3.14)$$

Examining the definition $\gamma_i^* = 2 - t_{2n+2-r-i}^*$, $i = 0, \ldots, 2n - r + 2$, we notice that the relations (3.14) are equivalent to the inequalities

$$1 < \gamma_{n-r+2}^* < \gamma_{n-r+3}^* < \cdots < \gamma_{2n-r+2}^* < 2. \qquad (3.15)$$

Therefore, $\bar{\gamma}_i^* = \min\{\gamma_i^*, 1\} = 1$ for $i = n - r + 2, \ldots, 2n - r + 2$, so by (1.16), the points $\{\bar{\gamma}_i^*\}_{i=n-r+2}^{2n-r+2}$ cannot be the points of sign change of $\widehat{K}_{s^*}$ on $[0, \tau_n^*] \subset [0, 1]$.

Property B. The kernel $\widehat{K}_{s^*}$ has precisely $n - r + 1$ sign changes on the interval $[0, \tau_n^*]$ at the distinct points $\{\bar{\gamma}_i^* = \gamma_i^*\}_{i=1}^{n-r+1}$,

$$0 < \gamma_1^* < \cdots < \gamma_{n-r+1}^* < \tau_n^* < 1. \qquad (3.16)$$

The definition $\{\gamma_i^* = 2 - t_{2n+2-r-i}^*\}_{i=0}^{2n-r+2}$ and (3.16) imply that

$$1 < t_{n+1}^* < t_{n+2}^* < \cdots < t_{2n-r+2}^* = 2. \qquad (3.17)$$

From the definition (1.16) of the kernel $\widehat{K}_{s^*}$ and the fact that $\widehat{K}_{s^*}$ changes its sign at $\{\gamma_i^*\}_{i=1}^{n-r+1}$ we infer that

$$\operatorname{sign} s_{2n-r+3-i}^* = (-1)^i \chi, \quad \chi \in \{-1, 1\}, \quad i = 1, \ldots, n - r + 2. \qquad (3.18)$$

From (3.13) we find that $\operatorname{sign} s_{n+1}^* = (-1)^{n+1+m}$, while relations (3.18) for $i = n - r + 2$ imply that $\operatorname{sign} s_{n+1}^* = (-1)^{n-r+2} \chi$. So, $\chi = (-1)^{r-1+m}$, and the combination of properties (3.13) and (3.18) leads us to the final conclusion that

$$\operatorname{sign} s_i^* = (-1)^{i+m}, \qquad i = 1, \ldots, 2n - r + 2. \qquad (3.19)$$

$\square$

At this point of our construction we remind the reader the dependence of all introduced functions and collections of points on ε. The following lemma shows that the points $\{z_i(\varepsilon) = \bar{t}_i^*(\varepsilon)\}_{i=0}^{n+1}$ of alternance of the function $H_{s^*(\varepsilon)}$ are uniformly separated.

LEMMA 6.3.2. *There exists such a constant* $\delta = \delta(\omega, r, n)$, *independent of* ε, *that for all* $\varepsilon > 0$,

$$|z_i(\varepsilon) - z_{i-1}(\varepsilon)| = |\bar{t}_i^*(\varepsilon) - \bar{t}_{i-1}^*(\varepsilon)| > \delta, \qquad i = 1, \ldots, n + 1. \qquad (3.20)$$

Proof. First, let us show the uniform boundedness of the norms $\|H_{s^*(\varepsilon)}\|_{C[0,1]}$ from below. By properties (3.19) and (3.4) for $\iota = (-1)^m$, and definitions (1.23)–(1.25) of the function $H_{s^*(\varepsilon)}$, we have

$$2(n+1)\|H_{s^*(\varepsilon)}\|_{C[0,1]} = \sum_{i=1}^{n+1}(-1)^{i+m}\left(H_{s^*(\varepsilon)}(z_i(\varepsilon)) - H_{s^*(\varepsilon)}(z_{i-1}(\varepsilon))\right) =$$

$$\sum_{i=1}^{n+1}\int_{z_{i-1}(\varepsilon)}^{z_i(\varepsilon)}(-1)^{i+m}\dot{H}_{s^*(\varepsilon)}(t)\,dt = \sum_{i=1}^{n+1}\int_{z_{i-1}](\varepsilon)}^{z_i(\varepsilon)}|U_{s^*(\varepsilon)}(t)|\,dt = \|U_{s^*(\varepsilon)}\|_{\mathbb{L}_1[0,1]}. \quad (3.21)$$

In the process of the proof of Lemma 6.3.1 we showed that the function $U_{s^*(\varepsilon)}^{(r)}$ has precisely $n - r$ points $\{\eta_i^r(\varepsilon)\}_{i=1}^{n-r}$ on the interval $[0,1]$.

Recall the definition of $G'_{s^*}\big|_{[0,\tau_n^*]} = U_{s^*}^{(r)}\big|_{[0,\tau_n^*]}$ as the derivative of the extremal function of the problem (1.18), and the definition (1.19) of the other restriction $G'_{s^*}\big|_{[\tau_n^*,1]} = U_{s^*}^{(r)}\big|_{[0,\tau_n^*]}$. From Corollary 4.4.2, applied to the restriction $U_{s^*}^{(r)}\big|_{[0,\tau_n^*]}$ and from the formula (1.19) and the property (3.19) it follows that

for all $i = 0, \ldots, n - r$, and a.e. $x \in [\eta_i^r, \eta_{i+1}^r]$, there exists an $e_x \in [0,1]$, such that $U_{s^*}^{(r)}(x) = g_{s^*}(x) = \operatorname{sign} s_{2n-r+2-i}^* \omega'(e_x) = (-1)^{r+i+m}\omega'(e_x)$.

Therefore, by (3.5), (iii),

$$(-1)^{r+i+m}U_{s^*(\varepsilon)}^{(r)}(x) \geq \inf_{t \in (0,1)} \omega'(t) = \omega_0, \qquad x \in [\eta_{i-1}^r(\varepsilon), \eta_i^r(\varepsilon)], \quad (3.22)$$

for all $i = 1, \ldots, n - r + 1$ (see (0.2)).

Thus, we showed the applicability of Corollary 4.4.2 to the function $U_{s^*(\varepsilon)}$: in notations of the corollary, $N = n - r + 1$ and $P = \omega_0$. Therefore, by Corollary 4.4.2,

$$\|U_{s^*(\varepsilon)}\|_{\mathbb{L}_1[0,1]} \geq \frac{\omega_0 d_r}{(n-r+1)^r}, \quad (3.23)$$

where the constant d_r is defined in Corollary 4.4.2.

Combining the identity (3.21) and the inequality (3.23) we conclude that for all $\varepsilon > 0$,

$$\|H_{s^*(\varepsilon)}\|_{C[0,1]} \geq \frac{\omega_0 d_r}{2(n+1)(n-r+1)^r} =: J(\omega, r, n). \quad (3.24)$$

On the other hand, since $U_{s^*(\varepsilon)}$ has at least n zeroes at the points $\{\tau^*(\varepsilon)\}_{i=1}^n$, we can use Proposition 1.1.5 to derive the inequalities

$$\|U_{s^*(\varepsilon)}\|_{C[0,1]} \leq \|U'_{s^*(\varepsilon)}\|_{C[0,1]} \leq \ldots \|U_{s^*(\varepsilon)}^{(r-1)}\|_{C[0,1]} \leq \omega(1). \quad (3.25)$$

The estimates (3.24) and (3.25) lead us to the inequalities

$$2J \leq 2\|H_{s^*(\varepsilon)}\|_{C[0,1]} = |H_{s^*(\varepsilon)}(z_i(\varepsilon)) - H_{s^*(\varepsilon)}(z_{i-1}(\varepsilon))| \leq$$

$$\leq \|\frac{d}{dt}H_{s^*(\varepsilon)}\|_{C[0,1]}(z_i(\varepsilon) - z_{i-1}(\varepsilon)) \leq \omega(1)(z_i(\varepsilon) - z_{i-1}(\varepsilon)), \quad (3.26)$$

for all $i = 1, \ldots, n+1$. Therefore,

$$z_i(\varepsilon) - z_{i-1}(\varepsilon) \geq \frac{J(\omega, r, n)}{2\omega(1)} =: \delta, \qquad i = 1, \ldots, n+1. \qquad \square$$

By the Rolle theorem, each of the derivatives $U_{s^*(\varepsilon)}^{(l)}$, $l = 0, \ldots, r$, has precisely $n - l$ points of sign change at some $\{\eta_i^l(\varepsilon)\}_{i=1}^{n-l}$. The Rolle theorem gives us the following inequalities between the points of sign change of the consecutive derivatives:

$$\eta_i^l(\varepsilon) < \eta_i^{l+1}(\varepsilon) < \eta_{i+1}^l(\varepsilon), \qquad i = 1, \ldots, n - l - 1. \qquad (3.27)$$

Notice that $\eta_i^0(\varepsilon) = \tau_i^*(\varepsilon)$, $i = 1, \ldots, n$.

COROLLARY 6.3.3. *Let $\{\eta_i^r(\varepsilon)\}_{i=1}^{n-r}$ be the points of sign change of the function $U_{s^*(\varepsilon)}^{(r)}$. Then, there exist such constants $\hat{\delta} > 0$ and $E > 0$ that for all $0 < \varepsilon < E$,*

$$\tau_i^*(\varepsilon) + \hat{\delta} < \eta_i^r(\varepsilon), \qquad i = 1, \ldots, n - r. \qquad (3.28)$$

Proof. By definitions (1.6)–(1.8) of points $\{\tau_i\}_{i=1}^n$ we have the following inequalities for $i = 1, \ldots, n$:

$$|\tau_i^*(\varepsilon) - t_i^*(\varepsilon)| = |\tau_i^*(\varepsilon) - \bar{t}_i^*(\varepsilon)| = \left| \frac{\bar{t}_i^*(\varepsilon) + \varepsilon(2i+1)}{1 + \varepsilon(2n - r + 2)} - \bar{t}_i^*(\varepsilon) \right| \leq (2n - r + 2)\varepsilon. \qquad (3.29)$$

Thus, for $\varepsilon < E := \dfrac{1}{4(2n - r + 2)}$, we infer from (3.29) and (3.20) that

$$\tau_i^*(\varepsilon) - \tau_{i-1}^*(\varepsilon) \geq \frac{1}{2}\delta, \qquad i = 1, \ldots, n. \qquad (3.30)$$

Let us show that we can apply Corollary 4.4.3 to the function $U_{s^*(\varepsilon)}$, its simple zeroes $\{\tau_i^*(\varepsilon)\}_{i=1}^n$ and points $\{\eta_i^r(\varepsilon)\}_{i=1}^{n-r}$ of sign change of $U_{s^*(\varepsilon)}^{(r)}$. Indeed, by inequalities (3.25) and (3.22),

$$\|U_{s^*(\varepsilon)}\|_{C[0,1]} \leq \omega(1), \qquad |U_{s^*(\varepsilon)}(t)| \geq \omega_0. \qquad (3.31)$$

Revoking also inequalities (3.30), we can apply Corollary 4.4.3 to conclude that

$$\eta_i^1(\varepsilon) - \tau_i^*(\varepsilon) \geq \frac{L\delta^r}{2^r\omega(1)} =: \hat{\delta}, \qquad i = 1, \ldots, n, \qquad (3.32)$$

where $L := \dfrac{\omega_0 d_r}{(n - r + 1)^r}$ (see (3.23)). $\qquad \square$

The following result lets us find inequalities between the zeroes $\{\tau_i^*(\varepsilon)\}_{i=1}^n$ of the function $U_{s^*(\varepsilon)}$ and the points $\{\gamma_i^*(\varepsilon)\}_{i=1}^{n-r+1}$ of sign change of the kernel $\widehat{K}_{s^*(\varepsilon)}$.

COROLLARY 6.3.4. *Let $\hat{\delta}, E$ be as in Corollary 6.3.3. Then, for all $0 < \varepsilon < E$,*

$$\gamma_i^*(\varepsilon) \geq \tau_{i-1}^*(\varepsilon) + \hat{\delta}, \qquad i = 1, \ldots, n - r + 1. \tag{3.33}$$

Proof. We defined the function $U_{s^*(\varepsilon)}^{(r)}\big|_{[0,\tau_n^*(\varepsilon)]}$ as the derivative of the function $U_{s^*(\varepsilon)}^{(r-1)}\big|_{[0,\tau_n^*(\varepsilon)]}$ extremal in the problem

$$\int_0^{\tau_n^*(\varepsilon)} h(t)\widehat{K}_{s^*(\varepsilon)}(t)\, dt \to \sup, \qquad h \in H^\omega[0, \tau_n^*(\varepsilon)]. \tag{3.34}$$

The points $\{\bar{\gamma}_i^*(\varepsilon) = \gamma_i^*(\varepsilon)\}_{i=1}^{n-r+1}$ are the points of sign change of the generating kernel $\widehat{K}_{s^*(\varepsilon)}$. By Remark 2.2.3 (use (2.7.18)), we have the following inequalities between $\{\gamma_i^*(\varepsilon)\}_{i=1}^{n-r+1}$ and the points of sign change of $U_{s^*(\varepsilon)}^{(r)}$:

$$\gamma_i^*(\varepsilon) < \eta_i^r(\varepsilon) < \gamma_{i+1}^*(\varepsilon), \qquad i = 1, \ldots, n - r. \tag{3.35}$$

Therefore, the inequalities (3.35) and the assertion (3.28) of Corollary 6.3.3 imply that

$$t_{i-1}^*(\varepsilon) + \hat{\delta} < \eta_{i-1}^r(\varepsilon) < \gamma_i^*(\varepsilon). \qquad i = 1, \ldots, n - r. \tag{3.36}$$

$\square$

6.4. Limiting procedure as $\varepsilon \to 0$

As we showed in (3.25), the families of functions $\{U_{s^*(\varepsilon)}^{(k)}\}_{\varepsilon>0}$ are uniformly bounded:

$$\|U_{s^*(\varepsilon)}^{(k)}\|_{C\,[0,1]} \leq \omega(1), \qquad k = 0, \ldots, r - 1. \tag{4.1}$$

Inequalities (4.1) also imply that the functions $U_{s^*(\varepsilon)}^{(k)}$ have a common majorizing modulus of continuity on the interval $[0, 1]$:

$$\omega\left(U_{s^*(\varepsilon)}^{(k)}; t\right) \leq \omega(1)t, \qquad 0 \leq t \leq 1, \qquad k = 0, \ldots, r - 2. \tag{4.2}$$

As for the $(r-1)^{\text{th}}$ derivative, $U_{s^*(\varepsilon)}^{(r-1)}(t) = G_{s^*(\varepsilon)}(t) + \lambda$, so $U_{s^*(\varepsilon)}^{(r-1)} \in H^\omega[0,1]$.

Then, the Arzela–Ascoli theorem enables us to extract such a subsequence $\varepsilon_k \downarrow 0$, as $k \uparrow \infty$, that

$$U_{s^*(\varepsilon_k)} \to U \quad \text{in } \mathbb{C}^{r-1}[0,1], \qquad \text{as } k \to \infty. \tag{4.3}$$

NOTATIONS. For $i = 0, \ldots, 2n - r + 2$, set

$$(i) \quad \begin{cases} t_i = \lim_{k \to \infty} t_i^*(\varepsilon_k); \\ \bar{t}_i = \min\{t_i, 1\}; \end{cases} \quad (ii) \quad \begin{cases} \gamma_i = \lim_{k \to \infty} \gamma_i^*(\varepsilon_k); \\ \bar{\gamma}_i = \min\{\gamma_i, 1\}. \end{cases} \quad (4.4)$$

Notice that by the definition of points $\{\tau_i(s)\}_{i=1}^n$ in (1.6)–(1.8),

$$\tau_i(s) = \begin{cases} \dfrac{\bar{t}_i(s) + \varepsilon(2i+1)}{1 + \varepsilon(2n-r+2)}, & i = 1, \ldots, n-r; \\[3mm] \dfrac{\bar{t}_i(s) + \varepsilon(i+n-r)}{1 + \varepsilon(2n-r+2)}, & i = n-r+1, \ldots, n. \end{cases} \quad (4.5)$$

Thus, by (4.4), (i) and (3.12), (3.16),

$$\lim_{k \to \infty} \tau_i^*(\varepsilon_k) = \lim_{k \to \infty} \bar{t}_i^*(\varepsilon_k) = \lim_{k \to \infty} t_i^*(\varepsilon_k) = \bar{t}_i, \quad i = 1, \ldots, n,$$
$$\bar{\gamma}_i = \gamma_i, \quad i = 1, \ldots, n-r+1. \quad (4.6)$$

By Lemma 6.3.2, the limiting points $\{\bar{t}_i\}_{i=0}^{n+1}$ also remain uniformly separated:

$$\bar{t}_i - \bar{t}_{i-1} \geq \delta, \quad i = 1, \ldots, n+1. \quad (4.7)$$

From Corollary 6.3.4 and (4.6) it follows that

$$\gamma_i = \bar{\gamma}_i \geq \bar{t}_{i-1} + \hat{\delta}, \quad i = 1, \ldots, n-r+1. \quad (4.8)$$

Recall the definition of the kernel K_s in (1.15):

$$K_{s^*(\varepsilon_k)} = \sum_{i=0}^n \alpha_i(\varepsilon_k)(\tau_i^*(\varepsilon_k)) - t)_+^{r-1}, \quad (4.9)$$

where $\{\alpha_i(\varepsilon_k)\}_{i=0}^n$ satisfy the equation (1.14) for $s = s^*(\varepsilon_k)$.
    According to (1.30), (i),

$$\operatorname{supp} K_{s^*(\varepsilon_k)} = [0, t_{j_k}^*(\varepsilon_k)], \quad \text{for some } j_k, \ r-1 \leq j_k \leq n, \quad k \in \mathbb{N}. \quad (4.10)$$

Without loss of generality (if necessary, by extracting a subsequence $\varepsilon_{k_l}, l \in \mathbb{N}$), we can assume that $j_k \equiv j, \ \forall k \in \mathbb{N}$, for some $j : r-1 \leq j \leq n$.

LEMMA 6.4.1. Let

$$\beta_i = \lim_{k \to \infty} \beta_i^*(\varepsilon_k), \quad i = 0, \ldots, n-r+1, \quad (4.11)$$

where points $\{\beta_i(s)\}_{i=0}^{n-r+1}$ are defined in (1.3)–(1.9). Then,

$$\beta_i = \gamma_i, \quad i = 1, \ldots, j-r+1. \quad (4.12)$$

Proof. By Corollary 3.1.3 and inequalities (4.7), we have the following relations between the knots $\{\tau_i^*(\varepsilon_k)\}_{i=1}^{j}$ and zeroes $\{\beta_i^*(\varepsilon_k)\}_{i=1}^{j-r+1}$ of the kernel $K_{s^*(\varepsilon)}(t)$ lying inside the support of $K_{s^*(\varepsilon)}$:

$$\beta_i^*(\varepsilon_k) < \tau_{i+r-2}^*(\varepsilon_k) < \tau_{i+r-1}^*(\varepsilon_k) - \delta, \qquad i = 1, \ldots, j - r + 1, \tag{4.13}$$

with the constant δ from (4.7), independent of $k \in \mathbb{N}$.

Taking the limit in (4.13) and using (4.6), (i) and (4.11), we find inequalities between points $\{\beta_i\}_{i=1}^{j-r+1}$ and $\{\bar{t}_i\}_{i=r}^{j}$:

$$\beta_i \leq \bar{t}_{i+r-1} - \delta, \qquad i = 1, \ldots, j - r + 1. \tag{4.14}$$

Recall the definition of points $\{\beta_i(s)\}_{i=1}^{n-r+1}$ in (1.3)–(1.9):

$$\beta_i^*(\varepsilon_k) = \frac{\bar{\beta}_i^*(\varepsilon_k) + 2i\varepsilon_k}{1 + \varepsilon_k(2n - r + 2)}, \qquad i = 1, \ldots, n - r + 1, \tag{4.15}$$

where $\bar{\beta}_i^*(\varepsilon_k) = \min\left\{\max\{\gamma_i^*(\varepsilon_k); \bar{t}_{i-1}^*(\varepsilon_k)\}; \bar{t}_{i+r-1}^*(\varepsilon_k)\right\}$, $i = 1, \ldots, n - r + 1$.

Taking the limit in (4.15) and using inequalities (4.8), we have

$$\beta_i = \lim_{k \to \infty} \beta_i^*(\varepsilon_k) = \min\left\{\max\{\gamma_i; \bar{t}_{i-1}\}; \bar{t}_{i+r-1}\right\} = \min\{\gamma_i; \bar{t}_{i+r-1}\} \tag{4.16}$$

for $i = 1, \ldots, n - r + 1$. Then, the combination of (4.14) and (4.16) proves the desired property (4.12). $\qquad\square$

The result of Lemma 6.4.1, (4.6) and inequalities (4.8) and (4.14) imply that

$$\bar{t}_{i-1} + \hat{\delta} < \gamma_i = \beta_i \leq \bar{t}_{i+r-1} - \delta, \qquad i = 1, \ldots, j - r + 1. \tag{4.17}$$

Let the coefficients $\{\alpha_i\}_{i=0}^{j}$ be derived from the system of linear equations

$$\begin{cases} \displaystyle\sum_{i=0}^{j} \alpha_i [\bar{t}_i]^p = m! \delta_{m,j}, & p = 0, \ldots, r - 1; \\[4mm] \displaystyle\sum_{i=0}^{j} \alpha_i (\bar{t}_i - \beta_l)_+^{r-1} = 0, & l = 1, \ldots, j - r + 1. \end{cases} \tag{4.18}$$

Inequalities (4.17) assure the unique solvability of the system (4.18). Put

$$K(t) = -\frac{1}{(r-1)!} \sum_{i=0}^{j} \alpha_i (\bar{t}_i - t)_+^{r-1}. \tag{4.19}$$

Set also

$$\widehat{K}(t) = (-1)^{i+r+m-1} |K(t)|, \qquad t \in [\gamma_{i-1}, \gamma_i], \quad i = 1, \ldots, j - r + 2, \tag{4.20}$$

Compare the definitions (1.14), (1.15) of the kernel K_{s^*} and (3.28) of $\widehat{K}_{s^*}$ with definitions (4.19), (4.20) of kernels $K(t)$ and $\widehat{K}(t)$. From the limiting relations

$$\lim_{k\to\infty} \beta_i^*(\varepsilon_k) = \beta_i = \gamma_i = \lim_{k\to\infty} \gamma_i^*(\varepsilon_k), \qquad i = 1, \dots, j - r + 1,$$

of Lemma 6.4.1 and (4.6) it follows that

$$\begin{aligned}
K &= \lim_{k\to\infty} K_{s^*(\varepsilon_k)} \quad \text{in } \mathbb{C}[0, t_n]; \\
\widehat{K} &= \lim_{k\to\infty} \widehat{K}_{s^*(\varepsilon_k)} \quad \text{in } \mathbb{C}[0, t_n].
\end{aligned} \tag{4.21}$$

By (3.1.6), (iv),

$$(-1)^{i+r+m-1} K(t) \geq 0, \quad t \in [\gamma_{i-1}, \gamma_i], \qquad i = 1, \dots, n - r + 1. \tag{4.22}$$

Juxtaposing (4.20) and (4.22), we finally infer that for $t \in [\gamma_{i-1}, \gamma_i]$, $i = 1, \dots, n - r + 1$,

$$\widehat{K}(t) := (-1)^{i+r+m-1} |K(t)| = (-1)^{r+i+m-1} \cdot (-1)^{r+i+m-1} K(t) = K(t), \tag{4.23}$$

In addition, by the inherited property (3.1), (iii), the limiting kernel $K(t) = \widehat{K}(t)$ has the zero mean on the interval $[0, t_j]$. Therefore, as we showed in Corollary 3.1.2, the support of the kernel K contains the interval $[0, \bar{t}_r] = [0, t_r]$.

Next, the function $U_{s^*(\varepsilon_k)}^{(r-1)}$ is extremal in the problem (3.18). Since the kernels $\widehat{K}_{s^*(\varepsilon_k)}$ converge to the kernel K in the uniform metrics on $[0, t_n]$, Corollary 2.2.6 guarantees that the limiting function $U^{(r-1)}(t)$ is extremal in the problem

$$\int_0^{t_n} h(t) K(t) \, dt \to \sup, \qquad h \in H^\omega[0, t_n]. \tag{4.24}$$

Put

$$H(t) = \lim_{k\to\infty} H_{s^*(\varepsilon_k)}(t), \qquad 0 \leq t \leq 1. \tag{4.25}$$

By (3.19),

$$\frac{d}{dt} H_{s^*(\varepsilon_k)}(t) = (-1)^{i+m} \left| U_{s^*(\varepsilon_k)} \right|, \quad t_{i-1}^*(\varepsilon_k) < t < t_i^*(\varepsilon_k), \quad i = 1, \dots, n+1. \tag{4.26}$$

On the other hand, $U_{s^*}^{(r)}$ has $n - r$ sign changes on $[0, \tau_n^*]$, so all zeroes $\{\tau_i^*(\varepsilon_k)\}_{i=1}^n$ of $U_{s^*(\varepsilon_k)}$ are simple:

$$\operatorname{sign} U_{s^*(\varepsilon_k)}(t) = (-1)^{i+m}, \qquad \tau_{i-1}^*(\varepsilon_k) < t < \tau_i^*(\varepsilon_k), \qquad i = 1, \dots, n+1. \tag{4.27}$$

From the limiting relations (4.6) and the comparison of (4.26) and (4.27) we deduce that

$$\frac{d}{dt}H(t) = U(t), \qquad 0 \le t \le t_{n+1}. \tag{4.28}$$

Finally, the limiting function $H(t)$ inherits the properties (3.17) for $\xi = (-1)^m$ and the norm (5.32) of functions $\{H_{s^*(\varepsilon_k)}\}_{k \in \mathbb{N}}$:

$$H(t_i) = (-1)^i \|H\|_{C[0,t_n]}, \qquad i = 0, \ldots, n+1. \tag{4.29}$$

It remains to rename the function H, the alternance points $\{t_i\}_{i=0}^{n+1}$, and the points $\{\gamma_i\}_{i=0}^{n-r+2}$:

$$\begin{aligned}
\mathcal{Z}_n(t) &:= H(t), \qquad 0 \le t \le t_n =: d, \\
\vartheta_i &:= \gamma_i, \qquad i = 0, \ldots, n-r+2, \\
\nu_i &:= t_i, \qquad i = 0, \ldots, n+1.
\end{aligned} \tag{4.30}$$

REMARK 6.4.1. By our choice (1.19) of the extension, the derivative of the continuous function $\mathcal{Z}_n^{(r)}(t)$ is expressed on the interval $[t_n, 1]$ by the formula

$$\mathcal{Z}_n^{(r+1)}(t) = (-1)^{n+1}\omega'(t), \qquad t \in [t_n, 1]. \tag{4.31}$$

Theorem 6.0.1 is proved completely for strictly concave modulii of continuity ω endowed with the property $\inf_{t \in (0,1)} \omega'(t) > 0$.

In the general case, given a concave modulus of continuity ω on $\mathbb{R}_+$, we introduce the sequence $\{\omega_l\}_{l \in \mathbb{N}}$ of modulii of continuity

$$\omega_l(t) = \omega(t) + \frac{1}{l}[t^{1/2} + t], \qquad t \in \mathbb{R}_+, \quad l \in \mathbb{N}. \tag{4.32}$$

Notice that each of the strictly concave modulii of continuity ω_l, $l \in \mathbb{N}$, enjoys the property (0.2):

$$\inf_{t \in (0,1)} \omega_l'(t) > \frac{1}{l}, \qquad l \in \mathbb{N}, \tag{4.33}$$

and $\omega_l \to \omega$ in $\mathbb{C}[0,1]$.

Then, for each $l \in \mathbb{N}$, Theorem 6.0.1 enables us to construct the family of functions $X_l(t) = \mathcal{Z}_{B,n,r,m,\omega_l}(t)$, $t \in [0,1]$, endowed with the properties (4.1) for $\omega = \omega_l$.

Arguing as in (4.2), we can show that

$$\omega\left(X_l^{(k)}; t\right) \le \omega_l(1)t \le \omega(1)t, \qquad 0 \le t \le 1, \quad k = 0, \ldots, r-2. \tag{4.34}$$

Therefore, by the Arzela–Ascoli theorem, we can extract such a subsequence of indices $\{l_k\}_{k \in \mathbb{N}}$ that $X_{l_k} \to \mathcal{Z}$ in $\mathbb{C}^r[0,1]$, as $k \to \infty$. Clearly, the limiting function $\mathcal{Z}(t)$ inherits the property (0.1), (i) of functions $\{X_{l_k}\}_{k \in \mathbb{N}}$, i.e. the complete alternance at some points $\{\nu_i\}_{i=0}^{n+1}$. Corollary 2.2.8 assures that the function $\mathcal{Z}(t)$ enjoys the property (0.1), (ii), as well. The proof of Theorem 6.0.1 for $0 < m < r$ is now complete. $\qquad \square$

REMARK 6.4.2. In (4.3) we obtained the function $H^{(r)}(t) = U^{(r-1)}(t)$ as the limit of functions $U^{(r-1)}_{s^*(\varepsilon_k)}(t)$ extremal in the problem (1.18) for $s = s^*(\varepsilon_k)$. The extremality of $U^{(r-1)}_{s^*(\varepsilon_k)}$ in (1.18) implies that the functions $U^{(r-1)}_{s^*(\varepsilon_k)}$ enjoy the property formulated in Corollary 2.2.5. Therefore, the limiting function $\mathcal{Z}^{(r)}_n(t) = U^{(r-1)}(t)$ also inherits this property on the interval $[0, \nu_n]$:

for a.e. $x \in [0, \nu_n]$ there exists such $r_x, v_x \in [0, \nu_n]$ that

$$|\mathcal{Z}^{(r+1)}_n(x)| = \omega'(r_x) \quad and \quad |\mathcal{Z}^{(r)}_n(x) - \mathcal{Z}^{(r)}_n(v_x)| = \omega(r_x). \tag{4.35}$$

REMARK 6.4.3. In the case $0 < m < r$ the definition (3.2), (3.3) of the kernel $K(t)$ involves $2n - r + 1$ parameters $\{\nu_i\}_{i=1}^n$ and $\{\vartheta_i\}_{i=1}^{n-r+1}$. This explains the use of a $(2n - r + 1)$-dimensional sphere in our proof of Theorem 6.0.1 for $0 < m < r$.

However, in the case $m = r$ the kernel $K(t)$ is defined by $2n - r$ parameters $\{\nu_i\}_{i=1}^n$ and $\{\vartheta_i\}_{i=1}^{n-r}$. Therefore, proving Theorem 6.0.1, we would construct an appropriate odd and continuous mapping $\varkappa$ on the $(2n - r)$-dimensional sphere $\mathbb{S}^{2n-r}$. This modification is the only difference in the proofs of Theorems 6.0.1 for $0 < m < r$ and $m = r$.

Chapter 7
Properties of Chebyshev Ω-Splines

Relying on the results of Theorem 6.0.1, we first introduce the discrete family of Chebyshev functions $\{R_n = R_{B,\omega_\alpha,r,m,n}\}_{n \geq r}$ extremal in the problem

$$f^{(m)}(0) \to \sup, \qquad f \in W^r H^\omega[0, d_n], \quad \|f\|_{C[0,d_n]} \leq B, \tag{0.0}$$

for $\omega(t) = t^\alpha$, $0 < \alpha \leq 1$, and some interval $[0, d_n]$, $d_n = d_n(B, \omega, r, m)$. Then, referring to the results of our paper [7] or [8], we describe the Chebyshev ω-splines of the problem (0.0) for arbitrary ω. Finally, we analyze various properties of Chebyshev ω-splines crucial in the construction of extremal functions in the Kolmogorov problem on the half-line $\mathbb{R}_+$.

7.1. Review of the structure of Chebyshev ω-splines on $[0, 1]$

For our convenience we consider the case $0 < m < r$. The proofs of the corresponding properties of functions R_n in the case $m = r$ are identical to those for $0 < m < r$.

Fix n, $m \in \mathbb{N}$, $n \geq r$, $m < 1$, and a concave modulus of continuity ω. In Theorem 6.0.1 we constructed the function $\mathcal{Z}_n = \mathcal{Z}_{\omega,r,m,n}$ extremal in the problem

$$f^{(m)}(0) \to \sup, \quad f \in W^r H^\omega[0, 1]; \quad \|f\|_{C[0,1]} \leq \rho_n, \tag{1.1}$$

where $\rho_n = \rho_{\omega,r,m,n}$. Each of the functions $\mathcal{Z}_n$ has a complete set of $n + 2$ points of alternance $\{\hat{\nu}_i^n\}_{i=0}^{n+1}$:

$$\mathcal{Z}_n(\hat{\nu}_i^n) = (-1)^{i+m}\|\mathcal{Z}_n\|_{C[0,1]}, \qquad i = 0, \ldots, n+1. \tag{1.2}$$

The function $\mathcal{Z}_n^{(r)}$ is generated by the kernel

$$K_n(t) = -\frac{1}{(r-1)!} \sum_{i=0}^{n} \hat{\alpha}_i^n (\hat{\nu}_i^n - t)_+^{r-1}, \tag{1.3}$$

whose coefficients $\{\hat{\alpha}_i^n\}_{i=0}^{n}$ solve equations

$$\begin{cases} \sum_{i=0}^{n} \hat{\alpha}_i^n [\hat{\nu}_i^n]^j = m! \, \delta_{m,j}, & j = 0, \ldots, r; \\[2mm] \sum_{i=0}^{n} \hat{\alpha}_i^n (\hat{\nu}_i^n - \hat{\vartheta}_l^n)_+^{r-1} = 0, & l = 1, \ldots, n-r+1. \end{cases} \tag{1.4}$$

for a collection of points $\{\hat{\vartheta}_i^n\}_{i=0}^{n-r+2}$ satisfying inequalities (3.1). More precisely, the function $\mathcal{Z}_n^{(r)}$ enjoys the property

$$\sup_{h\in H_0^\omega[0,1]} \int_0^1 h(t)K_n(t)\,dt = \int_0^1 [\mathcal{Z}_n^{(r)}(t) - \mathcal{Z}_n^{(r)}(0)]K_n(t)\,dt. \qquad (1.5)$$

Let also $\{\hat{\eta}_i^n\}_{i=1}^{n-r}$ be the $n-r$ monotonely ordered points of sign change of the function $\mathcal{Z}_n^{(r+1)}(t)$: $0 =: \hat{\eta}_0^n < \hat{\eta}_1^n < \ldots \hat{\eta}_{n-r}^n < \hat{\eta}_{n-r}^{n+1} := 1$.

In the following proposition we show that the norms $\rho_n = \|\mathcal{Z}_n\|_{C[0,1]}$ tend to zero as $n \to \infty$.

PROPOSITION 7.1.1. *For $n \geq r$, let $\rho_n = \rho_{n,r,m,\omega} = \|\mathcal{Z}_{n,r,m,\omega}\|_{C[0,1]}$. Then,*

$$\rho_n \leq \frac{\omega(1)}{2(n+1)}, \qquad n \geq r. \qquad (1.6)$$

Proof. By (1.2) and the simplicity of zeroes of $\dfrac{d}{dt}\mathcal{Z}_n(t)$, we have

$$2(n+1)\rho_n = \sum_{i=1}^{n+1}(-1)^{i+m}[\mathcal{Z}_n(\hat{\nu}_i) - \mathcal{Z}_n(\hat{\nu}_{i-1})] = \sum_{i=1}^{n+1}\int_{\hat{\nu}_{i-1}}^{\hat{\nu}_i}(-1)^{i+m}\frac{d}{dt}\mathcal{Z}_n(t)\,dt =$$

$$= \sum_{i=1}^{n+1}\int_{\hat{\nu}_{i-1}}^{\hat{\nu}_i}|\frac{d}{dt}\mathcal{Z}_n(t)|\,dt = \|\mathcal{Z}_n'\|_{\mathbb{L}_1[0,1]}. \qquad (1.7)$$

In addition, the derivative $\dfrac{d}{dt}\mathcal{Z}_n(t)$ has $n \geq r$ zeroes $\{\hat{\nu}_i\}_{i=1}^n$ on $[0,1]$. Thus, using (1.7) and applying Proposition 1.1.5 to $\dfrac{d}{dt}\mathcal{Z}_n(t) \in W^{r-1}H^\omega[0,1]$, we infer that

$$2(n+1)\rho_n = \|\mathcal{Z}_n'\|_{\mathbb{L}_1[0,1]} \leq \|\mathcal{Z}_n^{(r)}\|_{C[0,1]} \leq \omega(1). \qquad (1.8)$$

$\square$

7.2. Rescaled Chebyshev ω_α-splines of the fixed norm B

Fix α, $0 < \alpha \leq 1$, and $B > 0$. Let $\mathcal{Z}_n(t) = \mathcal{Z}_{\omega_\alpha,r,m,n}(t)$ be the Chebyshev ω_α-splines from Section 7.1 in the particular case of the Hölder modulus of continuity $\omega_\alpha(t) = t^\alpha$. The property

$$f(\cdot) \in W^r H^\alpha[0,1] \iff \gamma^{r+\alpha}f(\cdot/\gamma) \in W^r H^\alpha[0,\gamma], \qquad (2.1)$$

will enable us to choose γ in such a way that the rescaled function $\gamma^{r+\alpha} f(\cdot/\gamma) \in W^r H^\alpha[0, \gamma]$ has a given norm B on $[0, \gamma]$. We will also show that the transformation $f(\cdot) \mapsto \gamma^{r+\alpha} f(\cdot/\gamma)$ preserve the property of extremality in the Kolmogorov–Landau inequalities.

Indeed, set

$$d_n = d_{\omega_\alpha, r, m, n, B} := [B/\rho_n]^{\frac{1}{r+\alpha}}. \tag{2.2}$$

Let us define the function $R_n = R_{B, \omega_\alpha, r, m, B}$ in $W^r H^\alpha[0, d_n]$ by the formula

$$R_n(t) = d_n^{r+\alpha} \mathcal{Z}_n(t/d_n), \qquad 0 \le t \le d_n. \tag{2.3}$$

The normalization in (2.3) is chosen in such a way that

$$\|R_n\|_{\mathbb{C}[0, d_n]} = B, \qquad n \ge r. \tag{2.4}$$

Put

$$
\begin{aligned}
&\nu_i^n := d_n \hat{\nu}_i^n, && i = 0, \dots, n+1; \quad \vartheta_i^n = d_n \hat{\vartheta}_i^n, && i = 0, \dots, n-r; \\
&\eta_i^n := d_n \hat{\eta}_i^n, && i = 0, \dots, n-r+1; \quad \alpha_i^n = d_n^{-m} \hat{\alpha}_i^n, && i = 0, \dots, n,
\end{aligned}
\tag{2.5}
$$

where $\{\hat{\nu}_i^n\}_{i=0}^{n+1}, \{\hat{\vartheta}_i^n\}_{i=1}^{n-r}, \{\hat{\eta}_i^n\}_{i=1}^{n-r}$ are the points of alternance of $\mathcal{Z}_n$, the knots of the kernel K_n, and the points of sign change of the function $\mathcal{Z}_n^{(r+1)}(t)$, respectively.

From the property (1.2) and the definitions (2.3) of the function R_n and (2.5) of the points $\{\nu_i^n\}_{i=0}^{n-1}$ it follows that

$$R_n(\nu_i^n) = (-1)^{i+m} \|R_n\|_{\mathbb{C}[0, d_n]} = (-1)^{i+m} B, \qquad i = 0, \dots, n+1. \tag{2.6}$$

The equations (1.4) and definitions (2.5) imply that the coefficients $\{\alpha_i^n\}_{i=0}^n$ are the solutions of the system

$$
\begin{cases}
\displaystyle\sum_{i=0}^n \alpha_i^n [\nu_i^n]^j = m! \, \delta_{m,j}, & j = 0, \dots, r; \\
\displaystyle\sum_{i=0}^n \alpha_i^n (\nu_i^n - \vartheta_l^n)_+^{r-1} = 0, & l = 1, \dots, n-r+1;
\end{cases}
\tag{2.7}
$$

Let us introduce kernels $T_n(t)$ and $W_n(t)$:

$$
T_n(t) = -\frac{1}{(r-1)!} \sum_{i=0}^n \alpha_i^n (\nu_i^n - t)_+^{r-1},
$$

$$
W_n(t) = \frac{1}{r!} \sum_{i=0}^n \alpha_i (\nu_i^n - t)_+^r.
$$
$$\tag{2.8}$$

By the definitions (2.5) and (2.8),

$$\begin{aligned} T_n(t) &= d_n^{r-m-1} K_n\left(t/d_n\right), && 0 \le t \le d_n, \\ W_n(t) &= \sigma_n^{r-m} F_n\left(t/d_n\right), && 0 \le t \le d_n, \end{aligned} \qquad (2.9)$$

where the kernel $K_n(t)$ is introduced in (1.3), and $F_n(t) = \dfrac{1}{r!} \sum_{i=0}^{n} \hat{\alpha}_i^n (\hat{\nu}_i^n - t)_+^r$.

Then, the extremal property (1.5) of the function $\mathcal{Z}_n^{(r)}$ and Corollary 2.2.4, applied to the dilation T_n of the kernel K_n, guarantee that

$$\sup_{h \in H_0^\alpha[0,d_n]} \int_0^{d_n} h(t) T_n(t)\, dt = \int_0^{d_n} [R_n^{(r)}(t) - R_n^{(r)}(0)] T_n(t)\, dt. \qquad (2.10)$$

As in Chapter 5, we arrive at the formula for the mth derivative of the function $f \in W^r H^\alpha(\mathbb{R}_+)$ at the origin:

$$f^{(m)}(0) = \sum_{i=0}^{n} \alpha_i^n f(\nu_i^n) + \int_0^{d_n} f(t) T_n(t)\, dt. \qquad (2.11)$$

The formula (2.11) and properties (2.6) and (2.10) enable us to conclude that the following inequality (with $\omega = \omega_\alpha$)

$$f^{(m)}(0) \le R_n^{(m)}(0) = \sum_{i=0}^{n} |\alpha_i^n| B + \int_0^{d_n} \Re_\omega(W_n; t) \omega'(t)\, dt, \qquad (2.12)$$

holds for all functions $f \in W^r H^\alpha[0, d_n]$ with $\|f\|_{C[0,d_n]} \le B$.

7.3. Chebyshev ω-splines of the fixed norm B

In Section 7.2 we showed how to transform the Chebyshev ω_α-splines on the interval $[0,1]$ into the Chebyshev ω_α-splines of the norm B with the given number $n+2$ of alternance points on an appropriate interval $[0, d_n]$. In our papers [7], [8] we constructed the extremal Chebyshev ω-splines for all concave modulii of continuity ω with the property

$$\lim_{t \to +\infty} \omega(t) = +\infty. \qquad (\star)$$

THEOREM 7.3.1. *Let* $B > 0$, m, $r\,n \in \mathbb{N}$, $0 < m < r$, $n \geq r$, *and* ω *be a concave modulus of continuity on* $\mathbb{R}_+$ *satisfying* $(\star)$. *Then, there exist* $d_n = d(B,\omega,r,m,n) > 0$, *collections of points* $\{\hat{\nu}_i^n = \hat{\nu}_i(B,\omega,r,m,n)\}_{i=0}^{n+1}$ *and* $\{\hat{\vartheta}_i^n = \hat{\vartheta}_i(B,\omega,r,m,n)\}_{i=0}^{n-r+1}$, *and the function* $R_n = R_{B,\omega,r,m,n}$, *endowed with the properties*

(1) $R_n(\nu_i) = (-1)^{i+m}\|R_n\|_{\mathrm{C}[0,d_n]} = (-1)^{i+m}B,$ $i = 0,\ldots,n+1;$

$$\text{(2)}\quad \sup_{h\in H_0^\omega[0,d_n]} \int_0^{d_n} h(x)T_n(x)\,dx = \int_0^{d_n} [R_n^{(r)}(x) - R_n^{(r)}(0)]T_n(x)\,dx, \tag{3.1}$$

where the coefficients $\{\alpha_i\}_{i=0}^n$ *of the kernel* $K(t) = -\dfrac{1}{(r-1)!}\sum_{i=0}^n \alpha_i(\nu_i - t)_+^{r-1}$ *satisfy equations* (2.7).

Once again, the properties (3.1), (1), (2) lead us to the extremal inequality (2.12) for all functions $f \in W^r H^\omega[0,d_n]$: $\|f\|_{\mathrm{C}[0,d_n]} \leq B$.

In addition, the function R_n as the extremal function of the problem (3.1), (2) enjoys the analog of the property (6.4.35): *for any point* $x \in [0,\nu_n^n]$ *there exists such* r_x, $v_x \in [0,\nu_n^n]$ *that* $|x - v_x| = r_x$, and

$$R_n^{(r+1)}(x) = \omega'(r_x), \qquad |R_n^{(r)}(x) - R_n^{(r)}(v_x)| = \omega(r_x). \tag{3.2}$$

Notice that in the case of $\omega(t) = \omega_\alpha(t)$, the transformation (2.3) preserves (6.4.35), so the property (3.2) follows immediately from (6.4.35).

In order to implement the limiting procedure to the family of extremal functions $\{R_n(t)\}_{n\geq r}$, we need a series of results on the boundedness of the alternance points $\{\nu_i^n\}_{i=0}^{n+1}$, the points $\{\vartheta_i^n\}_{i=1}^{n-r}$ and the coefficients $\{\alpha_i^n\}_{i=0}^n$.

7.4. Properties of Chebyshev ω-splines of the fixed norm

First of all, applying Proposition 1.1.5 to the function $R_n(x)$, we obtain the estimate

$$2(n+1)B = \|R_n'\|_{\mathrm{L}_1[0,d_n]} \leq d_n\|R_n'\|_{\mathrm{C}[0,d_n]} \leq d_n^r \omega(d_n). \tag{4.1}$$

In view of our assumption $(\star)$, we infer from (4.1) that

$$\lim_{n\to+\infty} d_n = +\infty. \tag{4.2}$$

For $\omega(t) = \omega_\alpha(t)$, the property (4.2) follows immediately from the expression (2.2) for d_n and Proposition 7.1.1.

LEMMA 7.4.1. *Let* $\{\eta_i^n\}_{i=1}^{n-r}$ *be the set of points of sign change of the function* $R_n^{(r+1)}$, *and* $\eta_0^n = 0$. *There exists such a constant* $L_1 = L_1(\omega,r,B)$ *that*

$$\eta_{i+2r}^n - \eta_i^n \geq L_1.$$

for all $n \geq 4r$ *and* $i = 0,\ldots,n-3r$.

Proof. The inequalities

$$\nu_i^n < \eta_i^n < \nu_{i+r}^n, \qquad i = 1, \ldots, n-r, \tag{4.3}$$

between the points $\{\eta_i^n\}_{i=1}^{n-r}$ of sign change of $R_n^{(r+1)}$ and the zeroes $\{\nu_i^n\}_{i=1}^{n}$ of the derivative $R_n'(t)$ follow immediately from the Rolle theorem. From (4.3) we derive the inclusion

$$[\nu_{i+r}^n, \nu_{i+2r}^n] \subset [\eta_i^n, \eta_{i+2r}^n]. \tag{4.4}$$

Notice that $\dfrac{d}{dt} R_n(\nu_i) = 0, \quad i = 1, \ldots, n..$ Therefore, the function $\dfrac{d}{dt} R_n'(t)$ has at least r zeroes on the interval $[\nu_{i+r}^n, \nu_{i+2r}^n]$. Applying Proposition 1.1.5 to the function $R_n' \in W^{r-1} H^\omega[\nu_{i+r}^n, \nu_{i+2r}^n]$ and using the identity

$$\|R_n'\|_{L_1[\nu_{i+r}^n, \nu_{i+2r}^n]} = 2r\|R_n\|_{C[\nu_{i+r}^n, \nu_{i+2r}^n]} = 2rB,$$

(see (1.7), (1.8)), we obtain the chain of inequalities:

$$2rB = \|R_n'\|_{L_1[\nu_{i+r}^n, \nu_{i+2r}^n]} \le (\nu_{i+2r}^n - \nu_{i+r}^n)\|R_n'\|_{C[\nu_{i+r}^n, \nu_{i+2r}^n]} \le$$

$$\le (\nu_{i+2r}^n - \nu_{i+r}^n)^r\|R_n^{(r)}\|_{C[\nu_{i+r}^n, \nu_{i+2r}^n]} \le (\nu_{i+2r}^n - \nu_{i+r}^n)^r \omega(\nu_{i+2r}^n - \nu_{i+r}^n). \tag{4.5}$$

From inclusions (4.3) and inequalities (4.5) we obtain the desired estimate:

$$\eta_{i+2r}^n - \eta_i^n \ge \nu_{i+2r}^n - \nu_{i+r}^n \ge g^{-1}(2rB) =: L_1, \tag{4.6}$$

where $g^{-1}(x)$, $x \in \mathbb{R}_+$, is the inverse function of $g(x) = x^r \omega(x)$. $\qquad \square$

Now we can show that the number of active knots of kernels $\{T_n(t)\}_{n \ge r}$ on a compact $\mathcal{K} \in \mathbb{R}_+$ is bounded by a constant dependent only on r, B, ω, $\mathcal{K}$.

LEMMA 7.4.2. *For any compact $\mathcal{K}$, there exists such a constant $\mathcal{M}_\mathcal{K} = \mathcal{M}_\mathcal{K}(r, \omega, B)$ that for all $n \ge r$,*

$$T_n(t) = \frac{(-1)^{r-m}}{(r-1-m)!} t^{r-1} + \frac{(-1)^{r-1}}{(r-1)!} \sum_{i=0}^{\mathcal{M}_\mathcal{K}} \alpha_i^n (t - \nu_i^n)_+^{r-1}, \quad t \in \mathcal{K}. \tag{4.7}$$

Proof. By the definition,

$$x_+^{r-1} = x^{r-1} - (-1)^{r-1}(-x)_+^{r-1}, \qquad x \in \mathbb{R}. \tag{4.8}$$

Therefore, by (4.8) and equations (2.7) for the coefficients $\{\alpha_i^n\}_{i=0}^{n}$, we have

$$\sum_{i=0}^{n} \alpha_i^n (\nu_i^n - t)_+^{r-1} = \sum_{i=0}^{n} \alpha_i^n (\nu_i^n - t)^{r-1} - (-1)^{r-1} \sum_{i=0}^{n} \alpha_i^n (t - \nu_i^n)_+^{r-1} =$$

$$= \sum_{i=0}^{n} \alpha_i^n \sum_{j=0}^{r-1} \binom{r-1}{j} [\nu_i^n]^j (-t)^{r-1-j} + (-1)^r \sum_{i=0}^{n} \alpha_i^n (t - \nu_i^n)_+^{r-1} =$$

$$= \sum_{j=0}^{r-1} \left(\sum_{i=0}^{n} \alpha_i^n [\nu_i^n]^j \right) \binom{r-1}{j} (-t)^{r-1-j} + (-1)^r \sum_{i=0}^{n} \alpha_i^n (t - \nu_i^n)_+^{r-1} =$$

$$= \frac{(-1)^{r-1-m}(r-1)!}{(r-1-m)!} t^{r-1-m} + (-1)^r \sum_{i=0}^{n} \alpha_i^n (t - \nu_i^n)_+^{r-1}. \tag{4.9}$$

For any compact $\mathcal{K} \in \mathbb{R}_+$, there exists a $j = j(\mathcal{K}) \in \mathbb{N}$, such that $\mathcal{K} \subset [0, L_1 j]$, where the constant L_1 is defined in (4.6). Also by (4.6),

$$\nu_{rj}^n = \sum_{k=1}^{j} (\nu_{rk}^n - \nu_{r(k-1)}^n) \geq L_1 j. \tag{4.10}$$

Thus, $\mathcal{K} \subset [0, \nu_{rj}^n]$. Consequently,

$$(t - \nu_i^n)_+^{r-1} \equiv 0, \qquad t \in \mathcal{K}, \qquad \text{for all} \quad i > rj, \tag{4.11}$$

and, by (4.9), for all $n \geq (r+1)j$, we finally have

$$T_n(t) := -\frac{1}{(r-1)!} \sum_{i=0}^{n} \alpha_i^n (\nu_i^n - t)_+^{r-1} =$$

$$= \frac{(-1)^{r-m}}{(r-1-m)!} t^{r-1-m} + \frac{(-1)^{r-1}}{(r-1)!} \sum_{i=0}^{(r+1)j} \alpha_i^n (t - \nu_i^n)_+^{r-1}. \tag{4.12}$$

$\square$

7.5. General properties of extremal functions R_n

LEMMA 7.5.1. *Let $0 \leq m \leq r$, and the function $R_n = R_{B,\omega,r,m,n}$ be as in Theorem 7.3.1. Then, there exists a finite limit of the sequence $\{|R_n^{(m)}(0)|\}_{n \geq r}$:*

$$\lim_{n \to +\infty} |R_n^{(m)}(0)| < +\infty. \tag{5.1}$$

Proof. By (2.12), for each fixed $l \geq r$, the function $R_l(t)$ is extremal in the problem

$$|f^{(m)}(0)| \to \sup, \qquad f \in W^r H^\omega[0, d_l], \quad \|f\|_{C[0,d_l]} \leq B. \tag{5.2}$$

As we showed in (4.2), $\lim_{t \to +\infty} d_n = +\infty$. Therefore, for each fixed $n \geq r$, there exists an $\mathcal{I}_n \in \mathbb{N}$, such that

$$d_l \geq d_n, \qquad \text{for all} \quad l \geq \mathcal{I}_n. \tag{5.3}$$

From the extremality of the function R_l in the problem (5.2) and the inequality (5.3) it follows that

$$|R_i^{(m)}(0)| = |R_i|_{[0,d_n]}^{(m)}(0)| \leq |R_n^{(m)}(0)|, \qquad \text{for all} \quad i \geq \mathcal{I}_n. \tag{5.4}$$

From (5.4) we infer that

$$\limsup_{i \to +\infty} |R_i^{(m)}(0)| \leq \liminf_{i \to +\infty} |R_i^{(m)}(0)|. \tag{5.5}$$

$\square$

COROLLARY 7.5.2. *For $n \geq r$, let the coefficients $\{\alpha_i^n\}_{i=0}^n$ be derived from equations (1.7). Then,*

$$\limsup_{n \to +\infty} \sum_{i=0}^n |\alpha_i^n| < +\infty. \tag{5.6}$$

Proof. Recall the formula (5.12):

$$|R_n^{(m)}(0)| = \sum_{l=0}^n |\alpha_l^n| + \int_0^{d_n} \Re\left(W_n; t\right) \omega'(t)\, dt. \tag{5.7}$$

Now the property (5.6) follows from the result of Lemma 7.5.1. $\square$

At this point we remind the reader that the function R_n and its domain $[0, d_n]$ depend on m: $R_n(t) = R_{n,m}(t)$ and $d_n = d_{n,m}$. Put

$$d = \max_{0 \leq m \leq r} d_{r,m}. \tag{5.8}$$

COROLLARY 7.5.3. *Let m, $l \in \mathbb{Z}_+$, $0 \leq m$, $l \leq r$. There exists such a constant $\mathcal{L} = \mathcal{L}(r, \omega, B)$ that*

$$\|R_{n,m}^{(l)}\|_{\mathbb{L}_\infty[0,d_{n,m}]} \leq \mathcal{L}, \qquad for\ all \quad n \geq r. \tag{5.9}$$

Proof. Let the constant $d = d(r, \omega, B)$ be defined by (5.8).

First, let us consider such functions $R_{n,m}$ that $d_{n,m} \leq 2d$.

Recall that the function $R'_{n,m}(t) \in W^{r-1}H^\omega[0, d_{n,m}]$ has $n \geq r$ zeroes at the points $\{\nu_i^n = \nu_i^n(m)\}_{i=1}^n$. Applying Proposition 1.1.5 to $R'_n(t)$, we derive inequalities

$$\|R_{n,m}^{(l)}\|_{\mathbb{C}[0,d_{n,m}]} \leq d_{n,m}^{r-l}\omega(d_{n,m}) \leq (2d)^{r-l}\omega(2d), \qquad 1 \leq l \leq r. \tag{5.10}$$

It remains to estimate the $\mathbb{C}[0, d_{n,m}]$-norms of the functions $R_{n,m}$ with $\sigma_{n,m} > 2d$. For such a function $R_{n,m}$ and $\tau \in [0, d_{n,m}]$, put

$$G_\tau(x) = \begin{cases} R_{n,m}(x + \tau), & x \in [0, d], & \text{if } \tau < d; \\ R_{n,m}(\tau - x), & x \in [0, d], & \text{if } \tau > d. \end{cases} \tag{5.11}$$

The function G_τ belongs to the class $W^r H^\omega[0, d]$, and $\|G_\tau\|_{\mathbb{C}[0,d]} \leq B$. From the property of extremality of function $R_{r,l}$ in (5.2) for $d_l = d_{r,l}$ it follows that for all $1 \leq l \leq r$,

$$|R_{n,m}^{(l)}(\tau)| = |G_\tau^{(l)}(0)| = |G|_{[0,d_{r,l}]}^{(l)}(0)| \leq |R_{r,l}^{(l)}(0)|. \tag{5.12}$$

Since τ is an arbitrary point on $[0, d_{n,m}]$, from (5.12) we conclude that

$$\|R_{n,m}^{(l)}\|_{\mathbb{C}[0,d_{n,m}]} \leq |R_{r,l}^{(l)}(0)|, \qquad 1 \leq l \leq r, \tag{5.13}$$

as long as $d_{n,m} > 2d$.

Finally, in view of inequalities (5.10), (5.13) and the property $\|R_n\|_{\mathbb{C}[0,d_{n,m}]} = B$, it remains to set

$$\mathcal{L} := \max_{1 \leq l \leq r} \{(2d)^{r-l}\omega(2d) + |R_{r,l}^{(l)}(0)|\} + B \tag{5.14}$$

in order to obtain the constant in the inequality (5.9). $\square$

COROLLARY 7.5.4. *There exists such a constant $\mathcal{D} = \mathcal{D}(r, \omega, B)$ that*

$$|R_n^{(r+1)}(x)| \geq \mathcal{D}, \quad x \in [0, \nu_n^n], \quad \text{for all } n \geq r. \tag{5.15}$$

We use the property (3.2) of the function R_n that holds for all $x \in [0, \nu_n^n]$ except a finite number of points:

$$R_n^{(r+1)}(x) = \omega'(r_x) \quad \exists\, v_x : |R_n^{(r)}(x) - R_n^{(r)}(v_x)| = \omega(r_x). \tag{5.16}$$

However, by Corollary 7.5.3, $\|R_n^{(r)}\|_{\mathbb{C}[0,d_n]} \leq \mathcal{L}$, where the constant $\mathcal{L} = \mathcal{L}(r, \omega, B)$ is defined in (5.14). Thus, by (5.16),

$$\omega(r_x) \leq 2\|R_n^{(r)}\|_{\mathbb{L}_\infty[0,d_n]} \leq 2\mathcal{L}. \tag{5.17}$$

Consequently, by (5.16), (5.17),

$$|R_n^{(r+1)}(x)| = \omega'(r_x) \geq \omega'(\omega^{-1}(2\mathcal{L})) =: \mathcal{D}. \tag{5.18}$$

$\square$

Now we are in a position to establish the *uniform* boundedness of the points $\{\nu_i^n\}_{i=0}^n$, $\{\vartheta_i^n\}_{i=0}^{n-r+1}$ and $\{\eta_i^n\}_{i=1}^{n-1}$ from above.

LEMMA 7.5.5. *Let $\{\nu_i^n\}_{i=0}^{n+1}$ be the alternance points of R_n, $n \geq r$. Then for any $j \in \mathbb{N}$, there exists such a constant $c_r = c_{r,\omega,B}$ that*

$$\nu_j^n \leq c_r j, \quad \text{for all } n \geq j. \tag{5.19}$$

Proof. From inequalities (4.3) it follows that the function $R_n^{(r+1)}$ can have at most r points $\{\nu_i\}_{i=k-r+1}^{i=k}$ of sign change on each of the intervals $[\nu_k^n, \nu_{k+1}^n]$, $k = 0, \ldots, n-1$. In particular, there exist an index $l : 0 \leq l \leq j-1$, and an interval $(\alpha_n, \beta_n) \subset [\nu_l^n, \nu_{l+1}^n]$, such that

$$\beta_n - \alpha_n \geq \frac{\nu_j^n}{r \cdot j}, \quad \text{and} \quad (\alpha_n, \beta_n) \bigcap \{\eta_i^n\}_{i=1}^{n-r} = \varnothing. \tag{5.20}$$

Notice that by our choice of the interval (α_n, β_n) in (5.20), both functions $R_n'(t)$ and $R_n^{(r+1)}(t)$ maintain their sign on the interval (α_n, β_n).

Using the inequality (5.15) and applying Corollary 4.4.2 to estimate the norm $\|R_n\|_{\mathbb{L}_1[\alpha_n, \beta_n]}$, we arrive at the inequalities:

$$2B = |R_n(\nu_{l+1}^n - R_n(\nu_l^n)| = \int_{\nu_l^n}^{\nu_{l+1}^n} |R_n(t)|\, dt \geq$$

$$\geq \|R_n\|_{\mathbb{L}_1[\alpha_n, \beta_n]} \geq \mathcal{D}(\beta_n - \alpha_n)^{r+1} u_r \geq \mathcal{D}\left[\frac{\nu_j^n}{j \cdot r}\right]^{r+1} u_r, \tag{5.21}$$

where the constant u_r is introduced in the proof of Corollary 4.4.2. Therefore,

$$\nu_j^n \leq c_r j, \quad c_r := r\left[\frac{2B}{\mathcal{D}u_r}\right]^{\frac{1}{r+1}}. \tag{5.22}$$

$\square$

COROLLARY 7.5.6. *Let the points* $\{\eta_i^n\}_{i=1}^{n-1}$ *and* $\{\vartheta_i^n\}_{i=1}^{n-r+1}$ *be the points of sign change of* $R_n^{(r+1)}$ *and the alternance points of* R_n, *respectively, and the constant* c_r *be introduced in* (5.22). *Then, for any* $j \in \mathbb{N}$ *and all* $n \geq r + j$,

$$\vartheta_j^n \leq c_r(j + r - 1), \qquad \eta_j^n \leq c_r(j + r). \tag{5.23}$$

Proof. From inequalities (5.1.1), definitions (2.5) and inequalities (5.19) it follows that

$$\vartheta_j^n < \nu_{j+r-1}^n \leq c_r(j + r - 1), \tag{5.24}$$

while inequalities (5.25) and (5.53) imply that

$$\eta_j^n < \nu_{j+r}^n \leq c_r(j + r). \tag{5.25}$$

$\square$

7.6. The restricted action of the generating kernel $T_n(t)$

By the definition, the function $R_n^{(r)}(t)$ is extremal in the problem

$$\int_0^{d_n} h(t) T_n(t)\, dt \rightarrow \sup, \qquad h \in H^\omega[0, d_n]. \tag{6.1}$$

As we showed in Theorem 6.0.1, $\operatorname{supp} T_n(t) = [0, \nu_{I_n}^n]$, for some index I_n, $r \leq I_n \leq n$. Theorem X enables us to characterize the structure of the function R_n:

there exists a $V_0^{I_n}$-*partition* $\mathcal{V}[n] = \left\{ \{A_i[n]\}_{i=1}^{I_n-1}, \{B_{ij}[n], C_{ji}[n]\}_{(i,j)\in\mathcal{P}(I_n)} \right\}$ *of the interval* $[0, \nu_{I_n}^n]$ *with the following properties:*

$$(A)\ \nu_i^n \in A_i[n], \qquad i = 1, \ldots, I_n - 1;$$

$$(B)\ \int_{B_{ij}[n]\cup C_{ji}[n]} T_n(t)\, dt = 0, \qquad (i,j) \in \mathcal{P}(I_n); \tag{6.2}$$

$$(C)\ \int_{A_i[n]} T_n(t)\, dt = 0, \qquad i = 1, \ldots, I_n - 1;$$

For all $(i,j) \in \mathcal{P}(I_n)$, *the function* $R_n^{(r)}(t)$ *is a solution of the problem*

$$\int_0^{d_n} h(t) T_{ij}^n(t)\, dt \rightarrow \sup, \qquad h \in H^\omega[0, d_n],$$

$$T_{ij}^n(t) = T_n(t) \cdot \mathcal{X}(B_{ij}[n] \cup C_{ji}[n]; t), \qquad t \in [0, d_n]; \tag{6.3}$$

and for all $i = 1, \ldots, I_n - 1$, the function $R_n^{(r)}(t)$ solves the problem

$$
\int_0^{d_n} h(t) T_i^n(t)\, dt \to \sup, \qquad h \in H^\omega[0, d_n],
\tag{6.4}
$$

$$
T_i^n(t) = T_n(t) \cdot \mathcal{X}(A[n]; t), \qquad t \in [0, d_n].
$$

According to Korneichuk's Lemma 2.1.1, the function $R_n^{(r+1)}(t)$ is given by the following formulas on the interval $[0, \nu_{I_n}]$ for some $\chi \in \{\pm 1\}$.

$$
\chi R_n^{(r+1)}(t) = \begin{cases} (-1)^{i+1} \omega'(\rho_{ij}(t) - t), & t \in B_{ij}[n]; \\ (-1)^{i+1} \omega'(t - \rho_{ij}^{-1}(t)), & t \in C_{ji}[n]; \end{cases}
\tag{6.5}
$$

for all $(i, j) \in \mathcal{P}(I_n)$, where the function $\rho_{ij} = \rho_{i,j,n} : B_{ij}[n] \to C_{ji}[n]$ is determined from the equation

$$
F_{ij}^n(t) = F_{ij}^n(\rho_{ij}(t)), \quad t \in B_{ij}[n], \quad \rho_{ij}(t) \in C_{ji}[n],
\tag{6.6}
$$

where $F_{ij}^n(t) = \int_0^t T_{ij}^n(y)\, dy, \quad t \geq 0$.

On the intervals $A_i[n] =: [\alpha_{2i-1}[n], \alpha_{2i}[n]], \quad i = 1, \ldots, I_n - 1$, we have

$$
\chi R_n^{(r+1)}(t) = \begin{cases} (-1)^{i+1} \omega'(\rho_i(t) - t), & t \in [\alpha_{2i-1}[n], \vartheta_i^n]; \\ (-1)^{i+1} \omega'(t - \rho_i^{-1}(t)), & t \in [\vartheta_i^n, \alpha_{2i}]; \end{cases}
\tag{6.7}
$$

where $\rho_i : [\alpha_{2i-1}, \vartheta_i^n] \to [\vartheta_i^n, \alpha_{2i}]$ is determined from the equation

$$
F_i^n(t) = F_i^n(\rho_i(t)), \quad t \in [\alpha_{2i-1}, \vartheta_i^n],
\tag{6.8}
$$

where $F_i^n(t) = \int_0^t T_i^n(y)\, dy, \quad t \geq 0$.

LEMMA 7.6.1. *For $n \geq r$ and $(i, j) \in \mathcal{P}(I_n)$, let $B_{ij}[n] =: [\beta_{ij}^1, \beta_{ij}^2] \neq \square$, i.e. $\beta_{ij}^1 < \beta_{ij}^2$. Then, there exists a constant $\mathcal{E} = \mathcal{E}_{r, \omega, B}$ such that*

$$
j - i \leq \mathcal{E}, \qquad n \geq r.
\tag{6.9}
$$

Proof. First of all, from the formula (6.5) and the property (2.1.10), proved in Corollary 2.1.3, it follows that if $\beta_{ij}^1 < \beta_{ij}^2$, then

$$\left| R_n^{(r)}(\gamma_{ij}^2) - R_n^{(r)}(\beta_{ij}^1) \right| = \left| R_n^{(r)}(\rho_{ij}(\beta_{ij})) - R_n^{(r)}(\beta_{ij}^1) \right| = \omega(\gamma_{ij}^2 - \beta_{ij}^1). \qquad (6.10)$$

On the other hand, by the assertion of Corollary 7.5.3,

$$\| R_n^{(r)} \|_{C[0,d_n]} \leq \mathcal{L}, \qquad (6.11)$$

where the constant $\mathcal{L} = \mathcal{L}_{r,\omega,B}$ is defined in (5.14).

From (6.10) and (6.11) we infer that

$$\gamma_{ij}^2 - \beta_{ij}^1 \leq \omega^{-1}(2\mathcal{L}). \qquad (6.12)$$

By (6.2), (A), we have inclusions

$$\beta_{ij}^1 \in [\vartheta_{i-1}^n, \vartheta_i^n], \qquad \gamma_{ij}^2 \in [\vartheta_{j-1}^n, \vartheta_j^n]. \qquad (6.13)$$

Remark 2.2.3 provides inequalities (2.2.18) between the points $\{\vartheta_i\}_{i=1}^{I_n-r+1}$ of sign change of the kernel T_n and the points $\{\eta_i^n\}_{i=1}^{I_n-r}$ of sign change of the function $R_n^{(r+1)}(t)$:

$$\vartheta_i^n < \eta_i^n < \vartheta_{i+1}^n, \qquad i = 1, \ldots, I_n - r. \qquad (6.14)$$

Let

$$k = \left[\frac{j - i - 2}{2r} \right]. \qquad (6.15)$$

Then, from inclusions (6.13) we observe that $[\eta_i^n, \eta_{i+(2r+2)k}^n] \subset [\beta_{ij}^1, \gamma_{ij}^2]$. Then, by Lemma 7.4.1,

$$\gamma_{ij}^2 - \beta_{ij}^1 \geq \eta_{i+2rk}^n - \eta_i^n = \sum_{l=1}^{k} \left(\eta_{i+2rl} - \eta_{i+2r(l-1)} \right) \geq L_1 k. \qquad (6.16)$$

The combination of the estimates (6.12) and (6.13) and the definition (6.15) of k leads us to the conclusion that

$$\frac{j - i - 2}{2r} - 1 \leq k \leq \frac{\omega^{-1}(2\mathcal{L})}{L_1}. \qquad (6.17)$$

Finally, by (6.17),

$$j - i \leq \mathcal{E} := (2r + 2) \left[\frac{\omega^{-1}(2\mathcal{L})}{L_1} + 1 \right]. \qquad (6.18)$$

$\square$

Chapter 8

Chebyshev Ω-Splines of the Half-line $\mathbb{R}_+$

Relying on the results of the previous chapter, we apply the limiting procedure to construct the extremal function of the problem

$$\|f^{(m)}\|_{L_\infty(\mathbb{R}_+)} \to \sup, \qquad f \in W^r H^\alpha(\mathbb{R}_+), \quad \|f\|_{L_\infty(\mathbb{R}_+)} \le B, \qquad (0.0)$$

and describe the corresponding multiplicative Kolmogorov inequalities associated with the problem (0.0).

8.1. Limiting procedure

Lemma 7.5.5 and Corollaries 7.5.6 and 7.5.2 enable us to extract such a subsequence $\{n_k\}_{k=1}^\infty$ that

$$\lim_{k \to \infty} \alpha_j^{n_k} = \alpha_j, \quad j \in \mathbb{Z}_+; \qquad \lim_{k \to \infty} \nu_j^{n_k} = \nu_j, \quad j \in \mathbb{N};$$
$$\lim_{k \to \infty} \vartheta_j^{n_k} = \vartheta_j, \quad j \in \mathbb{N}; \qquad \eta_j^{n_k} = \eta_j, \quad j \in \mathbb{N}. \qquad (1.1)$$

Now we are in a position to define the function $\mathcal{Z}$ and kernels K and F on the whole half-line $\mathbb{R}_+$. For $t \in \mathbb{R}_+$, put

$$F(t) = \frac{(-1)^{r-m}}{(r-m)!} t^r + \frac{(-1)^{r-1}}{r!} \sum_{i=0}^\infty \alpha_i (t - \nu_i)_+^r;$$

$$K(t) = \frac{(-1)^{r-m}}{(r-1-m)!} t^{r-1} + \frac{(-1)^{r-1}}{(r-1)!} \sum_{i=0}^\infty \alpha_i (t - \nu_i)_+^{r-1}. \qquad (1.2)$$

By (7.4.10), $\lim_{i \to \infty} \nu_i = \infty$. Therefore, both kernels $F(t)$ and $K(t)$ have only a finite number of knots on any given compact $\mathcal{K}$. By Lemma 7.4.2, each of the kernels T_{n_k} and W_{n_k}, defined in (7.2.8), is a polynomial spline with a finite number of knots on $\mathcal{K}$. The limiting relations (1.1) for the points $\{\nu_i^n\}_{i=0}^{n+1}$ and the coefficients $\{\alpha_i^n\}_{i=0}^n$ imply that the families of kernels $\{T_{n_k}\}$ and $\{W_{n_k}\}_{k \in \mathbb{N}}$ converge uniformly on each compact $\mathcal{K} \in \mathbb{R}_+$ to the kernels $K(t)$ and $F(t)$, respectively.

Corollary 7.5.3 enables us to apply the Arzela–Ascoli theorem to each of the families of functions $\{R_{n_k}^{(l)}\}_{k \in \mathbb{N}}$, $0 \le l \le r$. Therefore, we can assume that the sequence of indices $\{n_k\}_{k \in \mathbb{N}}$ is chosen in such a way that the family of functions $\{R_{n_k}^{(l)}\}_{k \in \mathbb{N}}$ converges uniformly on each compact $\mathcal{K} \in \mathbb{R}_+$ to the lth derivative $\mathcal{Z}^{(l)}(t)$ of some function $\mathcal{Z} \in W^r H^\omega(\mathbb{R}_+)$. In other words,

$$R_{n_k} \rightrightarrows \mathcal{Z}, \quad \text{as } k \to \infty, \qquad (1.3)$$

where the convergence in (1.3) is understood in the sense of convergence in the metrics of $\mathbb{C}^r(\mathcal{K})$ for each compact $\mathcal{K} \subset \mathbb{R}_+$.

Our first goal is to prove the following extremal property of the function $\mathcal{Z}(t)$:

$$\sup_{h \in H_0^\nu(\mathbb{R}_+)} \int_{\mathbb{R}_+} h(t) K(t)\, dt = \int_{\mathbb{R}_+} [\mathcal{Z}^{(r)}(t) - \mathcal{Z}^{(r)}(0)] K(t)\, dt. \qquad (1.4)$$

Recall that for each $n \geq r$, $\operatorname{supp} T_n(t) = [0, \nu_{I_n}^n]$ for some $I_n \geq r$. Put

$$\mathcal{J} := \sup_{k \in \mathbb{N}} I_{n_k}. \qquad (1.5)$$

REMARK 8.1.1. We may assume that the subsequence $\{n_k\}_{k \in \mathbb{N}}$ is chosen in such a way that the sequence $\{I_{n_k}\}_{k \in \mathbb{N}}$ is nondecreasing, if $\mathcal{J} = +\infty$, and $I_{n_k} \equiv \mathcal{J}$, $k \in \mathbb{N}$, if $\mathcal{J} < +\infty$.

For all pairs $(i, j) \in \mathcal{P}(\mathcal{J})$: $j - i \leq \mathcal{E}$, define

$$B_{ij} = \lim_{k \to \infty} B_{ij}[n_k], \qquad C_{ji} = \lim_{k \to \infty} C_{ji}[n_k]. \qquad (1.6)$$

We do not consider atoms B_{ij}, C_{ji} with $j - i > \mathcal{E}$, because according to Lemma 7.6.1, $B_{ij}[n] = \square$, $C_{ji}[n] = \square$, for all $(i, j) \in \mathcal{P}(I_n)$ such that $j - i > \mathcal{E}$.

For $i = 1, \ldots, \mathcal{J} - 1$, put

$$A_i = \lim_{k \to \infty} A_i[n_k]. \qquad (1.7)$$

The convergence of intervals in (1.6) and (1.7) is understood in the sense of convergence of their respective endpoints.

From the properties (7.6.2) and the uniform convergence of the family of kernels $\{T_{n_k}\}_{k \in \mathbb{N}}$ to K it follows that

$$(A) \quad \nu_i \in A_i, \qquad i = 1, \ldots, \mathcal{J} - 1;$$

$$(B) \quad \int_{B_{ij} \cup C_{ji}} K(t)\, dt = 0, \qquad (i, j) \in \mathcal{P}(\mathcal{J}); \qquad (1.8)$$

$$(C) \quad \int_{A_i} K(t)\, dt = 0, \qquad i = 1, \ldots, \mathcal{J} - 1.$$

For $i = 1, \ldots, \mathcal{J} - 1$, put

$$K_i(t) := K(t) \mathcal{X}(A_i; t), \quad F_i(t) := \int_0^t K_i(y)\, dy, \quad t \in \mathbb{R}_+. \qquad (1.9)$$

For all $(i, j) \in \mathcal{P}(\mathcal{J})$, put

$$K_{ij}(t) := K(t)\mathcal{X}(B_{ij} \cup C_{ji}; t), \quad F_{ij}(t) := \int_0^t K_{ij}(y)\, dy, \quad t \in \mathbb{R}_+. \qquad (1.10)$$

By (1.8), each of the kernels in (1.9) and (1.10) is simple. Set

$$\Re_\omega(F; t) := \sum_{i=1}^{\mathcal{J}-1} \Re\left(F_i; t\right) + \sum_{(i,j)\in\mathcal{P}(\mathcal{J})} \Re\left(F_{ij}; t\right), \qquad t \in \mathbb{R}_+, \qquad (1.11)$$

where rearrangements $\Re(\Psi; t)$ of simple kernels Ψ are introduced in Definition 2.1.2.

By definitions in (1.9),

$$\operatorname{supp} F_i(t) = A_i =: [\alpha_{2i-1}, \alpha_{2i}], \quad i = 1, \ldots, \mathcal{J} - 1. \qquad (1.12)$$

By definitions in (6.10), for all pairs $(i, j) \in \mathcal{P}(\mathcal{J})$,

$$\operatorname{supp} F_{ij} = [\beta_{ij}^1, \gamma_{ij}^2], \quad \text{where } [\beta_{ij}^1, \beta_{ij}^2] := B_{ij}, \quad \text{and} \quad [\gamma_{ij}^1, \gamma_{ij}^2] := C_{ji}. \qquad (1.13)$$

By our estimates (7.6.12), the lengths of the supports in (1.13) are bounded by the constant $\omega^{-1}(2\mathcal{L})$ dependent only on r, ω, B. Since

$$\mathcal{Z}^{(r)}(\alpha_{2i}) - \mathcal{Z}^{(r)}(\alpha_{2i-1}) = \omega(\alpha_{2i} - \alpha_{2i-1}), \quad i = 1, \ldots, \mathcal{J} - 1,$$

the argument in (7.6.10)–(7.6.12) guarantees that the supports in (1.12) are majorized by the same constant. Therefore, the support of the rearrangement $\Re_\omega(F; t)$ is finite:

$$\operatorname{supp} \Re_\omega(F; t) = [0, \kappa], \qquad (1.14)$$

where

$$\kappa := \max\{ \sup_{1\leq i\leq \mathcal{J}-1} |\alpha_{2i} - \alpha_{2i-1}|; \sup_{(i,j)\in\mathcal{P}(\mathcal{J})} |\gamma_{ij}^2 - \beta_{ij}^1|\}. \qquad (1.15)$$

LEMMA 8.1.1. *Let the function* $\mathcal{Z}$, *the kernels* K, F, *and the rearrangement* $\Re_\omega(F; t)$ *be defined in* (1.3), (1.2) *and* (1.11), *respectively. Then,*

$$\sup_{h\in H_0^\omega(\mathbb{R}_+)} \int_{\mathbb{R}_+} h(t)K(t)\, dt = \int_{\mathbb{R}_+} [\mathcal{Z}^{(r)}(t) - \mathcal{Z}^{(r)}(0)]K(t)\, dt = \int_0^\kappa \Re_\omega\left(F; t\right)\omega'(t)\, dt.$$

$$(1.16)$$

Proof. Recall the definitions of kernels in (1.9) and (1.10). The uniform convergence of the family of kernels $\{T_{n_k}(t)\}_{k\in\mathbb{N}}$ to $K(t)$ and the limiting relations (1.11) imply that we can apply Corollary 2.2.6 to the following families of simple kernels:

(1) such simple kernels $\{T_i^{n_k}(t)\}_{k\in\mathbb{N}}$, $i = 1,\ldots,\mathcal{J}-1$, that $A_i \neq \square$;
(2) such simple kernels $\{T_{ij}^{n_k}(t)\}_{k\in\mathbb{N}}$, $(i,j) \in \mathcal{P}(\mathcal{J})$, that $B_{ij} \neq \square$.

Then, an application of Corollary 2.2.6 assures that

for all $(i,j) \in \mathcal{P}(\mathcal{J})$, the function $\mathcal{Z}^{(r)}(t)$ solves the problem

$$\int_{\mathbb{R}_+} h(t)K_{ij}(t)\,dt \to \sup, \qquad h \in H^\omega(\mathbb{R}_+), \tag{1.17}$$

and for all $i = 1,\ldots,\mathcal{J}-1$, the function $\mathcal{Z}^{(r)}(t)$ solves the problem

$$\int_{\mathbb{R}_+} h(t)K_i(t)\,dt \to \sup, \qquad h \in H^\omega(\mathbb{R}_+). \tag{1.18}$$

Then, from Korneichuk's Lemma 2.1.1 and the extremality of the function $\mathcal{Z}^{(r)}(t)$ in problems (1.17) and (1.18) we derive the following chain of inequalities for any function $h \in H_0^\omega(\mathbb{R}_+)$:

$$\sup_{h\in H_0^\omega(\mathbb{R}_+)} \int_{\mathbb{R}_+} h(t)K(t)\,dt =$$

$$= \sup_{h\in H_0^\omega(\mathbb{R}_+)} \int_{\mathbb{R}_+} h(t)\left[\sum_{i=1}^{\mathcal{J}-1} K_i(t) + \sum_{(i,j)\in\mathcal{P}(\mathcal{J})} K_{ij}(t)\right]dt \leq$$

$$\leq \sum_{i=1}^{\mathcal{J}-1} \sup_{h\in H_0^\omega(\mathbb{R}_+)} \int_{\mathbb{R}_+} h(t)K_i(t)\,dt + \sum_{(i,j)\in\mathcal{P}(\mathcal{J})} \sup_{h\in H^\omega(\mathbb{R}_+)} \int_{\mathbb{R}_+} h(t)K_{ij}(t)\,dt \leq$$

$$\leq \int_{\mathbb{R}_+} \left(\sum_{i=1}^{\mathcal{J}-1} \Re(F_i;t) + \sum_{(i,j)\in\mathcal{P}(\mathcal{J})} \Re(F_{ij};t)\right)\omega'(t)\,dt =$$

$$= \int_{\mathbb{R}_+} \Re_\omega(F;t)\omega'(t)\,dt = \int_{\mathbb{R}_+} [\mathcal{Z}^{(r)}(t) - \mathcal{Z}^{(r)}(0)]K(t)\,dt, \tag{1.19}$$

implying that we have equalities everywhere in (1.19). $\qquad\square$

By (1.14), (1.15), the rearrangement $\Re_\omega(F;t)$ has a compact support $[0,\kappa]$.

Recall the formula (7.5.7) for the value of the mth derivative of the functions R_n at the origin:

$$R_n^{(m)}(0) = \sum_{l=0}^{n} |\alpha_l^n| \cdot B + \int_0^{d_n} \Re_\omega(W_n;t)\,\omega'(t)\,dt; \tag{1.20}$$

Now, Lemma 7.5.1 and an application of the Fatou lemma yield the estimate for the value of $|\mathcal{Z}^{(m)}(0)|$ from below:

$$\mathcal{Z}^{(m)}(0) \geq \sum_{i=0}^{\infty} |\alpha_i| B + \int_{\mathbb{R}_+} \Re_\omega(F;t)\,\omega'(t)\,dt. \tag{1.21}$$

In particular,

$$\sum_{i=0}^{\infty} |\alpha_i| < \infty, \tag{1.22}$$

and by (1.14),

$$\int_{\mathbb{R}_+} \Re_\omega(F;t)\,\omega'(t)\,dt = \int_0^{\kappa} \Re_\omega(F;t)\,\omega'(t)\,dt < \infty. \tag{1.23}$$

Since $\omega'(t) \geq \omega'(\kappa) > 0$, $0 \leq t \leq \kappa$, from (1.23) and our observation (1.21) it follows that

$$\|F\|_{L_1(\mathbb{R}_+)} \leq \|\Re_\omega(F;\cdot)\|_{L_1(\mathbb{R}_+)} < +\infty. \tag{1.24}$$

Let us introduce the functional class $\Im_B = \Im_{r,\omega,B}$:

$$\Im_B := \left\{ f \in W^r H^\omega(\mathbb{R}_+) \ \middle|\ \|f\|_{L_\infty(\mathbb{R}_+)} \leq B \right\} \tag{1.25}$$

We claim that for any function $f \in \Im_B$,

$$\left| f^{(m)}(0) \right| \leq \sum_{i=0}^{\infty} |\alpha_i| B + \int_{\mathbb{R}_+} \Re_\omega(F;t)\,\omega'(t)\,dt, \tag{1.26}$$

(compare (1.26) with the inequality (1.21)).

In order to prove the inequality (1.26), let us consider some function $\chi \in C_0^\infty(\mathbb{R}_+)$ with the properties

$$\begin{aligned}
&(i)\ \ \chi(x) \equiv 1 && x \in [0,1];\\
&(ii)\ \ \chi(x) \equiv 0, && x \geq 2;\\
&(iii)\ -2 < \chi'(x) \leq 0, && x \in \mathbb{R}_+.
\end{aligned} \tag{1.27}$$

Given a function $f \in \Im_B$, we introduce the function

$$f_N(x) = f(x)\chi(x/N), \qquad N \in \mathbb{N}. \tag{1.28}$$

In the following lemma we give the estimate for the modulus of continuity of the rth derivative of the function f_N.

LEMMA 8.1.2. *For a given function $f \in \mathfrak{F}_B$, let the family of functions $\{f_N\}_{N \in \mathbb{N}}$ be defined in (1.28). Then, there exists such a constant $M = M(\omega, r, B)$ that for all $N \in \mathbb{N}$,*

$$\omega\left(f_N^{(r)}; t\right) \leq \omega(t) + \frac{M}{N}t, \qquad t \in \mathbb{R}_+. \tag{1.29}$$

Proof. By the Leibnitz's formula,

$$f_N^{(r)}(x) = \sum_{j=0}^{r} f^{(j)}(x)\chi^{(r-j)}(x/N)N^{j-r}, \qquad x \in \mathbb{R}_+. \tag{1.30}$$

Fix any index $N \in \mathbb{N}$ and any pair of points $x_1, x_2 \in \mathbb{R}_+$. Then, by (1.30),

$$f_N^{(r)}(x_2) - f_N^{(r)}(x_1) = \sum_{j=0}^{r} \left(f^{(j)}(x_2) - f^{(j)}(x_1)\right)\chi^{(r-j)}(x_2)N^{j-r} +$$

$$+ \sum_{j=0}^{r} f^{(j)}(x_1)\left(\chi^{(r-j)}(x_2) - \chi^{(r-j)}(x_1)\right)N^{j-r}. \tag{1.31}$$

Lemma 5.2.1 guarantees the existence of such constants $\{D_{r,i}\}_{i=1}^{r-1}$ and $\{E_{r,i,\omega}\}_{i=1}^{r-1}$ that

$$\|g^{(l)}\|_{L_\infty(\mathbb{R}_+)} \leq D_{r,l}\|g\|_{L_\infty(\mathbb{R}_+)} + E_{r,l,\omega}, \tag{1.32}$$

for all $l = 1, \ldots, r-1$ and all $g \in W^r H^\omega(\mathbb{R}_+)$.

Therefore, by (1.32), we have the estimates

$$\|f^{(l)}\|_{L_\infty(\mathbb{R}_+)} \leq \mathcal{C}_{r,\omega}, \qquad 0 \leq l \leq r, \tag{1.33}$$

where

$$\mathcal{C}_{r,\omega} := \max_{0 < l \leq r}\left\{D_{r,l}B + E_{r,l,\omega} + B\right\}, \tag{1.34}$$

and the constants $\{D_{r,l}, E_{r,l,\omega}\}_{l=1}^{r}$ are defined in Lemma 5.2.1.

Thus, the differences in (1.31) can be estimated as follows: for $j = 0, \ldots, r-1$,

$$|f^{(j)}(x_2) - f^{(j)}(x_1)| \leq \|f^{(j+1)}\|_{C[\vartheta_i, \vartheta_{i+1}]}|x_2 - x_1| \leq \mathcal{C}_{r,\omega}|x_1 - x_2|, \tag{1.35}$$

In addition, by the definition (1.25) of the class $\mathfrak{F}_B$,

$$|f^{(r)}(x_2) - f^{(r)}(x_1)| \leq \omega(|x_2 - x_1|). \tag{1.36}$$

Set

$$\Gamma = \Gamma(\chi) := \max_{0 \leq i \leq r+1}\|\chi^{(i)}\|_{L_\infty(\mathbb{R}_+)}. \tag{1.37}$$

Then, for any $j = 0, \ldots, r$,

$$\left| \chi^{(r-j)}(x_2) - \chi^{(r-j)}(x_2) \right| \leq \Gamma |x_2 - x_1|. \tag{1.38}$$

Let

$$M = M(\omega, r, B) := 2(r+1)\mathcal{C}_{r,\omega}\Gamma. \tag{1.39}$$

Finally, from the identity (1.31) and inequalities (1.33), (1.35), (1.36), and (1.38) it follows that

$$|f_N^{(r)}(x_2) - f_N^{(r)}(x_1)| \leq \omega(|x_2 - x_1|) + \frac{M}{N}|x_2 - x_1| \tag{1.40}$$

$\square$

For a fixed $n \in \mathbb{N}$, we introduce the concave modulus of continuity σ by the formula

$$\sigma(x) = \omega(x) + \frac{M}{N}x, \qquad x \in \mathbb{R}_+. \tag{1.41}$$

As we proved in Lemma 8.1.2, $f_N \in W^r H^\sigma(\mathbb{R}_+)$. Since $\omega(t) \leq \sigma(t)$, $t \in \mathbb{R}_+$, Corollary 2.2.9 leads us to the estimate

$$\int_{\mathbb{R}_+} \sigma'(x)\Re_\sigma (F; x) \, dx \leq \int_{\mathbb{R}_+} \sigma'(x)\Re_\omega (F; x) \, dx, \tag{1.42}$$

where $\Re_\sigma(F; t)$ and $\Re_\omega(F; t)$ are the extremal rearrangements associated with the modulii $\sigma(t)$ and $\omega(t)$, respectively.

Since each of the functions f_N, $n \in \mathbb{N}$, has a compact support, the standard application of Taylor's formula (as in Chapter 5) yields the following identity:

$$f^{(m)}(0) = f_N^{(m)}(0) = \sum_{i=0}^{\infty} \alpha_i f_N(\nu_i) + \int_0^\infty f_N^{(r)}(x)K(x) \, dx, \tag{1.43}$$

where the kernel K is defined in (1.2).

Thus, using the estimate (1.42), the identity (1.43) and Lemma 8.1.2, we obtain the chain of inequalities

$$|f^{(m)}(0)| \leq \sum_{i=0}^{\infty} |\alpha_i|B + \int_{\mathbb{R}_+} \sigma'(x)\Re_\sigma (F; x)]dx \leq$$

$$\leq \sum_{i=0}^{\infty} |\alpha_i|B + \int_{\mathbb{R}_+} \sigma'(x)\Re_\omega (F; x)]dx = \tag{1.44}$$

$$= \sum_{i=0}^{\infty} |\alpha_i|B + \int_{\mathbb{R}_+} \omega'(x)\Re_\omega (F; x) \, dx + \frac{M}{N}\sum_{i=0}^{\infty} \int_{\mathbb{R}_+} \Re_\omega (F; x) \, dx,$$

for all functions $f \in \mathfrak{S}_B$ and all natural N. By (1.23), the integral norm of the rearrangement $\mathfrak{R}_\omega\,(F;t)$ is finite, so the estimate (1.26) is proven.

The estimate (1.21) assures that the inequality (1.26) is sharp: it becomes the equality if $f(x) = \mathcal{Z}(x)$ or $f(x) = -\mathcal{Z}(x)$, $x \in \mathbb{R}_+$. Moreover, using properties (1.22) and (1.44), we can take the limit in (1.43) to obtain the following numerical differentiation formula for all $f \in W^r H^\omega(\mathbb{R}_+)$ with $\|f\|_{\mathrm{L}_\infty(\mathbb{R}_+)} \leq B$:

$$f^{(m)}(0) = \sum_{i=0}^{\infty} \alpha_i f(\nu_i) + \int_{\mathbb{R}_+} f^{(r)}(t)K(t)\, dt. \tag{1.45}$$

8.2. The structure of Chebyshev ω_α-splines

In this section we use the results of the previous section to describe the structure of extremal functions of the Kolmogorov problem in the Hölder classes $W^r H^\alpha(\mathbb{R}_+)$:

$$\|f^{(m)}\|_{\mathrm{L}_\infty(\mathbb{R}_+)} \to \sup, \qquad f \in W^r H^\alpha(\mathbb{R}_+), \quad \|f\|_{\mathrm{L}_\infty(\mathbb{R}_+)} \leq B. \tag{2.1}$$

The characteristic property

$$f(\cdot) \in W^r H^\alpha(\mathbb{R}_+) \iff \gamma^{r+\alpha} f(\cdot/\gamma), \quad \gamma > 0, \tag{2.2}$$

implies that it is sufficient to find the extremal function in (2.1) for $B = 1$. Setting $\omega(t) = \omega_\alpha(t)$ and $B = 1$, we summarize the results of Section 8.1 in connection with the problem (2.2).

We showed the existence of such collections $\{\nu_i = \nu_i(r,m,\alpha)\}_{i \in \mathbb{Z}_+}$ and $\{\vartheta_i = \vartheta_i(r,m,\alpha)\}_{i \in \mathbb{Z}_+}$ that

$$\nu_0 = 0; \quad \nu_{i-1} < \nu_i, \quad i \in \mathbb{N}; \quad \lim_{i \to +\infty} \nu_i = +\infty;$$

$$\vartheta_0 = 0; \quad \vartheta_{i-1} < \vartheta_i, \quad i \in \mathbb{N}; \quad \lim_{i \to +\infty} \vartheta_i = +\infty.$$

In (1.2) we defined the integrable (by (1.24)) kernel

$$F(t) = F_{\alpha,r,m}(t) = \frac{(-1)^{r-m}}{(r-m)!} t^r + \frac{(-1)^r}{r!} \sum_{i=0}^{\infty} \alpha_i (t - \nu_i)_+^r \tag{2.3}$$

with the properties

$$\begin{aligned} F^{(i)}(0) &= (-1)^{r-m}\delta_{r-m,i}, \qquad i = 0,\ldots,r-1; \\ F(\vartheta_i) &= 0, \qquad i \in \mathbb{N}. \end{aligned} \tag{2.4}$$

Any function $f \in W^r H^\alpha(\mathbb{R}_+)$ admits the numerical differentiation formula (1.45). The following description of the Chebyshev ω_α-spline follows from (1.4), (1.26).

THEOREM 8.2.1. *There exist a function $\mathcal{Z} = \mathcal{Z}_{\alpha,r,m}(t)$ endowed with the properties*

$$(i) \qquad \sup_{h \in H_0^\alpha(\mathbb{R}_+)} \int_{\mathbb{R}_+} h(t)F'(t)\,dt = \int_{\mathbb{R}_+} \left[\mathcal{Z}^{(r)}(t) - \mathcal{Z}^{(r)}(0)\right] F'(t)\,dt; \tag{2.5}$$

$$(ii) \quad \mathcal{Z}(\nu_i) = (-1)^{i+m}\|\mathcal{Z}\|_{\mathrm{L}_\infty(\mathbb{R}_+)} = (-1)^{i-m}, \qquad i \in \mathbb{Z}_+.$$

Corollary 8.2.2 is a consequence of inequality (1.26) and properties (2.5) of the Chebyshev ω_α-spline $\mathcal{Z}$.

COROLLARY 8.2.2. *Let $f \in W^r H^\alpha(\mathbb{R}_+)$ and $\|f\|_{\mathrm{L}_\infty(\mathbb{R}_+)} \leq 1$. Then,*

$$\|f^{(m)}\|_{\mathrm{L}_\infty(\mathbb{R}_+)} \leq \sum_{i \in \mathbb{Z}_+} |\alpha_i| \|f\|_{\mathrm{L}_\infty(\mathbb{R}_+)} + \int_{\mathbb{R}_+} \omega_\alpha'(t)\Re_{\omega_\alpha}(F;t)\,dt. \tag{2.6}$$

The function $\mathcal{Z}(t)$ transforms the inequality (2.3) into the equality.

The property (2.1) of Hölder classes implies that the function $B\mathcal{Z}(t/B^{\frac{1}{r+\alpha}})$ is extremal in the problem (2.2). Finally, relations (1.2.10)–(1.2.12) of Lemma 1.2.3 between the additive and multiplicative form of inequalities in Hölder classes $W^r H^\alpha(\mathbb{R}_+)$ enable us to find the sharp constant in the multiplicative inequality for intermediate derivatives.

COROLLARY 8.2.3. *If $f \in W^r H^\alpha(\mathbb{R}_+)$, then*

$$\|f^{(m)}\|_{\mathrm{L}_\infty(\mathbb{R}_+)} \leq C_{\alpha,r,m}\|f\|_{\mathrm{L}_\infty(\mathbb{R}_+)}^{\frac{r+\alpha-m}{r+\alpha}}, \tag{2.7}$$

with

$$C_{\alpha,r,m} = (r+\alpha)\left(\frac{A}{r+\alpha-m}\right)^{\frac{r+\alpha-m}{r+\alpha}}\left(\frac{B}{m}\right)^{\frac{m}{r+\alpha}},$$

$$A := \sum_{i \in \mathbb{Z}_+} |\alpha_i| \quad and \quad B := \int_{\mathbb{R}_+} \omega_\alpha'(t)\Re_{\omega_\alpha}(F;t)\,dt.$$

$\square$

Chapter 9

Maximization of Integral Functionals in $H^\omega[a_1, a_2]$, $\quad -\infty \le a_1 < a_2 \le +\infty$

9.0. Formulation of the extremal problem

We describe extremal functions and rearrangements of the problem

$$\int_{a_1}^{a_2} h(t)\psi(t)\, dt \to \sup, \qquad h \in H_0^\omega[a_1, a_2], \qquad\qquad (**)$$

where $a_1 < 0 < a_2$, and the kernel ψ has a finite number or a countable monotonely ordered set of points of sign changes on $[a_1, a_2]$, $-\infty \le a_1 < a_2 \le +\infty$. In particular, we give the solution of the problem $(**)$ in the case of the entire line $[a_1, a_2] = \mathbb{R}$.

9.1. Definitions

Recall Definition 1.2.1 of functional classes $\mathbb{M}_n[\alpha, \beta]$, $n \in \mathbb{N}$.

DEFINITION 9.1.1. Let $n_1 \in \mathbb{N} \cup \{\infty\}$, $\quad n_2 \in \mathbb{N} \cup \{\infty\}$, $\quad j \in \{-1, 0, +1\}$, and $a_1 < 0 < a_2$. Let $\psi \in \mathbb{L}_1[a_1, a_2]$, $\quad \psi_m(x) := \psi((-1)^m x)$, $\quad x \in [0, (-1)^m a_m]$, for $m = 1,\ 2$. Then,

$$\psi \in \mathcal{M}_{n_1,n_2}^j[a_1, a_2] \iff \begin{cases} \psi_m \in \mathbb{M}_{n_m}[0, (-1)^m a_m], \quad m = 1,\ 2; \\[2mm] \operatorname{sign} \displaystyle\int_{a_1}^{a_2} \psi(x)\, dx = j. \end{cases}$$

We also introduce the class

$$\mathbb{M}_{n_1,n_2}[a_1, a_2] := \bigcup_{j=-1}^{1} \mathcal{M}_{n_1,n_2}^j[a_1, a_2].$$

In addition to the sets of indices $\{J_i = J_i(N)\}_{i=1}^N$, $\{L_i\}_{i=1}^N$, and pairs of indices $\mathcal{P} = \mathcal{P}(N)$, introduced in Definition 1.2.2, we will need the collection of indices $\{K_i = K_i(N)\}_{i=1}^N$ and pairs of indices $\mathbb{P}(N_1, N_2)$.

DEFINITION 9.1.2. Let N, N_1, $N_2 \in \mathbb{N} \cup \{\infty\}$. Then,

(1) $\quad K_i(N) := \left\{ k = i \ (\mathrm{mod}\ 2) + 2l - 1, \quad l \in \mathbb{N} \ \middle| \ k \le N \right\};$

(2) $\quad \mathbb{P}(N_1, N_2) := \{(i,j) \in \mathbb{N} \times \mathbb{N} \mid 1 \le i \le N_1, \ j \in K_i(N_2) \}.$

DEFINITION 9.1.3. Let n_1, $n_2 \in \mathbb{N} \cup \{\infty\}$. We shall say that a partition $\mathcal{W}$ of the interval $[a_1, a_2]$ with atoms $\{A_i^m, B_i^m, C_i^m, E_i^m, D_i^m\}_{i=1}^{n_m}$, and subatoms $\{B_{ij}^m, C_{ji}^m\}_{(i,j) \in \mathcal{P}(n_m)}$ and $\{E_{ik}^m\}_{(i,k) \in \mathbb{P}(n_m, n_{3-m})}$ for $m = 1, 2$, is called *a W_{n_1, n_2}-partition of the interval* $[a_1, a_2]$, if the following conditions are satisfied for $m = 1, 2$.

(A) $C_i^m = [(-1)^m \gamma_{5i-5}^m, (-1)^m \gamma_{5i-4}^m], \quad D_i^m = [(-1)^m \gamma_{5i-4}^m, (-1)^m \gamma_{5i-3}^m],$
$E_i^m = [(-1)^m \gamma_{5i-3}^m, (-1)^m \gamma_{5i-2}^m], \quad B_i^m = [(-1)^m \gamma_{5i-2}^m, (-1)^m \gamma_{5i-1}^m],$
$A_i^m = [(-1)^m \gamma_{5i-1}^m, (-1)^m \gamma_{5i}^m], \quad \text{for } i = 1, \dots, n_m,$
and such $\gamma^m = \{\gamma_i^m\}_{i=0}^{5n_m}$ that $0 \rhd \gamma^m \blacktriangleright (-1)^m a_m;$

(B) $B_i^m = \square, \ i = n_m - 2, n_m - 1, n_m; \ C_i^m = \square, \ i = 1, 2, 3; \ A_{n_m}^m = \square;$

(C) $B_i^m = \bigcup_{j \in J_i(n_m)} B_{ij}^m, \quad 1 \le i \le n_m - 3, \text{ where}$
$B_{ij}^m = [(-1)^{m+1} \xi_i \left(\frac{j-i+1}{2}\right), (-1)^{m+1} \xi_i \left(\frac{j-i-1}{2}\right)], \qquad j \in J_i(n_m),$
for such $\xi_i^m = \{\xi_i^m(k)\}_{k=1}^{|J_i(n_m)|}$, that $\gamma_{5i-2}^m \blacktriangleleft \xi_i^m \lhd \gamma_{5i-1}^m;$

(D) $C_i^m = \bigcup_{l \in L_i} C_{il}^m, \quad 4 \le i \le n_m, \text{ where}$
$C_{il}^m = [(-1)^m \varkappa_i^m (\frac{i-l-1}{2}), (-1)^m \varkappa_i^m (\frac{i-l+1}{2})], \qquad l \in L_i,$
for such $\varkappa_i^m = \{\varkappa_i^m(k)\}_{k=1}^{|L_i|}$ that $\gamma_{5i-5}^m \rhd \varkappa_i^m \blacktriangleright \gamma_{5i-4}^m;$

(E) $E_i^m = \bigcup_{k \in K_i(n_{3-m})} E_{ik}^m, \quad 1 \le i \le n_m, \text{ where}$
$E_{ik}^m = [(-1)^{m+1} e_i(\frac{k+1-i \ (\mathrm{mod}\ 2)}{2}), (-1)^{m+1} e_i(\frac{k+3-i \ (\mathrm{mod}\ 2)}{2})],$
for such $e_i^m = \{e_i^m(l)\}_{l=1}^{|K_i(n_{3-m})|}$ that $\gamma_{5i-3}^m \rhd e_i^m \blacktriangleright \gamma_{5i-2}^m.$

9.2. Structure of perfect ω-splines

THEOREM Y. *Let n_1, $n_2 \in \mathbb{N} \cup \{\infty\}$, and $\psi \in \mathcal{M}_{n_1, n_2}^k[a_1, a_2]$, where $k \in \{-1, 0, 1\}$. Let $\alpha_0^m := 0$, $\alpha_{n_m}^m := a_i$, $m = 1, 2$, and $\alpha^m = \{\alpha_i^m\}_{i=1}^{n_m - 1}$ be such distinct points of sign change of ψ on $[0, a_m]$ that*

$$a_1 \blacktriangleleft \alpha^1 \lhd 0, \qquad 0 \rhd \alpha^2 \blacktriangleright a_2.$$

I. There exist a solution $y_{\omega, \psi}$ of the problem $(\ast\ast)$ and a W_{n_1, n_2}-partition $\mathcal{W}$ with the following properties holding for $m = 1, 2$.

(A) $\alpha_i^m \in A_i^m, \qquad i = 1, \dots, n_m - 1;$

(B) $\displaystyle\int_{B_{ij}^m \cup C_{ji}^m} \psi(t)\,dt = 0$, $(i,j) \in \mathcal{P}(n_m)$;,

(C) $\displaystyle\int_{A_i^m} \psi(t)\,dt = 0$, $i = 1, \ldots, n_m - 1$;

(D) $y_{\omega,\psi} = -\omega(|t|)$, $t \in D_{2k-1}^m \neq \square$, $k = 0, \ldots, \lceil n_m/2 \rceil$;

(E) $y_{\omega,\psi} = \omega(|t|)$, $t \in D_{2k}^m \neq \square$, $k = 0, \ldots, [n_m/2]$;

(F$_1$)   *for $(i,j) \in \mathcal{P}(n_m)$, the function $y_{\omega,\psi}$ is extremal in the problem*

$$\int_{a_1}^{a_2} h(t)\psi_{ij}^m(t)\,dt \to \sup, \qquad h \in H^\omega[a_1, a_2], \tag{2.1}$$

$$\psi_{ij}^m(t) = \psi(t) \cdot \mathcal{X}(B_{ij}^m \cup C_{ji}^m; t), \qquad t \in [a_1, a_2]; \tag{2.2}$$

(F$_2$)   *for $(i,j) \in \mathbb{P}(n_1, n_2)$, the function $y_{\omega,\psi}$ is extremal in the problem*

$$\int_{a_1}^{a_2} h(t)\phi_{ij}(t)\,dt \to \sup, \qquad h \in H^\omega[a_1, a_2], \tag{2.3}$$

$$\phi_{ij}(t) = \psi(t) \cdot \mathcal{X}(E_{ij}^1 \cup E_{ji}^2; t), \qquad t \in [a_1, a_2]; \tag{2.4}$$

(F$_3$)   *for $i = 1, \ldots, n_m - 1$, the function $y_{\omega,\psi}$ is extremal in the problem*

$$\int_{a_1}^{a_2} h(t)\psi_i^m(t)\,dt \to \sup, \qquad h \in H^\omega[a_1, a_2], \tag{2.5}$$

where

$$\psi_i^m(t) = \psi(t) \cdot \mathcal{X}(A_i^m; t), \qquad t \in [a_1, a_2]. \tag{2.6}$$

II.

$$k = 1 \implies D_{2i-1}^m = \square, \quad i = 1, \ldots, \lceil n_m/2 \rceil, \quad m = 1,\,2;$$
$$k = -1 \implies D_{2i}^m = \square, \quad i = 1, \ldots, [n_m/2], \quad m = 1,\,2;$$
$$k = 0 \implies D_i^m = \square, \quad k = 1, \ldots, n_m, \quad m = 1,\,2.$$

COROLLARY 9.2.1. *The extremal function $y_{\omega,\psi}$ and W_{n_1,n_2}-partitions of the problem* (**) *are unique.*

9.3 Kernels $\Psi(\cdot)$ and their rearrangements $\Re_\omega(\Psi; \cdot)$

Put

$$\Psi(t) = \int_{a_m}^{t} \psi(y)\, dy, \qquad t \in [0, a_m], \quad m = 1,\, 2. \tag{3.1}$$

and for $t \in [a_1, a_2]$ and $m = 1,\, 2$, let

$$\Psi_{ij}^{m}(t) = \int_{a_1}^{t} \psi_{ij}^{m}(y)\, dy, \qquad (i, j) \in \mathcal{P}(n_m),$$

$$\Psi_{i}^{m}(t) = \int_{a_1}^{t} \psi_{i}^{m}(y)\, dy, \qquad i = 1, \ldots, n_m - 1, \tag{3.2}$$

$$\Phi_{ij}(t) = \int_{a_1}^{t} \phi_{ij}(y)\, dy, \qquad (i, j) \in \mathbb{P}(n_1, n_2).$$

The argument, similar to the one used in Section 2.2 (see relations (2.2.10), (2.2.11)), combined with the results of Theorem Y implies that all kernels in (3.2) are simple. Figures 9.3.1–9.3.8 illustrate the graphs of functions $y_{\omega,\psi}$ for $\psi \in \mathbb{M}_{2,l}[a_1, a_2]$, $l = 2, 3$.

DEFINITION 9.3.1. Let $\psi \in \mathbb{M}_{n_1, n_2}[a_1, a_2]$, and $\Psi(t)$ be defined by (3.1). *The extremal rearrangement $\Re_\omega(\Psi; \cdot)$ of the kernel Ψ is defined as follows:*

$$\Re_\omega(\Psi; t) := \sum_{m=1}^{2} \sum_{(i,j) \in \mathcal{P}(n_m)} \Re\left(\Psi_{ij}^{m}; t\right) + \sum_{m=1}^{2} \sum_{i=1}^{n_m - 1} \Re\left(\Psi_{i}^{m}; t\right) +$$

$$+ \sum_{m=1}^{2} \sum_{(i,j) \in \mathbb{P}(n_1, n_2)} \Re\left(\Psi_{ij}^{m}; t\right) + |\Psi(|t|)| \cdot \mathcal{X}(D; t), \qquad t \in [0, a_2 - a_1], \tag{3.3}$$

where $D := \bigcup_{i=1}^{n_1} D_i^{1} \cup \bigcup_{i=1}^{n_2} D_i^{2}$, and rearrangements $\Re(\Phi; \cdot)$ of simple kernels Φ are introduced by the formula (1.5) in Definition 1.1.3.

The numerical value of the maximum in the problem $(**)$ is expressed in terms of the extremal rearrangement as follows:

$$\sup_{h \in H_0^{\omega}[a_1, a_2]} \int_{a_1}^{a_2} h(t)\psi(t)\, dt = \int_{0}^{a_2 - a_1} \Re_\omega(\Psi; t)\omega'(t)\, dt. \tag{3.4}$$

REMARK 9.3.1. If $\chi\psi(x) > 0$ for a.e. $x \in [a_1, a_2]$ and a fixed $\chi \in \{-1, 1\}$, then

$$y_{\omega,\psi}(t) = \chi\omega(|t|), \qquad t \in [a_1, a_2]. \tag{3.5}$$

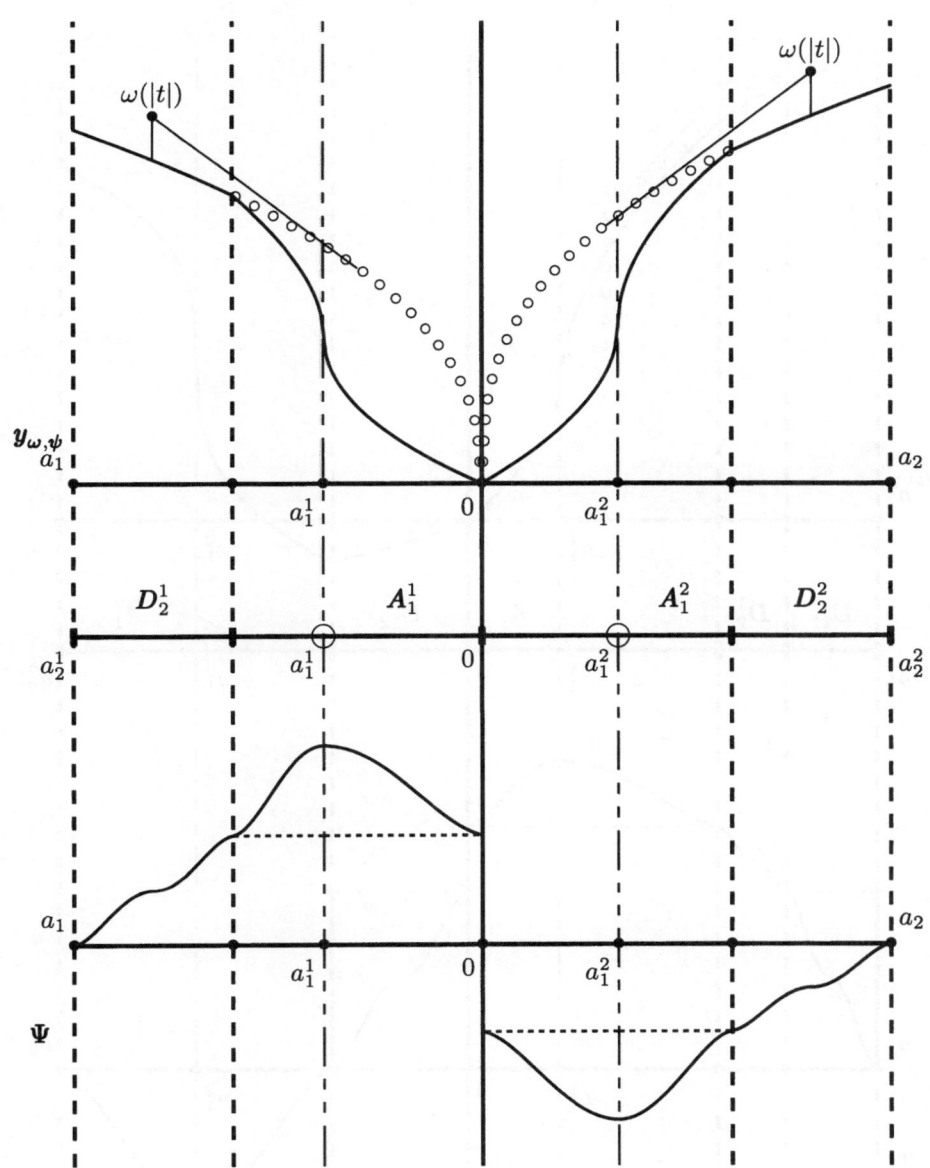

FIGURE 9.3.1. $V_{2,2}^{+1}$-partition, graphs of $y_{\omega,\psi}$, Ψ for $\psi \in \mathcal{M}_{2,2}^{+1}[a_1, a_2]$

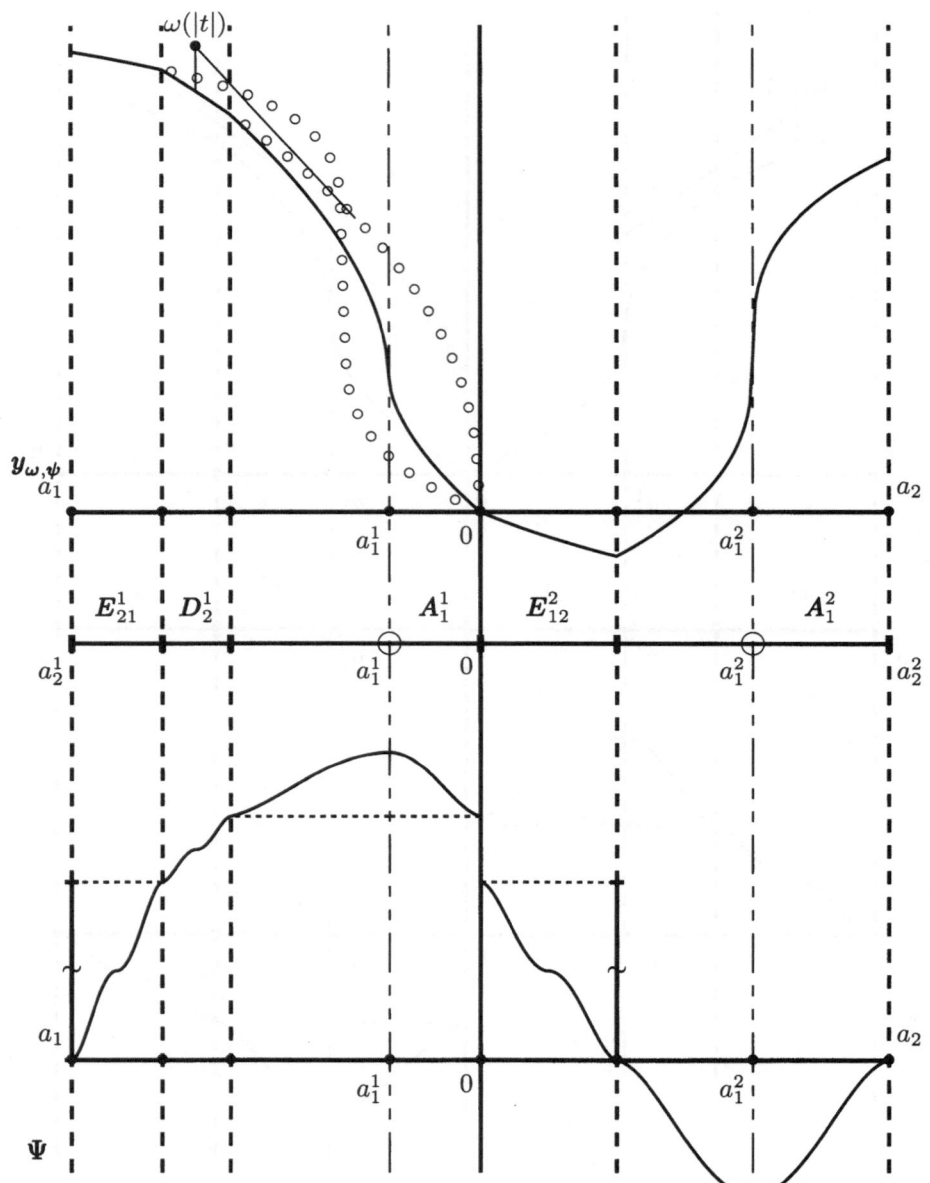

FIGURE 9.3.2. $V_{2,2}^{+1}$-partition, graphs of $y_{\omega,\psi}$, Ψ for $\psi \in \mathcal{M}_{2,2}^{+1}[a_1, a_2]$

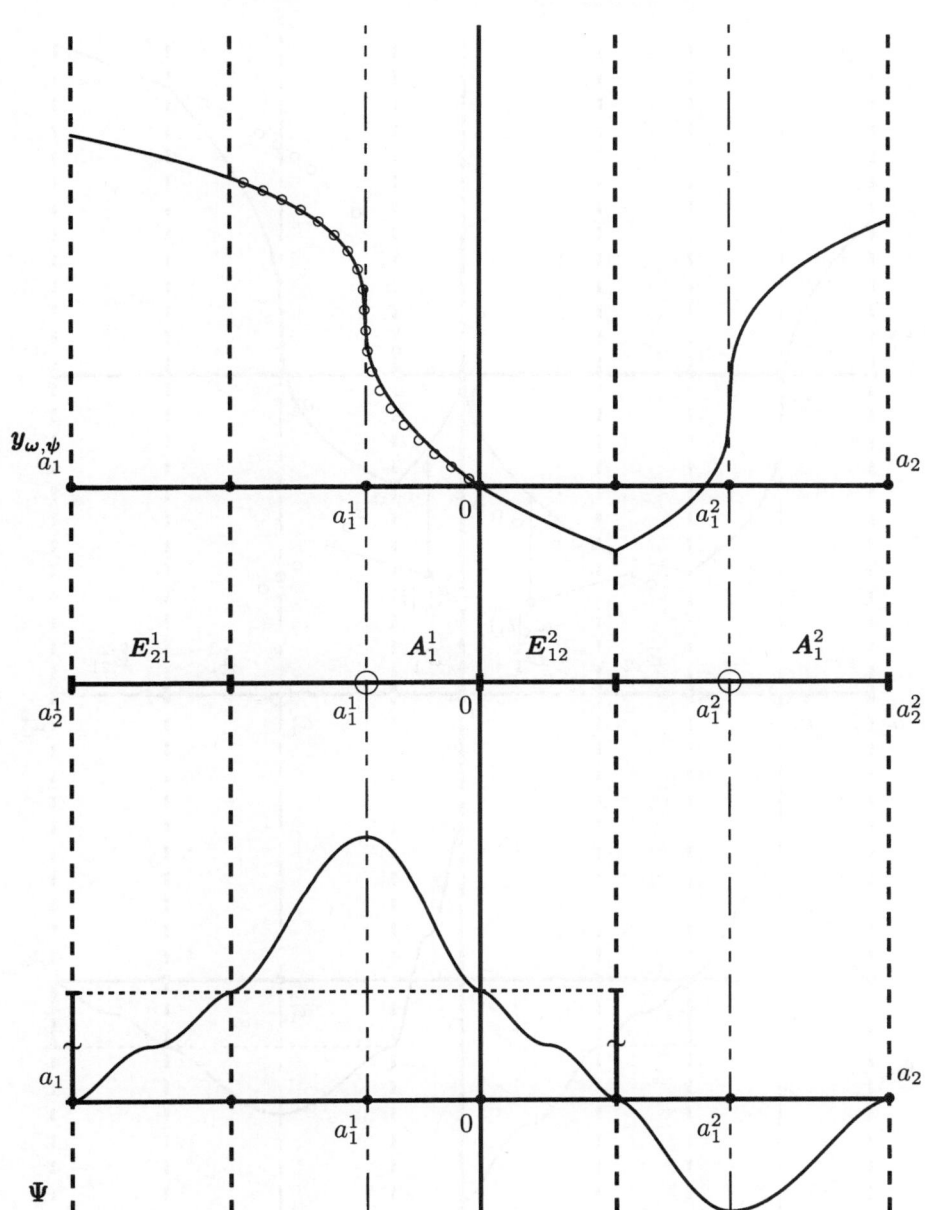

FIGURE 9.3.3. $V_{2,2}^0$-partition, graphs of $y_{\omega,\psi}$, Ψ for $\psi \in \mathcal{M}_{2,2}^0[a_1, a_2]$

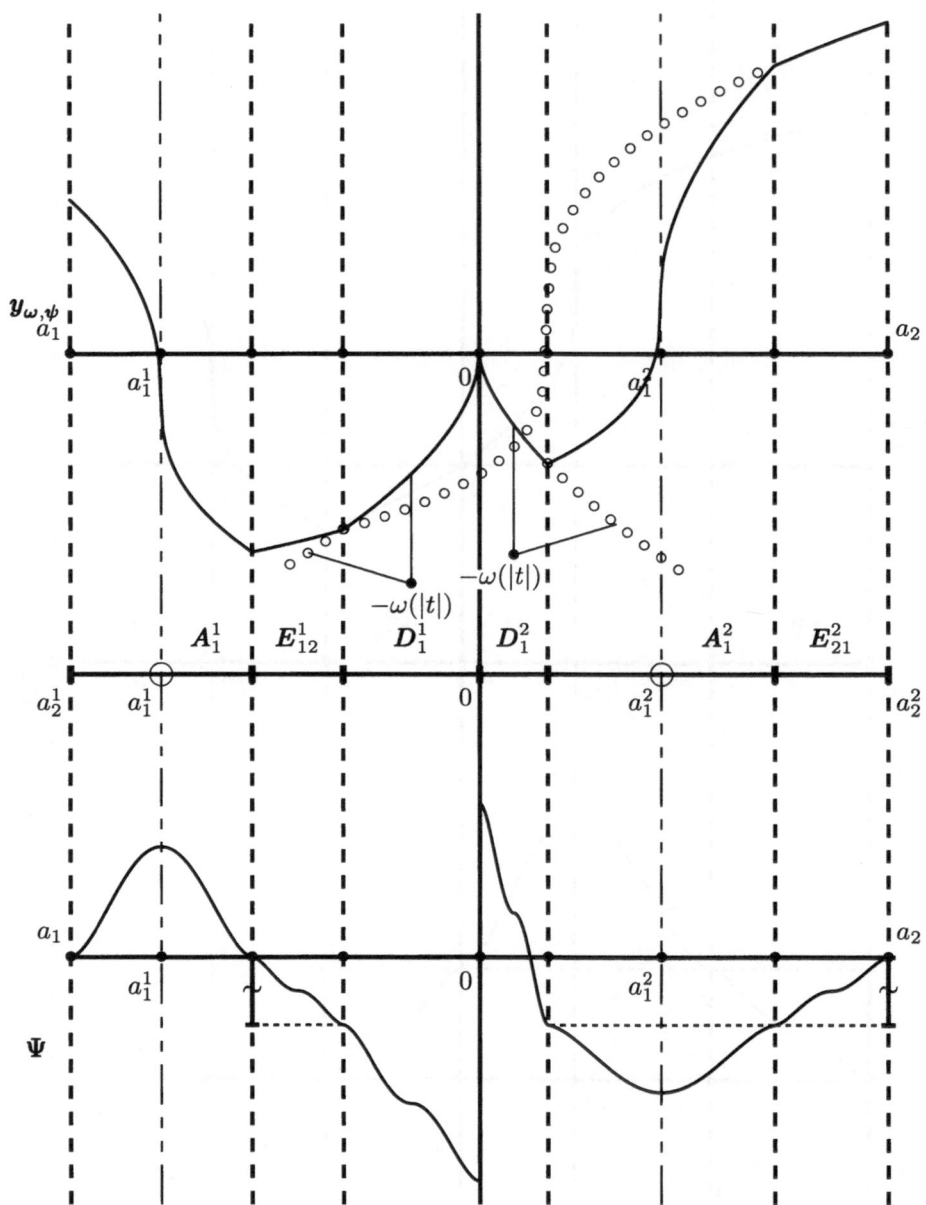

FIGURE 9.3.4. $V_{2,2}^{-1}$-partition, graphs of $y_{\omega,\psi}$, Ψ for $\psi \in \mathcal{M}_{2,2}^{-1}[a_1, a_2]$

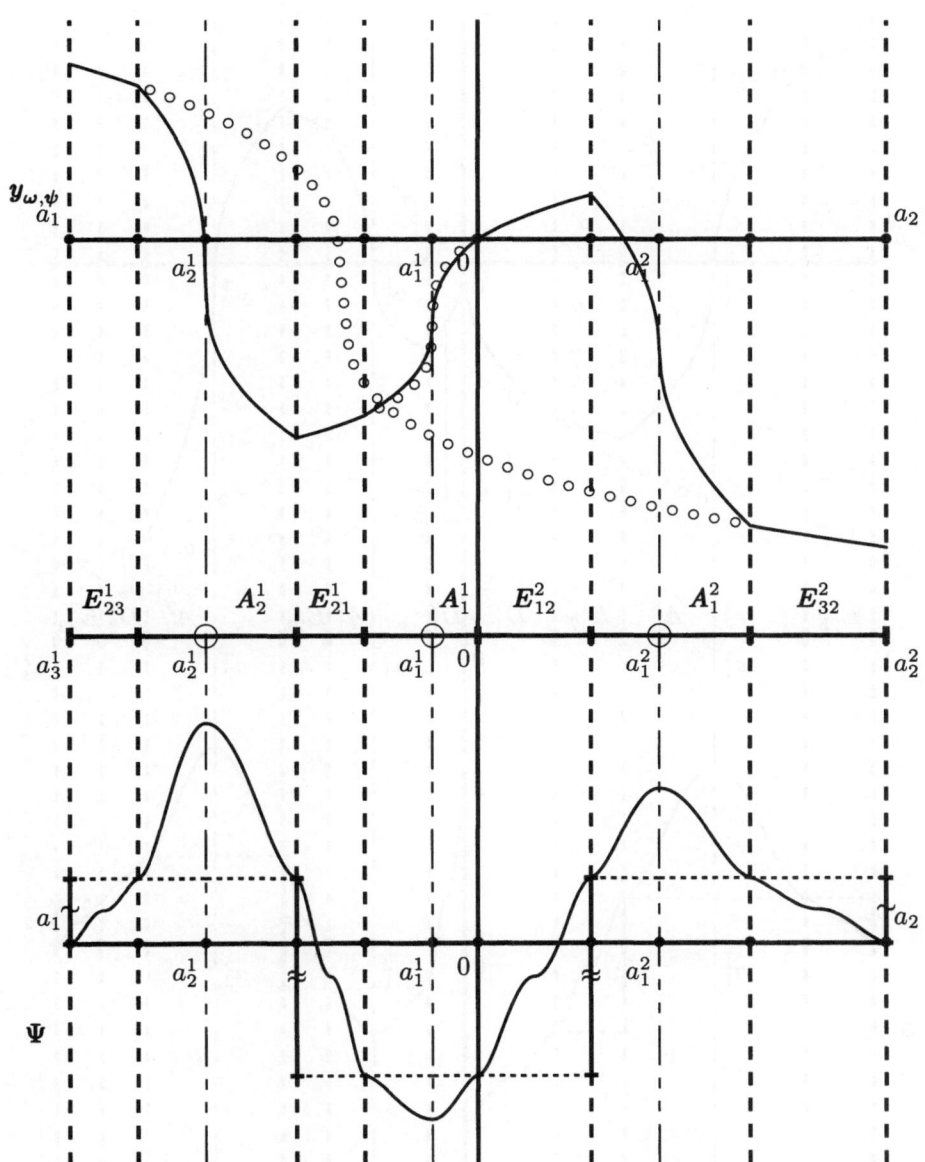

FIGURE 9.3.5. $V_{3,2}^0$-partition, graphs of $y_{\omega,\psi}$, Ψ for $\psi \in \mathcal{M}_{3,2}^0[a_1, a_2]$

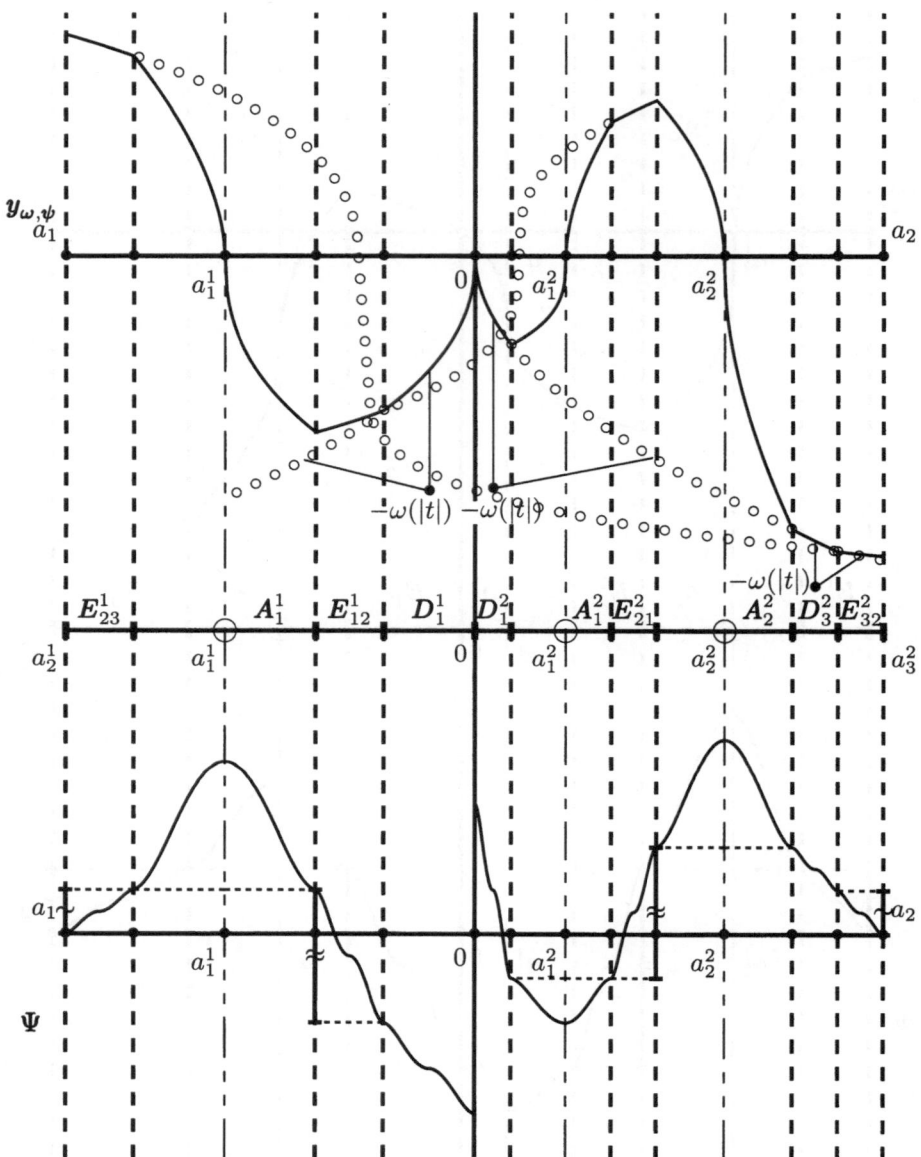

FIGURE 9.3.6. $V_{2,3}^{-1}$-partition, graphs of $y_{\omega,\psi}$, Ψ for $\psi \in \mathcal{M}_{2,3}^{-1}[a_1, a_2]$

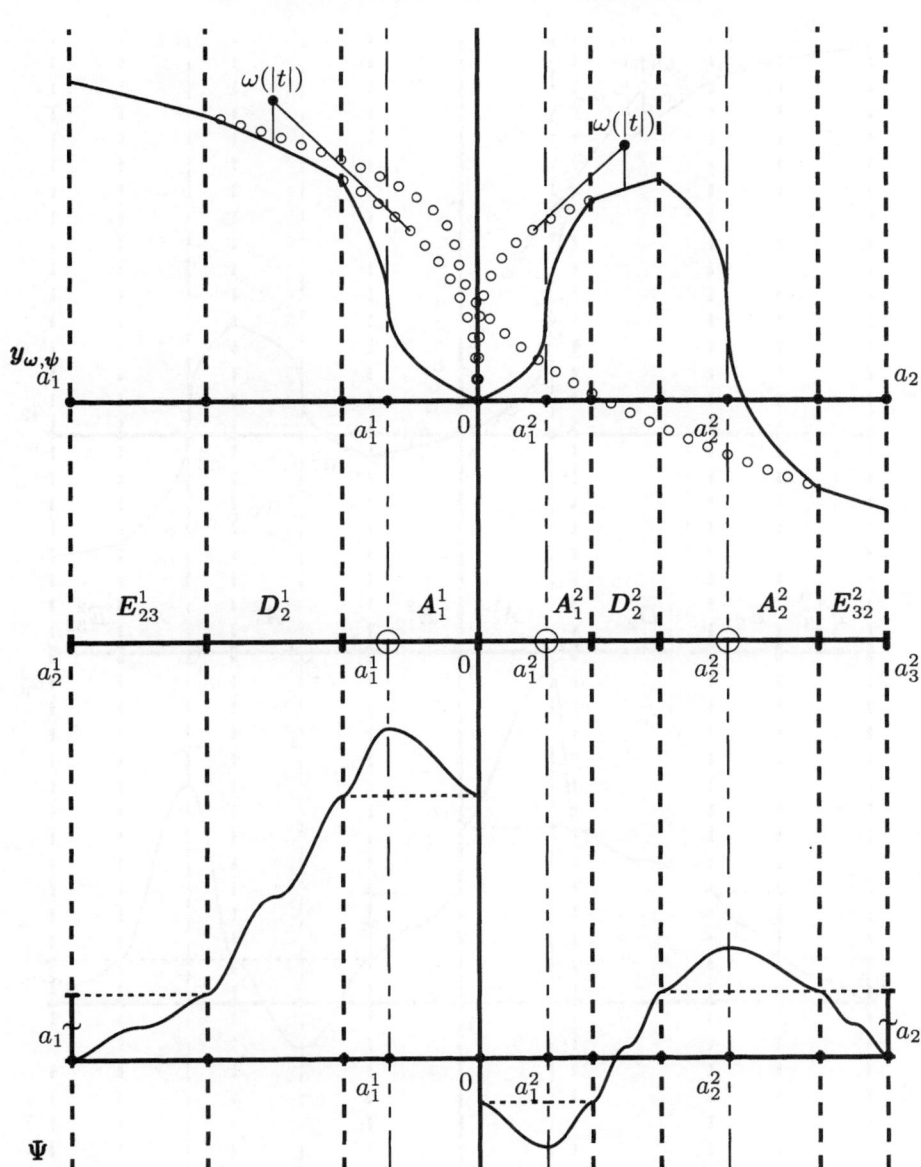

FIGURE 9.3.7. $V_{2,3}^{+1}$-partition, graphs of $y_{\omega,\psi}$, Ψ for $\psi \in \mathcal{M}_{2,3}^{+1}[a_1, a_2]$

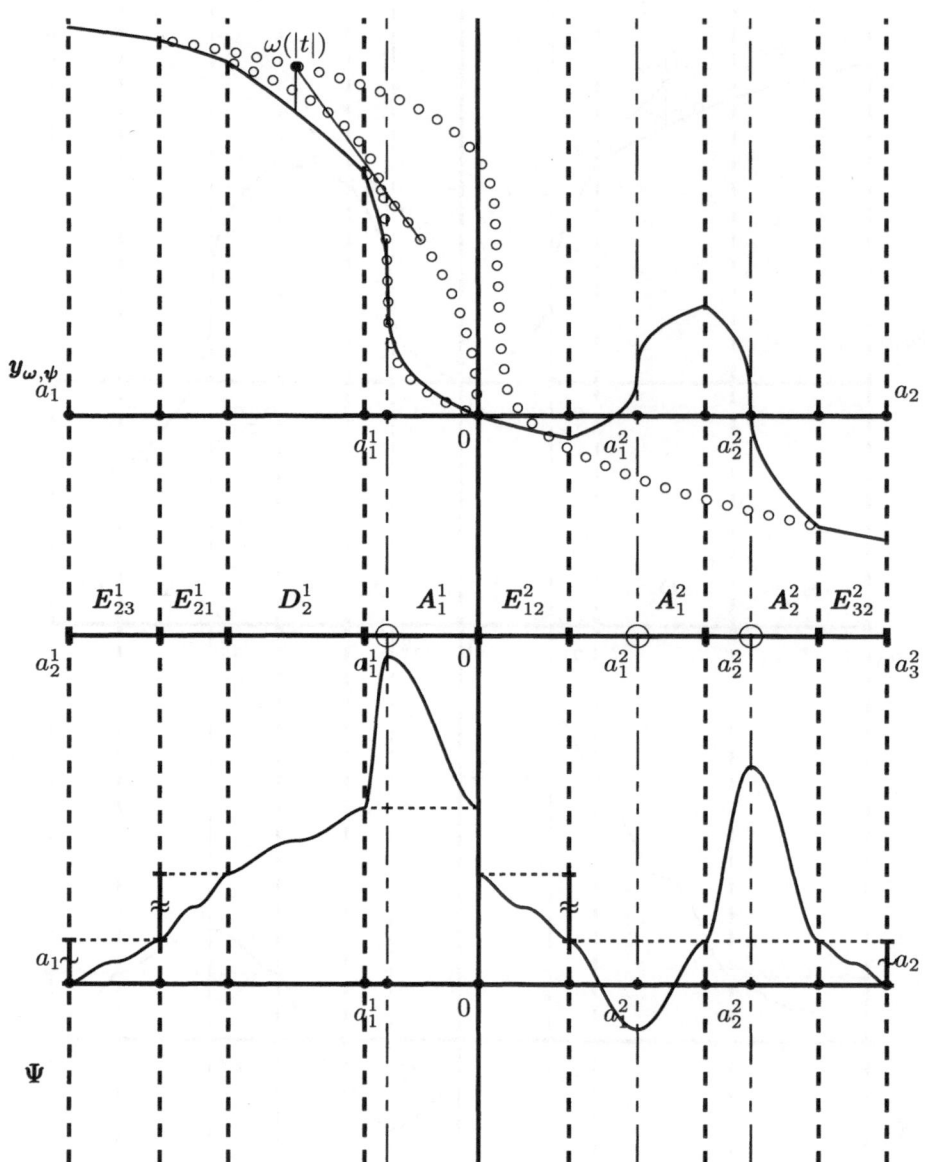

FIGURE 9.3.8. $V_{2,3}^{+1}$-partition, graphs of $y_{\omega,\psi}$, Ψ for $\psi \in \mathcal{M}_{2,3}^{+1}[a_1, a_2]$

Chapter 10

Sharp Kolmogorov Inequalities in $W^r H^\alpha(\mathbb{R})$

Let $m, r : 0 < m \leq r$, be integers. In this chapter we first describe the discrete family of Chebyshev ω-splines extremal in the problem

$$f^{(m)}(0) \to \sup, \qquad f \in W^r H^\omega[-1, 1], \quad \|f\|_{C[-1,1]} \leq B, \qquad (\text{P}.1)$$

for certain choices of B and all concave modulii of continuity ω. Then, we characterize the extremal functions in the problem

$$\|f^{(m)}\|_{L_\infty(\mathbb{R})} \to \sup, \qquad f \in W^r H^\alpha(\mathbb{R}), \quad \|f\|_{L_\infty(\mathbb{R})} \leq B, \qquad (\text{P}.2)$$

for all $B > 0$ and $\alpha \in (0, 1]$.

10.1. Euler ω-splines of the problem (P.1)

Recall the type of symmetry (1.2.9) of extremal functions $\mathcal{Z}$ in the problem (P.1):

$$\mathcal{Z}(-x) = (-1)^m \mathcal{Z}(x), \qquad x \in [-1, 1]. \qquad (1.1)$$

In particular, the extremal Euler ω-splines of the problem (P.0) are even and have an odd number of alternance points for even m, and they are odd and exhibit an even number of alternance points for odd m.

10.1.1. Numerical differentiation formulae

As in Chapter 5, in deriving the formulas for the mth derivative of a function from $W^r H^\omega[-1, 1]$, we consider two cases: $0 < m < r$ and $m = r$.

Case 1. $0 < \mathbf{m} < \mathbf{r}$. Fix $n \in \mathbb{N}, n \geq r$, such that n is odd for even m, and even for odd m.

Let the collections of points $\bar{\nu} = \{\nu_i\}_{i=0}^{n+1}$ and $\bar{\vartheta} = \{\vartheta_i\}_{i=1}^{n-r+1}$ on the interval $[-1, 1]$ be such that

(A) $-1 =: \nu_0 < \nu_1 < \cdots < \nu_{n+1} := 1; \quad \vartheta_1 < \vartheta_2 < \cdots < \vartheta_{n-r+1};$

(B) $\nu_i < \vartheta_i < \nu_{i+r-1}, \qquad i = 1, \ldots, n-r+1;$

(C) $\nu_i = \nu_{n+1-i}, \quad i = 0, \ldots, n+1; \quad \vartheta_i = -\vartheta_{n-r+2-i}, \quad i = 1, \ldots, n-r+1.$

$$(1.2)$$

In particular, the collections $\{\nu_i\}_0^{n+1}$ and $\{\vartheta_i\}_{i=1}^{n-r+1}$ are symmetric with respect to the origin.

Let the coefficients $\{\alpha_i\}_{i=0}^n$ be determined from the system of linear equations

$$
\begin{cases}
\displaystyle\sum_{i=0}^n \alpha_i(\nu_i+1)^j = 0, & j = 0,\ldots,m-1; \\[2ex]
\displaystyle\sum_{i=0}^n \alpha_i(\nu_i+1)^j = \dfrac{j!}{(j-m)!}, & j = m,\ldots,r-1; \\[2ex]
\displaystyle\sum_{i=0}^n \alpha_i(\nu_i-\vartheta_l)_+^{r-1} = \dfrac{(r-1)!}{(r-1-m)!}(-\vartheta_l)_+^{r-1-m}, & l = 1,\ldots,n-r+1.
\end{cases}
\tag{1.3}
$$

The inequalities (1.2), (B) between the points $\{\nu_i\}_{i=1}^n$ and $\{\vartheta_i\}_{i=1}^{n-r}$ guarantee that the system of equations (1.3) has a unique solution.

Let $f \in \mathbb{C}^r[-1,1]$. From the Taylor formula

$$
f(\tau) = \sum_{j=0}^{r-1} \beta_j(\tau+1)^j + \frac{1}{(r-1)!}\int_{-1}^1 f^{(r)}(x)(\tau-x)_+^{r-1}\,dx,
\tag{1.4}
$$

where $\beta_j = \dfrac{f^{(j)}(-1)}{j!}$, $\tau \in [-1,1]$, it follows that

$$
f^{(m)}(\tau) = \sum_{j=m}^{r-1} \frac{j!}{(j-m)!}\beta_j(\tau+1)^{j-m} + \frac{1}{(r-1-m)!}\int_{-1}^1 f^{(r)}(x)(\tau-x)_+^{r-1-m}\,dx.
\tag{1.5}
$$

Then, from the linear equations (1.3) and the identities (1.4) and (1.5) we derive the following numerical differentiation formula for the value of the mth derivative at the origin:

$$
f^{(m)}(0) = \sum_{i=0}^n \alpha_i f(\nu_i) + \int_{-1}^1 f^{(r)}(x)K_n(x)\,dx,
\tag{1.6}
$$

where the kernel $K_n(t)$ is defined as follows:

$$
K_n(t) = \frac{1}{(r-1-m)!}(-t)_+^{r-1-m} - \frac{1}{(r-1)!}\left[\sum_{i=1}^n \alpha_i(\nu_i-t)_+^{r-1} + \alpha_0(\nu_0-t)^{r-1}\right].
\tag{1.7}
$$

LEMMA 10.1.1. *Let the points $\{\nu_i\}_{i=0}^{n+1}$ and $\{\vartheta_i\}_{i=1}^{n-r+1}$ be as in (1.2), and the kernel $K_n(t)$ be defined in (1.3), (1.7). Then, $\alpha_0 = 0$, and*

$$
K_n(-t) = (-1)^{r-m}K_n(t).
\tag{1.8}
$$

Proof. Notice that the system (1.3) is equivalent to the following equations:

$$\begin{cases} K_n^{(j)}(-1) = K_n^{(j)}(1) = 0, & j = 0, \ldots, r-1; \\ K_n(\vartheta_l) = 0, & l = 1, \ldots, n-r+1. \end{cases} \tag{1.9}$$

Let us consider the kernel

$$\hat{K}_n(t) := (-1)^{r-m} K_n(-t). \tag{1.10}$$

In view of the symmetry (1.2), (C) of the knots $\{\nu_i\}_{i=1}^n$ of the kernel $K_n(t)$, the kernel $\hat{K}_n(t)$ has the same set of knots:

$$\hat{K}_n(t) = \frac{1}{(r-1-m)!}(-t)_+^{r-1-m} - \frac{1}{(r-1)!}[\sum_{i=1}^n \beta_i(\nu_i - t)_+^{r-1} + \beta_0(\nu_0 - t)^{r-1}],$$
$$\tag{1.11}$$

for some coefficients $\{\beta_i\}_{i=0}^n$. From the symmetry of $\{\vartheta_i\}_{i=1}^{n-r+1}$ with respect to 0 it follows that the kernel $\hat{K}_n(t)$ also satisfies equations (1.9) equivalent to the system (1.3). Therefore, coefficients $\{\beta_i\}_{i=0}^n$ solve (1.3), as well. Then, the uniqueness of the solution of the system (1.3) implies that

$$\alpha_i = \beta_i, \quad i = 0, \ldots, n, \quad \text{and} \quad \hat{K}_n(t) = K_n(t), \tag{1.12}$$

which proves (1.8). Next, notice that $K_n(t) \equiv 0$, $t \in (\nu_n, 1]$. But by (1.8),

$$0 = K_n(t) = (-1)^{r-m} K_n(-t) = -\frac{\alpha_0}{(r-1)!}(\nu_0 + t)_+^{r-1}, \quad \nu_n < t < 1. \tag{1.13}$$

Consequently, $\alpha_0 = 0$. $\qquad\square$

Let us assume that the points $\{\nu_i\}_{i=0}^n$ and $\{\vartheta_i\}_{i=1}^{n-r+1}$ are chosen in such a way that the kernel $K_n(t)$ has the zero mean on the interval $[-1, 1]$. These conditions are equivalent to an additional equation for the coefficients $\{\alpha_i\}_{i=0}^n$:

$$\sum_{i=0}^n \alpha_i(\nu_i + 1)^r = \frac{r!}{(r-m)!} \tag{1.14}$$

A. Pinkus [72] proved the following properties of the kernel $K(t)$:

(*i*) $\operatorname{supp} K = [\nu_k, \nu_{n+1-k}]$,   for some $k : 1 \leq k \leq [(n+2-r)/2]$;

(*ii*) the kernel K has precisely $n + 3 - 2k - r$ simple zeroes $\{\vartheta_i\}_{i=k}^{n+2-k-r}$
    on the interval (ν_k, ν_{n+1-k});

(*iii*) $\operatorname{sign} \alpha_i = (-1)^{i+m}$, $i = k+1, \ldots, n+1-k$;

(*iv*) $(-1)^{i+r+m} \operatorname{sign} K(t) \geq 0$, $\vartheta_i \leq t \leq \vartheta_{i+1}$, $i = k-1, \ldots, n+2-k-r$.
$$\tag{1.15}$$

By (1.15),

$$K(x) \in \pm\mathcal{M}^0_{n+4-2k-r}[\nu_k, \nu_{n+1-k}],$$

where classes $\mathcal{M}^j_l[a, b]$ for $l \in \mathbb{N}$ and $j \in \{-1, 0, +1\}$ are introduced in Definition 2.2.1.

Therefore, in the case of $0 < m < r$, we obtain the following estimate for the value of the mth derivative of the function $f \in W^r H^\omega[-1, 1]$ at the origin:

$$|f^{(m)}(0)| \leq \sum_{i=1}^n |\alpha_i| \cdot \|f\|_{\mathbb{C}[-1,1]} + \int_0^2 \Re_\omega\left(F_n; t\right) \omega'(t)\, dt, \tag{1.16}$$

where $\Re_\omega\left(F_n; t\right)$ is the rearrangement of the kernel $F_n(t) = \int_{\nu_n}^t K(y)\, dy$, as introduced in (2.2.20) of Definition 2.2.4.

Case 2. m=r. In this case, n is even for odd r and odd for even r.

Let the collections of points $\bar\nu = \{\nu_i\}_{i=0}^{n+1}$ and $\bar\vartheta = \{\vartheta_i\}_{i=1}^{n-r-1}$ on the interval $[-1, 1]$ be such that

(A) $-1 =: \nu_0 < \nu_1 < \cdots < \nu_{n+1} := 1;$ $\vartheta_1 < \vartheta_2 < \cdots < \vartheta_{n-r+1};$

(B) $\nu_i < \vartheta_i < \nu_{i+r},$ $i = 1, \ldots, n-r-1;$

(C) $\nu_i = \nu_{n+1-i},$ $i = 0, \ldots, n+1;$ $\vartheta_i = -\vartheta_{n-r-i},$ $i = 1, \ldots, n-r-1.$
$$\tag{1.17}$$

In particular, the collections $\{\nu_i\}_0^{n+1}$ and $\{\vartheta_i\}_{i=1}^{n-r-1}$ are symmetric with respect to the origin.

Let the coefficients $\{\alpha_i\}_{i=0}^n$ be determined from the system of linear equations

$$\begin{cases} \displaystyle\sum_{i=1}^n \alpha_i(\nu_i + 1)^j = 0, & j = 0, \ldots, r-1; \\[2mm] \displaystyle\sum_{i=1}^n \alpha_i(\nu_i - \vartheta_l)_+^{r-1} = 0, & l = 1, \ldots, n-r; \\[2mm] \displaystyle\sum_{i=1}^n \alpha_i(\nu_i + 1)_+^r = -r! \end{cases} \tag{1.18}$$

REMARK 10.1.1. The last equation in (1.18) is added for such a normalization of coefficients $\{\alpha_i\}_{i=1}^n$ that $\int_{-1}^1 K(x)\, dx = 1$.

Let

$$K_n(t) = -\frac{1}{(r-1)!} \sum_{i=1}^n \alpha_i(\nu_i - t)_+^{r-1}, \tag{1.19}$$

and

$$F_n(t) = \begin{cases} \displaystyle\int_{-1}^{t} K_n(y)\,dy, & t \in [-1,0]; \\[2mm] \displaystyle\int_{1}^{t} K_n(y)\,dy, & t \in [0,1]. \end{cases} \tag{1.20}$$

From the formula (1.4) and equations (1.8) we derive the numerical differentiation formula for the rth derivative of the function f at the origin:

$$f^{(r)}(0) = \sum_{i=1}^{n} \alpha_i f(\nu_i) + \int_{-1}^{1} [f^{(r)}(x) - f^{(r)}(0)]K(x)\,dx, \tag{1.21}$$

As in Lemma 10.1.1, we can show that $\alpha_0 = 0$ in (1.18) and the kernel $K_n(t)$ is even. The following properties of the kernel $K_n(t)$ are established in [72]:

(i) $\operatorname{supp} K_n = [\nu_k, \nu_{n+1-k}]$,   for some $k: 1 \le k \le [(n+1-r)/2]$;

(ii) the kernel K_n has precisely $n+1-2k-r$ simple zeroes $\{\vartheta_i\}_{i=k}^{n-k-r}$
on the interval (ν_k, ν_{n+1-k});

(iii) $\operatorname{sign} \alpha_i = (-1)^{i+m}$, $i = k+1, \ldots, n+1-k$;

(iv) $(-1)^{i+r+m} \operatorname{sign} K_n(t) \ge 0$, $\vartheta_i \le t \le \vartheta_{i+1}$, $i = k-1, \ldots, n+1-k-r$.
$$\tag{1.22}$$

Also, it follows from the last equation of the system (1.18) that

$$F_n(0-) = \frac{1}{2}, \qquad F_n(0+) = -\frac{1}{2}, \tag{1.23}$$

where $F_n(0+)$ and $F_n(0-)$ are the right-hand-side and the left-hand-side limit of F_n at the origin.

By (1.22), (1.23),

$$K_n(x) \in -\mathcal{M}_{l,l}^{-1}[\nu_k, \nu_{n+1-k}], \qquad l = \frac{1}{2}(n+3-2k-r), \tag{1.24}$$

where classes $\mathcal{M}_{l,l}^{j}[a,b]$ for $l \in \mathbb{N}$ and $j \in \{-1, 0+1\}$ are introduced in Definition 9.1.1.

Therefore, for any function $f \in W^r H^\omega[-1,1]$, the following inequality holds:

$$|f^{(r)}(0)| \le \sum_{i=1}^{n} |\alpha_i| \cdot \|f\|_{\mathbb{C}[-1,1]} + \int_{0}^{2} \Re_\omega(F_n;t)\,\omega'(t)\,dt, \tag{7.25}$$

where the rearrangement $\Re_\omega(F_n;t)$ is introduced in (9.3.3) of Definition 9.3.1.

10.1.2. Chebyshev ω-splines on the symmetric interval

We formulate a variant of Theorem 6.0.1 describing functions extremal in the problem (P.0) for $0 < m \le r$.

THEOREM 10.1.2. *Let ω be a concave modulus of continuity, $0 < m \le r$, and $n \in \mathbb{N}$ be such that $n \ge r$ and $n = m + 1 \pmod 2$. Then, there exist collections of points $\bar{\nu} = \bar{\nu}(n, r, m, \omega)$ and $\bar{\vartheta} = \bar{\vartheta}(n, r, m, \omega)$ as in (1.2) for $0 < m < r$ or as in (1.17) for $m = r$, and the function $T_n = T_{n,r,m,\omega}$ endowed with the properties:*

$$(A) \qquad \sup_{h \in H_0^\omega[-1,1]} \int_{-1}^{1} h(x) K_n(x)\, dx = \int_{-1}^{1} [T_n^{(r)}(x) - T_n^{(r)}(0)] K_n(x)\, dx,$$

where the kernel K_n is defined by (7.4), (7.8), (7.9) $\qquad$ (1.26)

for $0 < m < r$ or by (7.19) for $m = r$;

$$(B) \quad T_n(\nu_i) = (-1)^{i+m} \|T_n\|_{\mathrm{C}[-1,1]}, \qquad i = 0, \ldots, n+1.$$

10.2. Kolmogorov inequalities in $W^r H^\alpha(\mathbb{R})$

The results of this section can be proved in exactly the same manner as Theorem 8.2.1.

Fix $m, r \in \mathbb{N} : 0 < m \le r$. Let also

$$I(l) = \mathbb{Z}, \quad \text{if } l \text{ is even}; \qquad I(l) = \mathbb{Z} \setminus \{0\}, \quad \text{if } l \text{ is odd}. \qquad (2.1)$$

Let $\nu_0 = \vartheta_0 := 0$, and the collections of points $\{\nu_i\}_{i \in I(m)}$ and $\{\vartheta_i\}_{i \in I(r-m)}$ on the entire line $\mathbb{R}$ satisfy the conditions

$$\nu_{-i} = -\nu_i, \quad i \in I(m); \qquad \vartheta_{-j} = -\vartheta_j, \quad j \in I(r-m);$$

$$\lim_{i \to +\infty} \nu_i = +\infty; \quad \lim_{i \to +\infty} \vartheta_i = +\infty; \qquad (2.2)$$

$$\nu_i < \nu_{i+1}, \quad \vartheta_i < \vartheta_{i+1}, \quad i \in \mathbb{Z}.$$

The collections of points $\{\nu_i\}_{i \in I(m)}$, $\{\vartheta_i\}_{i \in I(r-m)}$ and the coefficients $\{\alpha_i\}_{i \in I(m)}$ can be chosen in such a way that the kernel

$$F(t) = \frac{1}{(r-m)!}(-t)_+^{r-m} + \frac{1}{r!} \sum_{i \in I(m)} \alpha_i (\nu_i - t)_+^r \qquad (2.3)$$

is integrable on the entire line $\mathbb{R}$ and $\sup\limits_{h \in H_0^\omega(\mathbb{R})} \int_{\mathbb{R}_+} h(t) F'(t)\, dt \le \infty$. In addition, $F(t)$ has the following properties:

$$F(-t) = (-1)^{r-m+1} F(t), \qquad t > 0;$$

$$\lim_{t \to \infty} F^{(j)}(t) = 0, \qquad j = 0, \ldots, r-1; \qquad (2.4)$$

$$F'(\vartheta_l) = 0, \qquad l \in I(r-m).$$

Also, any function from $f \in W^r H^\alpha(\mathbb{R})$ with a finite $\mathbb{L}_\infty(\mathbb{R})$ norm admits the following expression of the mth derivative of f at the origin:

$$f^{(m)}(0) = \sum_{i \in I(m)} \alpha_i f(\nu_i) + \int_{\mathbb{R}} [f^{(r)}(x) - f^{(r)}(0)] F'(x)\, dx. \qquad (2.5)$$

THEOREM 10.2.1. *Let $\omega_\alpha(t) = t^\alpha$, $0 < \alpha \leq 1$, and $m, r \in \mathbb{N}$, $0 < m \leq r$. There exist collections $\{\nu_i = \nu_i(r, m, \omega)\}_{i \in I(m)}$ and $\{\vartheta_i = \vartheta_i(r, m, \omega)\}_{i \in I(r-m)}$ as in (2.2), and the function $T = T_{r,m,\alpha}$ with the following properties:*

(A) $\displaystyle \sup_{h \in H_0^\alpha(\mathbb{R})} \int_{\mathbb{R}} h(x) F'(x)\, dx = \int_{\mathbb{R}} [T^{(r)}(x) - T^{(r)}(0)] F'(x)\, dx,$

   where the kernel F with properties (2.4), (2.5) is defined in (2.3);

(B) $T(\nu_i) = (-1)^{i+m} \|T\|_{\mathbb{L}_\infty(\mathbb{R})} = (-1)^{i+m}, \qquad i \in I(m);$

(C) *if $f \in W^r H^\alpha(\mathbb{R})$ and $\|f\|_{\mathbb{L}_\infty(\mathbb{R})} \leq 1$, then*

   $\|f^{(m)} \leq \mathcal{A}\|f\|_{\mathbb{L}_\infty(\mathbb{R})} + \mathcal{B},$

   where $\mathcal{A} = \mathcal{A}_{\alpha,r,m} := \displaystyle\sum_{i \in I(m)} |\alpha_i|$ and $\mathcal{B} = \mathcal{B}_{\alpha,r,m} := \displaystyle\int_{\mathbb{R}_+} \omega_\alpha'(t) \Re_{\omega_\alpha}(F; t)\, dt.$

$$(2.6)$$

The proof of Corollary 10.2.2 follows immediately from Theorem 10.2.1 and Corollary 2.2.4.

COROLLARY 10.2.2. *Let $0 < \alpha \leq 1$, $r, m \in \mathbb{N} : 0 < m \leq r$. Let the function $T = T_{r,m,\alpha}$ be as in Theorem 10.2.1. For any fixed $\gamma > 0$, the function $T_\gamma(t) = \gamma^{r+\alpha} T(t/\gamma)$ is a solution of the problem*

$$\|f^{(m)}\|_{\mathbb{L}_\infty(\mathbb{R})} \to \sup, \qquad f \in W^r H^\alpha(\mathbb{R}), \qquad \|f\|_{\mathbb{L}_\infty(\mathbb{R})} \leq \gamma. \qquad (2.7)$$

The relations (1.2.10)–(1.2.12) between additive and multiplicative inequalities in $W^r H^\alpha(\mathbb{R})$ lead us to the sharp multiplicative inequality for $\|f^{(m)}\|_{\mathbb{L}_\infty(\mathbb{R})}$.

COROLLARY 10.2.3. *Let the constants $\mathcal{A} = \mathcal{A}_{\alpha,r,m}$ and $\mathcal{B} = \mathcal{B}_{\alpha,r,m}$ be as in (2.6),*

(C), *and $C = (r + \alpha) \left(\dfrac{\mathcal{A}}{r + \alpha - m} \right)^{\frac{r+\alpha-m}{r+\alpha}} \left(\dfrac{\mathcal{B}}{m} \right)^{\frac{m}{r+\alpha}}$. Then,*

$$\|f^{(m)}\|_{\mathbb{L}_\infty(\mathbb{R})} \leq C \|f\|_{\mathbb{L}_\infty(\mathbb{R})}^{\frac{r+\alpha-m}{r+\alpha}}, \qquad (2.8)$$

for all $f \in W^r H^\alpha(\mathbb{R})$.

We compute the constants C in the cases $r = m = 1$ and $r = 2$, $m = 1$ or $m = 2$ in Chapter 11 and Chapter 12, respectively.

Chapter 11

Landau and Hadamard Inequalities in $W^1 H^\omega(\mathbb{R}_+)$ and $W^1 H^\omega(\mathbb{R})$

In this chapter we describe the extremal functions of the Kolmogorov–Landau problem

$$\|f'\|_{\mathrm{L}_\infty(I)} \to \sup, \qquad f \in W^1 H^\omega(I), \quad \|f\|_{\mathrm{L}_\infty(I)} \leq B, \tag{0.0}$$

for *all concave moduli of continuity* ω and $I = \mathbb{R}$ or $\mathbb{R}_+$. These results generalize the solution of the problem (0.0) for $\omega(t) = t$ by E. Landau [54] in the case $I = \mathbb{R}_+$ and J. Hadamard [31] in the case $I = \mathbb{R}$. A number of other elementary cases of the Kolmogorov–Landau problem for $\omega(t) = t$ are discussed by I. J. Schoenberg in [72].

11.1. Landau inequalities in $W^1 H^\omega(\mathbb{R}_+)$

Let us consider the problem

$$f'(0) \to \sup, \qquad f \in W^1 H^\omega(\mathbb{R}_+), \quad \|f\|_{\mathrm{L}_\infty(\mathbb{R}_+)} \leq B. \tag{1.1}$$

As we showed in Proposition 1.2.2, the extremal functions of the problem (1.1) also solve (0.0) for $I = \mathbb{R}_+$ and vice versa.

Let $f \in W^1 H^\omega(\mathbb{R}_+)$. Then,

$$f'(0) = \frac{1}{\tau}[f(\tau) - f(0)] - \frac{1}{\tau} \int_0^\tau [f'(x) - f'(0)]\, dx. \tag{1.2}$$

Therefore,

$$|f'(0)| \leq \frac{2}{\tau}\|f\|_{\mathrm{L}_\infty(\mathbb{R}_+)} + \frac{1}{\tau} \int_0^\tau \omega(x)\, dx. \tag{1.3}$$

Put

$$A_\omega^* = \lim_{t \to +\infty} \omega(t). \tag{1.4}$$

We adopt the convention $\omega(+\infty) := A_\omega^*$. Let ξ_A be defined as follows:

$$\xi_A = \min\{t > 0 \mid \omega(t) = A\}, \qquad A \in (0, A_\omega^*]. \tag{1.5}$$

Let

$$N_A(\omega) := \frac{1}{2} \int\limits_0^{\xi_A} (A - \omega(t))\, dt. \tag{1.6}$$

Then, we introduce the function $x_{\omega,A}(t)$ on the half-line $\mathbb{R}_+$ by the formulae

$$\frac{d}{dt} x_{\omega,A}(t) = \begin{cases} A - \omega(t), & 0 \le t \le \xi_A; \\ 0, & t > \xi_A, \end{cases}$$

$$x_{\omega,A}(y) = \int\limits_0^y \frac{d}{dt} x_{\omega,A}(t)\, dt - N_A(\omega), \quad y \in \mathbb{R}_+. \tag{1.7}$$

By definitions (1.6) and (1.7),

$$x_{\omega,A}(\xi_A) = -x_{\omega,A}(0) = \|x_{\omega,A}\|_{\mathrm{L}_\infty(\mathbb{R}_+)} = N_A(\omega). \tag{1.8}$$

Therefore, by (1.7), (1.8), the function $x_{\omega,A}$ transforms the inequality (1.3) into the equality for $\tau = \xi_A$. In our description of extremal functions in the problem (1.1), we distinguish two sets of concave modulii of continuity ω on $\mathbb{R}_+$:

$$\mathcal{N}_1 := \{\omega \mid \lim_{A \to A_\omega^*} N_A(\omega) = +\infty \}, \quad \mathcal{N}_2 := \{\omega \mid \lim_{A \to A_\omega^*} N_A(\omega) < +\infty \}. \tag{1.9}$$

Case 1. $\omega \in \mathcal{N}_1$.

Clearly, if $A_\omega^* = \infty$, then $\omega \in \mathcal{N}_1$. The Hölder modulii $\omega_\alpha(t) = t^\alpha$, $0 < \alpha \le 1$ provide a typical example of such functions.

If $A_\omega^* < \infty$, then

$$\omega \in \mathcal{N}_1 \quad \Longleftrightarrow \quad \int\limits_0^\infty (A_\omega^* - \omega(t))\, dt = \infty. \tag{1.10}$$

Among $\omega \in \mathcal{N}_1$ with a finite A_ω^* we mention $\omega(t) = \arctan(t)$ or $\omega(t) = c^p - (t+c)^p$ for all $t \ge 0$, $c > 0$ and $p \in [-1, 0)$.

If $\omega \in \mathcal{N}_1$, then $N_A(\omega)$ is a monotone function strictly increasing with A from 0 to $+\infty$. Thus, in view of the extremality of the functions $\{x_{\omega,A}\}_{A \in (0, A_\omega^*)}$ in the inequality (1.3) for the specified choice of τ, we arrive at the following complete (i.e., for all $B > 0$) description of extremal functions in the problem (1.1).

PROPOSITION 11.1.1. *If $\omega \in \mathcal{N}_1$, then the following sharp inequality holds for all functions $f \in W^1 H^\omega(\mathbb{R}_+)$ with $\|f\|_{\mathrm{L}_\infty(\mathbb{R}_+)} \le N_A(\omega)$ for some $A \in (0, A^*)$:*

$$\|f'\|_{\mathrm{L}_\infty(\mathbb{R}_+)} \le x'_{\omega,A}(0) = \frac{2}{\xi_A} N_A(\omega) + \frac{1}{\xi_A} \int\limits_0^{\xi_A} \omega(t)\, dt,$$

where ξ_A are defined in (1.5).

Recall the relations (1.2.10)–(1.2.12) of Lemma 1.2.3 describing the connection between multiplicative and additive inequalities in $W^r H^\alpha(\mathbb{R}_+)$. Then, the computation of constants ξ_A and $N_A(\omega_\alpha)$ in the case of Hölder modulii of continuity $\omega_\alpha(t) = t^\alpha$ and an application of Lemma 1.2.3 lead us to the following result.

COROLLARY 11.1.2. *For any function $f \in W^1 H^\alpha(\mathbb{R}_+)$, $\alpha \in (0,1]$, the following inequality holds:*

$$\|f'\|_{L_\infty(\mathbb{R}_+)} \leq \left[\frac{2(1+\alpha)}{\alpha}\right]^{\frac{1}{1+\alpha}} \|f\|_{L_\infty(\mathbb{R}_+)}^{\frac{1}{1+\alpha}}. \tag{1.11}$$

REMARK 11.1.1. By the definition, $W^1 H^1(\mathbb{R}_+) = W_\infty^2(\mathbb{R}_+)$, so, we have the original Landau inequality

$$\|f'\|_{L_\infty(\mathbb{R}_+)} \leq 2\|f\|_{L_\infty(\mathbb{R}_+)}^{\frac{1}{2}},$$

in (1.11) for $\alpha = 1$.

Case 2. $\omega \in \mathcal{N}_2$.

By (1.10),

$$\omega \in \mathcal{N}_2 \quad \Longleftrightarrow \quad A_\omega^* < \infty \quad \text{and} \quad \int_0^{\xi_{A_\omega^*}} (A_\omega^* - \omega(t))\, dt < \infty. \tag{1.12}$$

In particular, there exists the finite limit

$$C_\omega^* := \lim_{\tau \to \xi_{A_\omega^*}} \frac{1}{\tau} \int_0^\tau \omega(t)\, dt. \tag{1.13}$$

Notice that if ω is a strictly increasing modulus of continuity from $\mathcal{N}_2$, then $\xi_{A_\omega^*} = +\infty$ and $C_\omega^* = A_\omega^*$. Among such concave modulii of continuity we mention $\omega(t) = c^p - (t+c)^p$ for all $c > 0$ and $p < -1$.

The value of $\xi_{A_\omega^*}$ is finite, if and only if ω is *a stabilizing modulus of continuity* strictly increasing on $[0, \xi_{A_\omega^*}]$ and constant on $[\xi_{A_\omega^*}, +\infty)$.

As before, if $A \in (0, A_\omega^*]$, then the function $x_{\omega,A}$ is extremal in the problem (1.1) for $B = N_A(\omega)$, and the inequality (1.10) holds.

Taking the limit in (1.3) as $\tau \to \xi_{A_\omega^*}(\omega)$, we come to the conclusion that the following sharp inequality holds for any function $f \in W^1 H^\omega(\mathbb{R}_+)$ with $N_{A_\omega^*} < \|f\|_{L_\infty(\mathbb{R}_+)} < \infty$:

$$\|f'\|_{L_\infty(\mathbb{R}_+)} \leq A_\omega^* = x'_{\omega, A_\omega^*}(0). \tag{1.14}$$

There is another reason why the value $|f'(0)|$ of a function with a finite norm $\|f\|_{L_\infty(\mathbb{R}_+)}$ cannot exceed A_ω^*. Indeed, if $\chi f'(0) > A_\omega^* + c$, for some $\chi \in \{\pm 1\}$ and $c > 0$, then the inclusions $f' \in H^\omega(\mathbb{R}_+)$ and the definition (1.4) of A_ω imply that

$$\chi f'(t) \geq c, \qquad t \in \mathbb{R}_+.$$

Consequently, $\|f\|_{L[0,\gamma]} \geq \frac{1}{2}c\gamma$ for all $\gamma > 0$, so that $\|f\|_{L_\infty(\mathbb{R}_+)} = \infty$.

11.2. Hadamard inequalities in $W^1 H^\omega(\mathbb{R})$

By Proposition 1.2.2, the problem

$$f'(0) \to \sup, \qquad f \in W^1 H^\omega(\mathbb{R}), \quad \|f\|_{L_\infty(\mathbb{R})} \leq B. \tag{2.1}$$

is equivalent to the problem (0.0) for $I = \mathbb{R}$.

Let $f \in W^r H^\omega(\mathbb{R})$. Then,

$$f'(0) = \frac{1}{2\tau}[f(\tau) - f(-\tau)] - \frac{1}{2\tau} \int\limits_{-\tau}^{\tau} [f'(x) - f'(0)]\, dx. \tag{2.2}$$

Therefore,

$$|f'(0)| \leq \frac{1}{\tau}\|f\|_{L_\infty(\mathbb{R})} + \frac{1}{\tau} \int\limits_{0}^{\tau} \omega(x)\, dx. \tag{2.3}$$

For $A \in (0, A_\omega^*)$, let $\xi_A(\omega)$ and $N_A(\omega)$ be introduced in (1.5) and (1.6), respectively. Put

$$y_{\omega,A}(t) = \begin{cases} x_{\omega,A}(t) + N_A(\omega), & t > 0; \\ - x_{\omega,A}(-t) - N_A(\omega). & t \leq 0. \end{cases} \tag{2.4}$$

By the definition and (1.7),

$$y_{\omega,A}(\xi_A) = -y_{\omega,A}(-\xi_A) = \|y_{\omega,A}\|_{L_\infty(\mathbb{R})} = 2N_A(\omega). \tag{2.5}$$

Therefore, by (2.4), (2.5), the function $y_{\omega,A}$ transforms the inequality (2.3) into the equality for $\tau = \xi_A$. The following result is proved as an analog of Proposition 11.1.1.

PROPOSITION 11.2.1. *Let ω be a concave modulus of continuity on $\mathbb{R}_+$.*

I. For any function $f \in W^1 H^\omega(\mathbb{R})$ such that $\|f\|_{L_\infty(\mathbb{R})} \leq 2N_A(\omega)$ for some $A \in (0, A_\omega^*)$, the following inequality holds:

$$\|f'\|_{L_\infty(\mathbb{R})} \leq x_{\omega,A}'(0) = \frac{2}{\xi_A(\omega)} N_A(\omega) + \frac{1}{\xi_A(\omega)} \int\limits_{0}^{\xi_A(\omega)} \omega(t)\, dt. \tag{2.6}$$

II. If $\omega \in \mathcal{N}_2$, and $f \in W^1 H^\omega(\mathbb{R})$ is such that $2N_A(\omega) \leq \|f\|_{L_\infty(\mathbb{R})} < +\infty$, then

$$|f'(0)| \leq A_\omega^* = x_{\omega,A_\omega^*}'(0). \tag{2.7}$$

Computing $N_A(\omega_\alpha)$ and $\xi_A(\omega_\alpha)$ for $\omega_\alpha(t) = t^\alpha$ and applying Lemma 1.2.3, we find the extremal multiplicative inequality in the problem (0.0) for $I = \mathbb{R}_+$.

COROLLARY 11.2.2. *For any function* $f \in W^1H^\alpha(\mathbb{R}_+)$, $\alpha \in (0,1]$, *the following inequality holds:*

$$\|f'\|_{\mathrm{L}_\infty(\mathbb{R})} \le \left[\frac{(1+\alpha)}{\alpha}\right]^{\frac{1}{1+\alpha}} \|f\|_{\mathrm{L}_\infty(\mathbb{R})}^{\frac{1}{1+\alpha}}. \tag{2.8}$$

REMARK 11.2.1. For $\alpha = 1$, (2.8) is the original Hadamard inequality

$$\|f'\|_{\mathrm{L}_\infty(\mathbb{R})} \le \sqrt{2}\|f\|_{\mathrm{L}_\infty(\mathbb{R})}^{\frac{1}{2}}.$$

11.3. Specific feature of the Hölder classes $W^1H^\alpha(\mathbb{R})$ and $W^1H^\alpha(\mathbb{R}_+)$

For $B > 0$, let $X_{B,\omega,I}$ be an extremal function in the problem (1.1) for $I = \mathbb{R}_+$ and (2.1) for $I = \mathbb{R}$. In the following proposition we show that unless $\omega(t)$ is a Hölder's modulus of continuity $\tilde{\omega}_\gamma(t) = Ct^\gamma$, $0 < \gamma \le 1$, there exists no exact multiplicative inequality of the form

$$\|x'\|_{\mathrm{L}_\infty(I)} \le K_{\omega,I}\|x\|_{\mathrm{L}_\infty(I)}^\alpha, \qquad 0 < \alpha \le 1, \tag{3.1}$$

for which the functions $\{X_{B,\omega,I}\}_{B>0}$ comprise the set of extremal functions. The extremality of these functions in the inequality (3.1) is equivalent to the property

$$\|X'_{B,\omega,I}\|_{\mathrm{L}_\infty(I)} \equiv K_{\omega,I}\|X_{B,\omega,I}\|_{\mathrm{L}_\infty(I)}^\alpha, \qquad B > 0. \tag{3.2}$$

Before stating the corresponding result, we note that the property (3.2) cannot be satisfied for $\omega \in \mathcal{N}_2$. Indeed, for all $B > N(A_\omega^*)$ (if $I = \mathbb{R}_+$) or $B > 2N(A_\omega^*)$ (if $I = \mathbb{R}$), $X'_{B,\omega,I}(0) \equiv A_\omega^*$.

Consequently, in our analysis we can restrict ourselves only to $\omega \in \mathcal{N}_1$. For such modulii of continuity ω, the function ξ_A from (1.5) is a differentiable and strictly increasing function of A.

PROPOSITION 11.3.1. *Let* $\omega \in \mathcal{N}_1$. *Suppose that for any* $B > 0$, *there exists an* $\alpha = \alpha(\omega) : 0 < \alpha \le 1$, *such that the property* (3.2) *holds. Then,* $\alpha \in (0, 1/2]$, *and* $\omega(t)$ *is the Hölder's modulus of continuity*

$$\omega(t) = Ct^{\frac{\alpha}{1-\alpha}}, \qquad t \ge 0. \tag{3.3}$$

Proof. We consider the case $I = \mathbb{R}_+$. The proof in the case $I = \mathbb{R}$ is analogous.

By Proposition 11.3.1, $\|X'_{B,\omega,\mathbb{R}_+}\|_{\mathrm{L}_\infty(\mathbb{R}_+)} = B$, while $X_{B,\omega,\mathbb{R}_+}(0) = A$, where $A = A_B$ is determined from the equations

$$\omega(\xi_A) = A, \tag{3.4}$$

$$\int_0^{\xi_A} (A - \omega(t))\,dt = 2B. \tag{3.5}$$

Therefore, by (3.2) and (3.4), (3.5),

$$\frac{\left[X_{B,\omega,\mathbb{R}_+}(0)\right]^{\frac{1}{\alpha}}}{\left\|x_{B,\omega,\mathbb{R}_+}\right\|_{L_\infty(R_+)}} = \frac{A^{\frac{1}{\alpha}}}{\frac{1}{2}\int\limits_0^{\xi_A}(A-\omega(t))\,dt} = \text{const}, \quad A \in (0, A_\omega^*). \tag{3.6}$$

Differentiating (3.6) with respect to A, we obtain the identity

$$\frac{1}{\alpha}A^{\frac{1}{\alpha}-1}\int\limits_0^{\xi_A}(A-\omega(t))\,dt - A^{\frac{1}{\alpha}}\xi_A + A^{\frac{1}{\alpha}}\left(\omega(\xi_A)-A\right)\frac{d\xi_A}{dA} = 0. \tag{3.7}$$

By the definition (3.4) of ξ_A, the identity (3.7) is equivalent to the following equation:

$$\frac{1}{\alpha}\int\limits_0^{\xi_A}(A-\omega(t))\,dt = A\xi_A. \tag{3.8}$$

Another differentiation of (3.30) with respect to A produces the equation

$$\frac{1}{\alpha}\xi_A = \xi_A + A\frac{d\xi_A}{dA},$$

or, in a more compact form,

$$\left(\frac{1}{\alpha}-1\right)\frac{dA}{A} = \frac{d\xi_A}{\xi_A}. \tag{3.9}$$

Since $\xi_0 = 0$, the solutions of the differential equation (3.9) are of the form

$$\xi_A = C^{\frac{1-\alpha}{\alpha}}A^{\frac{1}{\alpha}-1}, \qquad C > 0. \tag{3.10}$$

The formula (3.10) in combination with the definition $\xi_A = \omega^{-1}(A)$ in (3.4) implies that

$$\omega(t) = Ct^{\frac{\alpha}{1-\alpha}}, \qquad t \geq 0. \tag{3.11}$$

Since ω is *concave*, we finally have $0 < \dfrac{\alpha}{1-\alpha} \leq 1$, or $0 < \alpha \leq \dfrac{1}{2}$. $\qquad\square$

11.4. Extrapolation problem in $W^1H^\omega(-\infty, \tau]$

Let ω be a concave modulus of continuity on $\mathbb{R}_+$ and $\tau > 0$. In this section we describe the extremal functions in the problem

$$f(\tau) \to \sup, \qquad f \in W^1H^\omega(-\infty, \tau], \quad \|f\|_{L_\infty(-\infty,0]} \leq B, \tag{4.1}$$

for all $B \geq 0$. As usual, we first consider the case $\omega \in \mathcal{N}_1$.

Fix $\beta > 0$, and consider the kernel

$$\rho_\beta(t) = \begin{cases} \beta, & 0 \le t \le \tau; \\ -\tau, & -\beta \le t < 0. \end{cases} \tag{4.2}$$

Then, the following identity holds for any function $f \in W^r H^\omega(-\infty, \tau]$:

$$f(\tau) = \frac{\tau + \beta}{\beta} f(0) - \frac{\tau}{\beta} f(-\beta) + \frac{1}{\beta} \int\limits_{-\beta}^{\tau} f'(t) \rho_\beta(t)\, dt. \tag{4.3}$$

By Korneichuk's Lemma 2.1.1,

$$\sup_{h \in H^\omega[-\beta, \tau]} \int\limits_{-\beta}^{\tau} h(t) \rho_\beta(t) = \frac{\tau \beta}{\tau + \beta} \int\limits_{0}^{\tau + \beta} \omega(t)\, dt, \tag{4.4}$$

Therefore, by (4.3), (4.4), for any function $f \in W^r H^\omega(-\infty, \tau]$ we have the inequality

$$|f(\tau)| \le \frac{2\tau + \beta}{\beta} \|f\|_{\mathbb{L}_\infty(-\infty, 0]} + \frac{\tau}{\tau + \beta} \int\limits_{0}^{\tau + \beta} \omega(t)\, dt. \tag{4.5}$$

The derivative $h_\beta(t)$ of the extremal function in (4.4) is given by the formula

$$h_\beta(x) = \begin{cases} \omega'(\dfrac{\beta + \tau}{\tau} x), & 0 < x \le \tau; \\[2mm] \omega'(-\dfrac{\beta + \tau}{\beta} x), & -\beta \le x < 0. \end{cases} \tag{4.6}$$

Put

$$E_{\tau, \beta} = \omega(\tau + \beta)\beta + \frac{\tau - \beta}{\tau + \beta} \int\limits_{0}^{\tau + \beta} \omega(t)\, dt,$$

$$L_{\tau, \beta} = \frac{\beta^2}{2(\tau + \beta)} \left[\omega(\tau + \beta)\beta - \frac{1}{\tau + \beta} \int\limits_{0}^{\tau + \beta} \omega(t)\, dt \right]. \tag{4.7}$$

For any fixed $\tau > 0$, the function $L_{\tau, \beta}$ increases with β from 0 to $+\infty$, while the function $E_{\tau, \beta}$ increases from $\int\limits_{0}^{\tau} \omega(t)\, dt$ to $+\infty$, as β increases from 0 to $+\infty$.

We define the functions $q_\beta(t)$ by

$$q_\beta(t) = \begin{cases} \displaystyle\int\limits_{-b}^{t} \int\limits_{-b}^{y} h_b(x)\, dx\, dy - L_{\tau, \beta}, & -b \le t \le \tau; \\[4mm] -L_{\tau, \beta}, & t < -\beta. \end{cases} \tag{4.8}$$

The function $q_\beta(t)$ is defined so that $q_\beta \in W^1 H^\omega(-\infty, \tau]$ and

$$(i) \quad q_\beta(0) = -q_\beta(-b) \doteq \|q_\beta\|_{\mathbb{L}_\infty(-\infty,0]} = L_{\tau,\beta}; \quad q_\beta(\tau) = E_{\tau,\beta};$$

$$(ii) \quad \sup_{h \in H^\omega[-b,\tau]} \int_{-b}^{\tau} h(t)\rho_\beta(t)\, dt = \int_{-b}^{\tau} q_\beta'(t)\rho_\beta(t)\, dt. \tag{4.9}$$

Therefore, the function $q_\beta(t)$ transforms the inequality (4.5) into the equality.

PROPOSITION 11.4.1. *Let the constants $E_{\tau,\beta}$ and $L_{\tau,\beta}$ be defined in (4.7) for $\beta \geq 0$. Then, for any function $f \in W^1 H^\omega(-\infty, \tau]$ with $\|f\|_{\mathbb{L}_\infty(-\infty,\tau]} \leq L_{\tau,\beta}$, the following inequality holds:*

$$|f(\tau)| \leq E_{\tau,\beta} = q_\beta(\tau). \tag{4.10}$$

Let us assume now that ω belongs to $\mathcal{N}_2$. The inequalities (4.10) still hold for all $\beta \geq 0$. But

$$\lim_{\beta \to +\infty} L_{\tau,\beta} = N_{A_\omega^*}(\omega) =: L_{\tau,\infty}, \quad \lim_{\beta \to +\infty} E_{\tau,\beta} = N_{A_\omega^*}(\omega) + A_\omega^* \tau =: E_{\tau,\infty}. \tag{4.11}$$

Also, as $\beta \to +\infty$ the functions q_β converge poinwise to the function $q_\infty(t)$, whose derivative is given by the formula

$$q_\infty'(t) = \begin{cases} A_\omega^* - \omega(-t), & t < 0; \\ A_\omega^*, & t \in [0, \tau]. \end{cases} \tag{4.12}$$

Taking the limit in the inequality (4.5) as $\beta \to +\infty$, we infer that

$$|f(\tau)| \leq E_{\tau,\infty} = q_\infty(\tau), \tag{4.13}$$

for all $f \in W^1 H^\omega(-\infty, \beta]$ such that $\|f\|_{\mathbb{L}_\infty(-\infty,0]} \geq L_{\tau,\infty}$.

Chapter 12

Sharp Kolmogorov–Landau Inequalities in $W^2 H^\omega(\mathbb{I})$, $\mathbb{I} = \mathbb{R} \vee \mathbb{R}_+$

In this chapter we describe extremal functions and sharp Kolmogorov inequalities in the problem

$$\|f^{(m)}\|_{\mathbb{L}_\infty(I)} \to \sup, \qquad f \in W^2 H^\omega(I), \quad \|f\|_{\mathbb{L}_\infty(I)} \le B,$$

for $m = 1, 2$, and $I = \mathbb{R}$ or $\mathbb{R}_+$. We also give the corresponding optimal numerical differentation formulae for $f'(x)$ and $f''(x)$.

12.1. Kolmogorov–Stechkin inequalities in $W^2 H^\omega(\mathbb{R})$

12.1.1. Estimates of the first derivative

The following problem is under our consideration:

$$\|f'\|_{\mathbb{L}_\infty(\mathbb{R})} \to \sup, \qquad \|f\|_{\mathbb{L}_\infty(\mathbb{R})} \le B, \quad f \in W^2 H^\omega(\mathbb{R}). \tag{1.0}$$

Let $b > 0$. Consider the kernel

$$\psi_b(t) = \begin{cases} t + b, & -b \le t \le 0. \\ t - b, & 0 \le t \le b. \end{cases} \tag{1.1}$$

Fix a function $f \in W^2 H^\omega(\mathbb{R})$. Then, we have the following formula for the first derivative at the origin:

$$f'(0) = \frac{1}{2b}[f(b) - f(-b)] + \frac{1}{2b} \int_{-b}^{b} f''(t)\, \psi_b(t)\, dt. \tag{1.2}$$

Therefore,

$$|f'(0)| \le \frac{1}{b}\|f\|_{\mathbb{L}_\infty(\mathbb{R})} + \frac{1}{2b} \sup_{h \in H^\omega[-b,b]} \int_{-b}^{b} h(t)\, \psi_b(t)\, dt. \tag{1.3}$$

Let

$$h_{\omega,b}(t) = \begin{cases} -\dfrac{1}{2}\omega(2t), & 0 \le t \le b. \\[2mm] \dfrac{1}{2}\omega(-2t), & -b \le t \le 0. \end{cases} \tag{1.4}$$

Since $\psi_b(t)$ is a simple odd kernel, by Corollary 2.1.2 of the Korneichuk lemma, the function $h_{\omega,b}$ is extremal in the problem

$$\int_{-b}^{b} h(t)\psi_b(t)\,dt \to \sup, \quad h \in H^\omega[-b,b].$$

We extend $h_{\omega,b}(t)$ to the entire line by the equations

$$h_{\omega,b}(t) = (-1)^n h_{\omega,b}(t - 2nb), \quad t \in [b(2n-1), b(2n+1)], \quad n \in \mathbb{Z}, \tag{1.5}$$

and put

$$g_{\omega,b}(t) = \int_0^t \int_b^\xi h_{\omega,b}(\tau)\,d\tau\,d\xi. \tag{1.6}$$

Notice that $g_{\omega,\pi/2n}(t)$ is the Euler spline $f_{n,2}(t)$ introduced in Definition 0.3.2. Let us summarize properties of the function $g_{\omega,b}(t)$:

(i) $g_{\omega,b}(t) \in W^2 H^\omega(\mathbb{R})$:

$$\omega\left(g''_{\omega,b}; t\right) = \begin{cases} \omega(t), & 0 \le t \le 2b. \\ \omega(2b), & t > 2b. \end{cases}$$

(ii) $g_{\omega,b}(b) = -g_{\omega,b}(-b) = \|g_{\omega,b}\|_{L_\infty(\mathbb{R})}.$ \hfill (1.7)

(iii) $\displaystyle\sup_{h \in H^\omega(\mathbb{R})} \int_{-b}^{b} h(t)\psi_b(t)\,dt = \int_{-b}^{b} g''_{\omega,b}(t)\psi_b(t)\,dt.$

Let $I_\omega(b) := \|g_{\omega,b}\|_{L_\infty(\mathbb{R})} = \int\limits_0^b \int\limits_0^t \frac{1}{2}\omega(2x)\,dx\,dt.$

One can easily verify that $I_\omega(b)$ is continuous and strictly increases from 0 to $+\infty$ as b increases from 0 to $+\infty$.

Properties (1.7), (i)–(iii) of the function $g_{\omega,b}(t)$ along with the identity (1.2) enable us to conclude that $g_{\omega,b}(t)$ is extremal in the problem (1.0) for $B = I_\omega(b)$.

By (1.2), (1.3), the sharp inequality is as follows:

$$\|f'\|_{L_\infty(\mathbb{R})} \le \frac{1}{b}\|f\|_{L_\infty(\mathbb{R})} + \frac{1}{8b}\int_0^{2b} \omega(u)(2b - u)\,du, \tag{1.8}$$

for functions $f \in W^2 H^\omega(\mathbb{R})$: $\|f\|_{L_\infty(\mathbb{R})} \le I_\omega(b)$. The following proposition summarizes these results.

PROPOSITION 12.1.1. *Let ω be a concave modulus of continuity on $\mathbb{R}_+$. For any $b > 0$, let*

$$I_\omega(b) := \int_0^b \int_t^b \frac{1}{2}\omega(2x)\,dx\,dt := \|g_{\omega,b}\|_{\mathrm{L}_\infty(\mathbb{R})};$$

$$A_\omega(b) := \int_0^b \frac{1}{2}\omega(2x)\,dx := \|g'_{\omega,b}\|_{\mathrm{L}_\infty(\mathbb{R})}. \tag{1.9}$$

Then, for any function $f \in W^2H^\omega(\mathbb{R})$ with $\|f\|_{\mathrm{L}_\infty(\mathbb{R})} \le I_\omega(b)$, the following sharp inequality holds:

$$\|f'\|_{\mathrm{L}_\infty(\mathbb{R})} \le A_\omega(b) = g'_{\omega,b}(0).$$

The computation of constants in (1.9) for $\omega_\alpha(t) = t^\alpha$ and Lemma 1.2.3 enable us to find sharp Kolmogorov inequalities in (1.0).

COROLLARY 12.1.1. *If $\omega(t) = t^\alpha$, $0 < \alpha \le 1$, and $f \in W^2H^\alpha(\mathbb{R})$, then*

$$\|f'\|_{\mathrm{L}_\infty(\mathbb{R})} \le C_\alpha \|f\|_{\mathrm{L}_\infty(\mathbb{R})}^{\frac{\alpha+1}{\alpha+2}}, \tag{1.10}$$

where $C_\alpha := 2^{\frac{\alpha-1}{\alpha+2}}(\alpha+2)^{\frac{\alpha+1}{\alpha+2}}(\alpha+1)^{-1}$.

REMARK 12.1.1. Notice that $C_\alpha \to 1$, as $\alpha \to 0$, and the inequality (1.10) for $\alpha = 0$ has the form

$$\|f'\|_{\mathrm{L}_\infty(\mathbb{R})} \le \|f\|_{\mathrm{L}_\infty(\mathbb{R})}^{\frac{1}{2}}.$$

This inequality is precisely the original *Hadamard inequality* for functions from the Sobolev class $\frac{1}{2}W^2_\infty(\mathbb{R}) = \{\frac{1}{2}f \mid f \in W^2_\infty(\mathbb{R})\}$. On the other hand, $W^2H^1(\mathbb{R}) = W^3_\infty(\mathbb{R})$, so (1.10) is the Kolmogorov inequality (0.1.5) in $W^3_\infty(\mathbb{R})$ for $r = 3$, $m = 1$:

$$\|f'\|_{\mathrm{L}_\infty(\mathbb{R}_+)} \le \frac{1}{2}3^{\frac{2}{3}}\|f\|_{\mathrm{L}_\infty(\mathbb{R})}.$$

12.1.2. Estimates of the second derivative

Let $b > 0$. Consider the kernel

$$\phi_b(t) = \begin{cases} -t - b, & -b \le t \le 0, \\ -b + t, & 0 \le t \le b. \end{cases} \tag{1.11}$$

Then,

$$f''(0) = \frac{1}{b^2}\left(f(b) - 2f(0) + f(-b)\right) + \frac{1}{b^2}\int_{-b}^b [f''(t) - f''(0)]\phi_b(t)\,dt. \tag{1.12}$$

Therefore,

$$|f''(0)| \leq \frac{4}{b^2}\|f\|_{\mathrm{L}_\infty(\mathbb{R})} + \frac{2}{b^2}\int_0^b \omega(t)(b-t)\,dt. \tag{1.13}$$

The function $\phi_b(t)$ is negative on $(-b, b)$. Therefore, a function $h^* \in H^\omega[-b, b]$ is extremal in the problem

$$\int_{-b}^b h(t)\phi_b(t)\,dt \to \sup, \qquad h \in H_0^\omega[-b, b], \tag{1.14}$$

if and only if

$$h^*(t) = \begin{cases} -\omega(t), & 0 \leq t \leq b; \\ -\omega(-t), & -b \leq t \leq 0. \end{cases} \tag{1.15}$$

Put

$$s_{\omega,b}(t) = h^*(t) + \frac{1}{b}\int_0^b \omega(\xi)\,dx. \tag{1.16}$$

We extend the function $s_{\omega,b}$ onto the whole line $\mathbb{R}$ periodically:

$$s_{\omega,b}(t) = s_{\omega,b}(t-2nb), \qquad t \in [b(2n-1), b(2n+1)], \quad n \in \mathbb{Z}. \tag{1.17}$$

Put

$$p_{\omega,b}(t) = \int_0^t \int_0^\xi s_{\omega,b}(\tau)\,d\tau\,d\xi - \frac{1}{2}\int_0^b \int_0^\xi s_{\omega,b}(\tau)\,d\tau\,d\xi. \tag{1.18}$$

The function $p_{\omega,b}$ is defined so that

$$p_{\omega,b}(nb) = (-1)^n \|p_{\omega,b}\|_{\mathrm{L}_\infty(\mathbb{R})}. \tag{1.19}$$

Consequently, the property (1.17) and the extremality of the function

$$h^*(x) = \frac{d^2}{dx^2}p_{\omega,b}(x) - \frac{d^2}{dx^2}p_{\omega,b}(0), \quad -b \leq x \leq b,$$

in the problem (1.14) imply that $p_{\omega,b}(t)$ transforms the inequality (1.13) into the equality.

Put

$$\begin{aligned}
I_{\omega,b} &= \frac{1}{4}b\int_0^b \omega(\tau)\,d\tau - \frac{1}{2}\int_0^b \int_0^\xi \omega(x)\,dx\,d\xi := \|p_{\omega,b}\|_{\mathrm{L}_\infty(\mathbb{R})}, \\
J_{\omega,b} &= \frac{1}{b}\int_0^b \omega(t)\,dt := \|p''_{\omega,b}\|_{\mathrm{L}_\infty(\mathbb{R})}.
\end{aligned} \tag{1.20}$$

One can easily verify that both functions $I_{\omega,b}$ and $J_{\omega,b}$ strictly increase from 0 to $+\infty$, as b increases from 0 to $+\infty$. In view of the inequality (1.13) and the form of extremal functions in (1.14), we have the following result.

PROPOSITION 12.1.2. *Let ω be a concave modulus of continuity on $\mathbb{R}_+$.*

Let $I_{\omega,b}$ and $J_{\omega,b}$ be defined in (1.20). Then, for any function $f \in W^2 H^\omega(\mathbb{R})$:
$\|f\|_{L_\infty(\mathbb{R})} \le J_{\omega,b}$, *the following sharp inequality holds:*

$$\|f''\|_{L_\infty(\mathbb{R})} \le I_{\omega,b}. \tag{1.21}$$

The version of Proposition 12.1.2 for $\omega_\alpha(t) = t^\alpha$ and an application of Lemma 1.2.3 leads us to the sharp multiplicative inequality in Hölder classes.

COROLLARY 12.1.3. *For any function $f \in W^r H^\alpha(\mathbb{R})$, $0 < \alpha \le 1$,*

$$\|f''\|_{L_\infty(\mathbb{R})} \le D_\alpha \|f\|_{L_\infty(\mathbb{R})}^{\frac{\alpha}{\alpha+2}}, \tag{1.22}$$

where $D_\alpha := 4^{\frac{\alpha}{\alpha+2}} \alpha^{-\frac{\alpha}{\alpha+2}} (\alpha+1)^{-\frac{2}{\alpha+2}} (\alpha+2)^{\frac{\alpha}{\alpha+2}}$.

REMARK 12.1.2. Notice that $D_\alpha \to 1$, as $\alpha \to 0$. Taking the limit in (1.22) as $\alpha \to 0$, we obtain the characteristic inequality of the class $W^2_\infty(\mathbb{R})$: $\|f''\|_{L_\infty(\mathbb{R})} \le 1$. If $\alpha = 1$ in (1.22), we have the Kolmogorov inequality (0.1.5) in $W^3_\infty(\mathbb{R})$ for $r = 3$, $m = 2$:

$$\|f''\|_{L_\infty(\mathbb{R}_+)} \le 3^{\frac{1}{3}} \|f\|_{L_\infty(\mathbb{R})}^{\frac{1}{3}}.$$

Summarizing the results of Section 12.1, we give the numerical differentiation formula for the first and second derivative in the class $W^2 H^\omega(\mathbb{R})$ and the error of approximation.

Consulting the formula (1.2) and the inequality (1.8), we find the numerical formula for $f'(x)$ with the step 2τ,

$$f'(x) \approx \mathcal{F}_1(x, \tau, \omega) = \frac{1}{2\tau}[f(x + \tau) - f(x - \tau)],$$

and the (maximal) error $E_1(\omega)$ of approximation of $f'(x)$ by $\mathcal{F}_1(x, \tau)$:

$$E_1(\omega) = \frac{1}{8\tau} \int_0^{2\tau} \omega(u)(2\tau - u)\, du.$$

Notice that for $\omega_\alpha(t) = t^\alpha$, $0 < \alpha \le 1$, $E_1(\omega_\alpha) = \dfrac{2^{\alpha-3}}{(\alpha+1)(\alpha+2)} \tau^{\alpha+1}$.

By the formula (1.12) and the inequality (1.13), the numerical formula for $f''(x)$ with the step 2τ is as follows:

$$f''(x) \approx \mathcal{F}_2(x, \tau, \omega) = \frac{1}{\tau^2}[f(x + \tau) - 2f(x) + f(x - \tau)].$$

The error $E_2(\omega)$ of approximation of $f''(x)$ by $\mathcal{F}_2(x, \tau)$ has the following expression:

$$E_2(\omega) = \frac{2}{\tau^2} \int_0^\tau \omega(u)(\tau - u)\, du.$$

For $\omega_\alpha(t) = t^\alpha$, $0 < \alpha \le 1$, $E_2(\omega_\alpha) = \dfrac{2}{(\alpha+1)(\alpha+2)} \tau^\alpha$.

12.2. Sharp estimates of derivatives in $W^2 H^\omega(\mathbb{R}_+)$, $W^2 H^\omega[0,1]$

12.2.1. Extensions of functions from $W^2 H^\omega[0, A]$ to $W^2 H^\omega(\mathbb{R}_+)$

Fix $A > 0$. We make the following observation on the possibility of functional extensions from the class $W^2 H^\omega[0, A]$ to the class $W^2 H^\omega(\mathbb{R}_+)$ without increasing the norm of $\mathbb{L}_\infty(\mathbb{R}_+)$.

Suppose that the derivative $\dfrac{d}{dt} g(t)$ of a function $g \in W^2 H^\omega[0, A]$ has two zeroes t_1, t_2, $0 \le t_1 < t_2 \le A$. Let $\delta = t_2 - t_1$. Then, the extension $E(g; t_1, t_2; \cdot)$ of the function $g(\cdot)$ from the interval $[0, t_2]$ to the entire half-line $\mathbb{R}_+$ is given by the formula

$$
E(g; t_1, t_2; t) = \begin{cases} g(t), & 0 \le t \le t_2; \\ g(t - 2n\delta), & t_2 + (2n-1)\delta \le t \le t_2 + 2n\delta, \quad n \in \mathbb{N}; \\ g(2t_2 + 2(n+1)\delta - t), & t_2 + 2(n-1)\delta \le t \le t_2 + (2n-1)\delta. \end{cases}
$$
$$(2.1)$$

The properties $\dfrac{d}{dt} g(t_1) = \dfrac{d}{dt} g(t_2) = 0$ assure the continuity of $\dfrac{d}{dt} g(t)$ on $\mathbb{R}_+$. In addition,

$$
\omega(E(g; t_1, t_2; \cdot); t) = \begin{cases} \omega(g; t), & 0 \le t \le t_2; \\ \omega(g; t_2), & t > t_2. \end{cases} \qquad (2.2)
$$

Thus, $E(g; t_1, t_2; t) \in W^2 H^\omega(\mathbb{R}_+)$. Also notice that

$$
\| E(g; t_1, t_2; \cdot) \|_{\mathbb{L}_\infty(\mathbb{R}_+)} = \| g \|_{C[0, t_2]}. \qquad (2.3)
$$

Therefore, we extended the function g to the entire half-line $\mathbb{R}_+$ without leaving the class $W^2 H^\omega(\mathbb{R}_+)$ and increasing the $\mathbb{L}_\infty$ norm.

12.2.2. Stechkin–Matorin inequalities in $W^2 H^\omega[0,1]$, $W^2 H^\omega(\mathbb{R}_+)$

In this subsection we first describe extremal functions of the problem

$$
f^{(m)}(0) \to \sup; \quad \| f \|_{C(I)} \le B, \quad f \in W^2 H^\omega(I), \quad (m = 1, 2), \qquad (2.4)
$$

in the case $m = 1, 2$, $I = [0, A]$ and any $B > 0$. This enables us to characterize the solution of the problem (2.4) for $m = 1, 2$, and $I = \mathbb{R}_+$, $B > 0$.

In order to avoid technical difficulties and emphasize the principal features of extremal functions in the problem (2.4), we restrict ourselves to a detailed consideration of the concave modulii of continuity ω on $\mathbb{R}_+$ with the property

$$
\lim_{t \to +\infty} \omega(t) = +\infty. \qquad (\star)
$$

Modulii ω, endowed with the property $(\star)$, are strictly concave and invertible: there exist a strictly convex inverse function ω^{-1}.

Let $0 =: \tau_0 < \tau_1 < \tau_2 \leq A$.
Determine $\{\alpha_i = \alpha_i(m)\}_{i=0}^2$ from the following system of linear equations:

$$\sum_{i=0}^2 \alpha_i \tau_i^k = m! \cdot \delta_{m,k}, \qquad k = 0, 1, 2, \quad m = 1 \text{ or } 2. \tag{2.5}$$

REMARK 12.2.1. $\tau_0{}^0 := 1$ in (2.5). Also, one can easily check that

$$
\begin{aligned}
&\alpha_0(1) = -\frac{1}{\tau_1} + \frac{1}{\tau_2}, \quad \alpha_1(1) = \frac{\tau_2}{\tau_1(\tau_2 - \tau_1)}, \quad \alpha_2(1) = -\frac{\tau_1}{\tau_2(\tau_2 - \tau_1)}; \\
&\alpha_0(2) = \frac{2}{\tau_1\tau_2}, \quad \alpha_1(2) = \frac{2}{\tau_1(\tau_1 - \tau_2)}, \quad \alpha_2(2) = \frac{2}{\tau_2(\tau_2 - \tau_1)}.
\end{aligned}
\tag{2.6}
$$

In particular, $\operatorname{sign}\alpha_i = (-1)^{i+m}$, $i = 0,\ 1,\ 2$.

Let $f \in W^2H^\omega[0, A]$. Then, an application of Taylor's formula to a $\mathbb{C}^2$-function $f(x)$ produces the identity

$$f^{(m)}(0) = \sum_{i=0}^2 \alpha_i f(\tau_i) + \int_0^b [f''(u) - f''(0)] K(u)\, du, \tag{2.7}$$

where

$$K(u) = -\sum_{i=0}^2 \alpha_i(\tau_i - u)_+, \tag{2.8}$$

Therefore,

$$|f^{(m)}(0)| \leq \left(\sum_{i=0}^2 |\alpha_i|\right) \|f\|_{C[0,b]} + \sup_{h \in H_0^\omega[0,b]} \int_0^b h(u)K(u)\, du. \tag{2.9}$$

The identity (2.7) and inequality (2.9) imply that if a function $g \in W^2H^\omega[0, A]$ enjoys two properties

$$
\begin{aligned}
&(i) \ \ g(\tau_i) = (-1)^{i+m} \|g\|_{C[0,b]}, \qquad i = 0, 1, 2; \\
&(ii) \quad \sup_{h \in H_0^\omega[0,\tau_2]} \int_0^{\tau_2} h(t)K(t)\, dt = \int_0^{\tau_2} [g''(t) - g''(0)]K(t)\, dt,
\end{aligned}
\tag{2.10}
$$

then $g(t)$ is a solution to the problem:

$$f'(0) \to \sup, \qquad f \in W^2H^\omega[0, A], \quad \|f\|_{C[0,\tau_2]} \leq \|g\|_{C[0,\tau_2]}. \tag{2.11}$$

In Theorem 2.1 of [8], we describe the family $\{Z_B\}_{B>0}$ of *Zolotarev ω-polynomials* in $W^rH^\omega[0, A]$. The following result is a variant of Theorem 2.1 in [8] for $r = 2$.

THEOREM 12.2.1. *Let $A > 0$. For any $B > 0$, there exist points*

$$\{\tau_i = \tau_i(B, A, m, \omega)\}_{i=0}^{2}, \quad 0 = \tau_0 < \tau_1 < \tau_2 \leq A,$$

and the function $Z_B(t) = Z_{B,A,\omega,m} \in W^2 H^\omega[0, A]$ with the properties

$$
\begin{aligned}
&(i) \quad \inf_{h \in H_0^\omega[0,A]} \int_0^{\tau_2} h(t)K(t)\,dt = \int_0^{\tau_2} [Z_B''(t) - Z_B''(0)]K(t)\,dt, \\
&\qquad \text{where } K \text{ is defined in (2.8) for } \{\tau_i = \tau_i(B, A, m, \omega)\}_{i=0}^{2}; \\
&(ii) \quad Z_B(\tau_i) = (-1)^{i+m}\|Z_{B,A}\|_{C[0,\tau_2]} = (-1)^{i+m}B; \\
&(iii) \quad \text{if } \tau_2 < A, \ \text{then } \frac{d}{dt}Z_{B,A}(\tau_2) = 0.
\end{aligned}
\tag{2.12}
$$

REMARK 12.2.2. In the case $m = 1$ the kernel $K(t)$ is simple, and the Korne-ichuk's lemma provides the formula for $\dfrac{d^3}{dt^3}Z_B(t)$ on the interval $[0, \tau_2]$:

$$
\frac{d^3}{dt^3}Z_B(t) =
\begin{cases}
\omega'(\rho(t) - t), & 0 \leq t \leq c, \\
\omega'(t - \rho^{-1}(t)), & c \leq t \leq \tau_2,
\end{cases}
\tag{2.13}
$$

where c is the only zero of K on the open interval $(0, \tau_2)$, and the function $\rho : [0, c] \to [c, \tau_2]$ is determined from the equation

$$
\int_0^t K(x)\,dx = \int_0^{\rho(t)} K(x)\,dx, \qquad 0 \leq t \leq c.
\tag{2.14}
$$

If $m = 2$, then $K(t) < 0$, $0 \leq t \leq \tau_2$. Therefore, the function $Z_B(t)$ enjoys the property (2.10), (ii), if and only if

$$
Z_B''(t) - Z_B''(0) = -\omega(t), \qquad 0 \leq t \leq \tau_2.
\tag{2.15}
$$

Now we can describe solutions of the problem

$$
f^{(m)}(0) \to \sup, \qquad f \in W^r H^\omega[0, A], \quad \|f\|_{C[0,A]} \leq B,
\tag{2.16}
$$

for $m = 1$, 2, and all $B > 0$ and $A > 0$. Let $Z_{B,A} = Z_{B,A,m,\omega}$ be the function from Theorem 12.2.1 with the alternance points $\{\tau_i = \tau_i(B, A, m, \omega)\}_{i=0}^{2}$. Then, if $\tau_2 = A$, then the function $Z_{B,A}(t)$ is defined on the entire interval $[0, A]$. Consequently, $Z_{B,A}(t)$ is extremal in the problem (2.16).

If $\tau_2 < A$, then by ((2.12), (iii)), $\dfrac{d}{dt}Z_{B,A}(\tau_2) = 0$. In addition, the derivative vanishes at the interior point of alternance τ_1: $\dfrac{d}{dt}Z_{B,A}(\tau_1) = 0$. Therefore, using the formula (2.1), we can extend the function $Z_{B,A}(t)$ from the interval $[0, \tau_2]$ to the entire half-line $\mathbb{R}_+$ without leaving the class $W^2 H^\omega(\mathbb{R}_+)$ and increasing the $\mathbb{L}_\infty$-norm: $\|E(Z_{B,A}; \tau_1, \tau_2; \cdot)\|_{\mathbb{L}_\infty(\mathbb{R}_+)} = \|Z_{A,B}\|_{\mathbb{C}[0, \tau_2]} = B$. Consequently, the function $E(Z_{B,A}; \tau_1, \tau_2; t)$ is extremal in the problem (2.16) for $A = +\infty$, i.e. for $[0, A] = \mathbb{R}_+$. Also, the restriction $E(Z_{B,A}; \tau_1, \tau_2; t)\big|_{[0,A]}$ solves the problem (2.16).

These observations show that in order to describe the *complete set of solutions* of the problem (2.16) on the half-line $[0, A] = \mathbb{R}_+$, it suffices to show that

$$\tau_2 = \tau_2(B, A, m, \omega) < A, \tag{2.17}$$

for each fixed $B > 0$ and all sufficiently large $A > 0$.

MICROLEMMA 12.2.2. *Let* $\{\tau_i(B, A)\}_{i=0}^2$ *be the points of alternance of the function* $Z_{B,A}(t) = Z_{B,A,m.\omega}(t)$, $A, B > 0$. *Then, for each* $B > 0$, *there exists an* $\mathcal{A}(B)$, *such that for all* $A > \mathcal{A}(B)$,

$$\tau_2(B, A) < A.$$

Proof. First of all, notice that the quantitative solutions of the family of problem

$$f''(0) \to \sup, \qquad f \in W^2 H^\omega[0, A], \quad \|f\|_{\mathbb{C}[0,A]} \leq B,$$

for $A \geq 1$ are bounded from below by some constant $\mathcal{D}$, dependent on B, ω, but independent of A. Indeed, set $\tau_1 = \dfrac{1}{2}$, $\tau_2 = 1$, and define the coefficients $\{\alpha_i(2)\}_{i=0}^2$ by (2.5) and the kernel $K(t)$ by (2.8) for the aforementioned choice of τ_1 and τ_2. Then, by the equation (2.7) for $m = 2$,

$$|f''(0)| \leq \mathcal{D} := \sum_{i=0}^2 \alpha_i(2)B + \int_0^1 \omega(t)|K(t)|\, dt, \tag{2.18}$$

for all functions $f \in W^2 H^\omega[0, A]$: $\|f\|_{\mathbb{C}[0,A]} \leq B$, and all $A \geq 1$. Also notice that by considering the function $f(A - t)$, we proved the boundedness of the values $|f(A)|$, as well:

$$|f''(0)| + |f''(A)| \leq 2\mathcal{D}, \tag{2.19}$$

for all functions $f \in W^2 H^\omega[0, A]$: $\|f\|_{\mathbb{C}[0,A]} \leq B$, and all $A \geq 1$.

Let us assume now that $\tau_2(B, A) = A$. Then, by Corollary 2.1.3 of the Kornaichuk lemma, the function $Z''_{B,A}(t)$ has the full modulus of continuity on the interval $[0, A]$ as an extremal function in (2.12), (ii): $\omega(Z''_{B,A}; t) = \omega(t)$, $0 \leq t \leq A$. In particular,

$$|Z''_{B,A}(A) - Z''_{B,A}(0)| = \omega(A). \tag{2.20}$$

Thus, if $\tau_2(B, A) = A$, then the juxtaposition of (2.19) and (2.20) leads us to the conclusion that $A < \omega^{-1}(2\mathcal{D})$. $\qquad\square$

Thus, taking $\hat{A} = \omega^{-1}(2\mathcal{D})$, we conclude that $\tau_2(B, \hat{A}) < A$, and the extension $E(Z_{\hat{A},B}; \tau_1, \tau_2; t)$ is extremal in the problem

$$\|f^{(m)}\|_{L_\infty(\mathbb{R}_+)} \to \sup, \qquad f \in W^2 H^\omega(\mathbb{R}_+), \quad \|f\|_{L_\infty(\mathbb{R}_+)} \le B.$$

In a conclusion of this chapter, we give more explicit inequalities in the problem (2.4) for $m = 2$.

12.2.3. Inequalities for $\|f''\|_{L_\infty(\mathbb{R}_+)}$ in $W^2 H^\omega(\mathbb{R}_+)$

Relying on the formula (2.5), we obtain the explicit description of extremal functions in (2.4) for $m = 2$.

For any $b > 0$, let $q_{\omega,b}$ be the function uniquely characterized by the following properties:

$$
\begin{aligned}
&(i) \quad \frac{d^3}{dt^3} q_{\omega,b}(t) = -\omega'(t), \qquad 0 \le t \le b; \\
&(ii) \quad q'_{\omega,b}(b) = 0; \\
&(iii) \quad q_{\omega,b} \text{ has three points of alternance } 0 =: \tau_0 < \tau_1(b) < \tau_2 := b \\
&\qquad \text{on the interval } [0, b].
\end{aligned}
\tag{2.21}
$$

We can give the formula for $q_{\omega,b}(x)$, $\quad 0 \le x \le b$:

$$q_{\omega,b}(x) = \int_0^x \int_b^t (A(b) - \omega(\xi))\, d\xi dt - \frac{1}{2} \int_0^{\tau_1(b)} \int_b^t (A(b) - \omega(\xi))\, d\xi\, dt, \tag{2.22}$$

where

$$
(i) \quad A(b) := \frac{2}{b^2} \int_0^b \int_t^b \omega(x)\, dx dt;
$$

$$
(ii) \quad \tau_1(b) \in [0, b] \text{ is derived from the equation}
\tag{2.23}
$$

$$\int_b^{\tau_1(b)} (A(b) - \omega(t))\, dt = 0.$$

Put

$$I(b) := \frac{1}{2} \int_0^{\tau_1(b)} \int_t^b (A(b) - \omega(x))\, dx\, dt, \tag{2.24}$$

One can easily verify that functions $A(b) := \|\tilde{q}''_{\omega,b}\|_{L_\infty(\mathbb{R}_+)}$ and $I(b) := \|\tilde{q}_{\omega,b}\|_{L_\infty(\mathbb{R}_+)}$ strictly increase from 0 to $+\infty$, as b increases from 0 to $+\infty$.

By our observation in Subsection 12.2.1 and Subsection 12.2.2, the extension $E(q_{\omega,b}; \tau_1, \tau_2; t)$ of the functions $q_{\omega,b}$ to the entire half-line $\mathbb{R}_+$ is extremal in the problem

$$\|f''\|_{L_\infty(\mathbb{R}_+)} \to \sup, \qquad f \in W^2 H^\omega(\mathbb{R}_+), \quad \|f\|_{L_\infty(\mathbb{R}_+)} \le I(b),$$

and the sharp inequality is as follows:

$$\|f''\|_{L_\infty(\mathbb{R}_+)} \le A(b). \tag{2.25}$$

Computations in the Hölder classes $W^2 H^\alpha(\mathbb{R}_+)$ lead to the following values of $A(b)$ and $I(b)$:

$$A(b) = \frac{2}{\alpha+2} b^\alpha, \qquad \tau_1(b) = k_\alpha b,$$

where $k_\alpha \in [0, b]$ is derived from the nonlinear equation

$$\frac{1 - [k_\alpha]^\alpha}{1 - k_\alpha} = \frac{2(\alpha+1)}{\alpha+2}. \tag{2.26}$$

Summarizing the results of Section 12.2, we describe the numerical differentiation formulae for $f'(x)$ and $f''(x)$. By the formulas (2.6)–(2.9), the approximating formulae for $f'(x)$ and $f''(x)$ with the step $h > 0$ are as follows:

$$f^{(k)}(x) \approx \frac{(-1)^k}{h^k} \{ af(x) - (a+b)f(x+ch) + bf(x+h) \}, \qquad k = 1, 2, \tag{2.27}$$

for some positive $a(\omega, k, h), b(\omega, k, h)$, and $c(\omega, k, h) < 1$. All these constants can be computed numerically.

MICROLEMMA 12.2.3. *The constants a, b, c_1 in (2.17) are independent of the step h in the case of the Hölder classes $W^2 H^\alpha(\mathbb{R}_+)$.*

Proof. Let $Z_{\hat{B}} = Z_{\hat{B},r,m,\omega_\alpha}$ be the function from Theorem 12.2.1 for $A = 1$, such that $\frac{d}{dt} Z_{\hat{B}}(1) = 0$. By Theorem 12.2.1, the function $Z_{\hat{B}}(t)$ has three alternance points $\{\tau_i\}_{i=0}^2$. Let the coefficients $\{\alpha_i = \alpha_i(m)\}_{i=0}^2$ be as in (2.6). As we showed in Chapter 8, the function $Q_h(t) = h^{2+\alpha} Z_{\hat{B}}(t/h)$ is extremal in the problem

$$f^{(m)}(0) \to \sup, \qquad f \in W^2 H^\alpha[0, h], \quad \|f\|_{C[0,h]} \le h^{2+\alpha} \hat{B}, \quad h > 0. \tag{2.28}$$

If $\{\tau_i[h]\}_{i=0}^2$ are the alternance points of Q_h, and $\{\alpha_i[h]\}_{i=0}^2$ are derived from (2.6) for $\{\tau_i[h]\}_{i=0}^2$, then the numerical differentiation formula (2.27) has the form

$$f^{(m)}(0) \approx \sum_{i=0}^2 \alpha_i[h] f(\tau_i[h]). \tag{2.29}$$

But $\alpha_i[h] = h^{-m} \alpha_i$, $\tau_i[h] = h\tau_i$, $i = 0, 1\, 2$. Therefore, from (2.28) we obtain the formula (2.27) with $c_1 = \tau_1$, $a = (-1)^m \alpha_0(m)$ and $b = (-1)^m \alpha_2(m)$, $\square$

Chapter 13

Chebyshev Ω-Splines and
N-Widths of $W^r H^\omega[0,1]$

In this chapter we review the Tihomirov's result that identifies the extremal functional and optimal approximating subspaces in the problem of N-widths of Sobolev classes $W_\infty^r[0,1]$. Then, we describe analogs of Chebyshev ω-splines in the problem of N-widths of functional classes $W^r H^\omega[0,1]$ for nonlinear ω.

13.1. N-widths of Sobolev classes $W_\infty^{r+1}[0,1]$

NOTATION 13.1.1. Let Y and Z be two subsets in the normed space X. We put

$$\mathrm{E}(Y,Z) := \sup_{y \in Y} \inf_{x \in X} \|y - x\|_X.$$

V. M. Tihomirov [87] computed the n-widths of the Sobolev class $W^r H^1[0,1]$ in terms of the norms $\{T_{k,r}\}_{\mathbb{C}[0,1]}$, $k \geq r$ and described the optimal approximating spline subspace. The following scheme of the proof of his result is borrowed from [61].

13.1.1. Estimates of $d_N \left(W_\infty^{r+1}[0,1], \mathbb{C}[0,1] \right)$ from below

Let $n, r \in \mathbb{N} : n \geq r$. Let $T_{n,r}$ be the Chebyshev perfect spline of degree $r+1$ with $n + 2$ alternance points $\{\nu_i = \nu_i(n,r)\}_{i=0}^{n+1}$ and $n - r$ knots $\{\vartheta_i = \vartheta_i(n,r)\}_{i=1}^{n-r}$, whose existence is assured by Lemma 4.1.1. Put

$$D_{N,r} := \|T_{N-1,r}\|_{\mathbb{C}[0,1]}, \qquad N \geq r + 1. \tag{1.1}$$

Given a natural $N \geq r + 1$, we introduce the $(N + 1)$-dimensional space $F_{N,r}$ as follows:

$$F_{N,r} := \mathrm{span}\,\{1, t, \dots, t^{r+1}, (t - \zeta_1)_+^{r+1}, \dots, (t - \zeta_{N-1-r})_+^{r+1}\}, \tag{1.2}$$

where $\{\zeta_i = \vartheta_i(N - 1, r)\}_{i=1}^{N-1-r}$ are the knots of the Chebyshev perfect spline $T_{N-1,r}$ arranged in the increasing order. Let

$$\mathcal{B}_{N,r} := \{s \in F_{N,r} \mid \|S\|_{\mathbb{C}[0,1]} \leq D_{N,r}\}, \qquad N \geq r + 1, \tag{1.3}$$

be the set of splines from $F_{N,r}$ whose norm does not exceed $D_{N,r}$. The proof of the following property of the set $\mathcal{B}_{n,r}$ involves a standard application of the zero counting technique based on the Rolle theorem.

LEMMA 13.1.1. *Let $N, r \in \mathbb{N} : N \geq r + 1$, and the set $\mathcal{B}_{n,r}$ be introduced by (1.3). Then,*

$$\mathcal{B}_{N,r} \subset W_\infty^{r+1}[0,1]. \tag{1.4}$$

Proof. Let us assume that there exists a spline S_* with the properties

$$\|S_*\|_{\mathbb{C}[0,1]} \le D_{N,r}, \qquad \|S_*\|_{\mathrm{L}_\infty[0,1]} > 1. \tag{1.5}$$

Let (ζ_{j-1}, ζ_j), for some $j \in \{1, \ldots, N-r\}$, be an interval, on which the piecewise constant function $S_*^{(r+1)}$ attains its maximal value, that is

$$|S_*^{(r+1)}(t)| = \|S_*^{(r+1)}\|_{\mathrm{L}_\infty[0,1]}, \qquad t \in (\vartheta_{j-1}, \vartheta_j).$$

Consider the difference

$$\Lambda(t) = \beta S_*(t) - T_{N-1,r}(t), \qquad t \in [0,1], \tag{1.6}$$

where $\beta = \dfrac{\alpha}{\|S_*^{(r+1)}\|_{\mathrm{L}_\infty[0,1]}}$, and $\alpha = \mathrm{sign}\left(S_*^{(r+1)}(t) \cdot T_{N-1,r}^{(r+1)}(t)\right)$, $t \in (\vartheta_{j-1}, \vartheta_j)$.
The choice of β guarantees that $\Lambda^{(r+1)}(t) \equiv 0$, $t \in (\vartheta_{j-1}, \vartheta_j)$. Also, $|\beta| < 1$, so $\|\beta S_*^{(r+1)}\|_{\mathrm{L}_\infty[0,1]} < \beta < 1$. On the other hand, $T_{N-1,r}$ is a perfect spline, so for some $\xi \in \{-1, 1\}$,

$$(-1)^i \xi T_{N-1,r}^{(r+1)}(t) \equiv 1, \qquad t \in (\zeta_{i-1}, \zeta_i), \qquad i = 1, \ldots, N-r.$$

Consequently,

$$(-1)^i \xi \Lambda^{(r+1)}(t) > 0, \qquad t \in (\zeta_{i-1}, \zeta_i), \qquad i = 1, \ldots, N-r, \tag{1.7}$$

with the equality in (1.7) holding at least on the interval $(\vartheta_{j-1}, \vartheta_j)$. Therefore, we showed that $\Lambda^{(r-1)}$ changes the sign on $[0,1]$ at most $N-1-r$ times.

On the other hand, from the inclusions $T_{N-1,r} \in \mathcal{B}_{N,r}$, $S_* \in \mathcal{B}_{N,r}$, the inequality $\beta < 1$, and the alternance property

$$T_{N-1,r}(\nu_i) = (-1)^i \sigma D_{n,r}, \qquad i = 0, \ldots, N+1, \qquad \sigma \in \{-1, 1\},$$

we deduce that

$$(-1)^i \sigma \Lambda(\nu_i) > 0, \qquad i = 0, \ldots, N+1. \tag{1.8}$$

In particular, Λ exhibits at least $N+1$ zeroes, so $\Lambda^{(r+1)}$ changes the sign at least $N-r$ times. This contradiction with the previous finding proves the result. $\square$

Now, an application of Theorem 1.1.6 leads us to the desired estimates

$$d_N\left(W_\infty^{r+1}, \mathbb{C}[0,1]\right) \ge D_{N,r}, \qquad N \ge r+1. \tag{1.9}$$

13.1.2. Estimates of $d_N\left(W_\infty^{r+1}[0,1], \mathbb{C}[0,1]\right)$ **from above**

Fix $N, r \in \mathbb{N} : N \geq r+1$. Let $\{z_i = z_i(N,r)\}_{i=1}^N$ be the zeroes of the Chebyshev spline $T_{N-1,r}$ arranged in the increasing order. In accordance with the inequalities (4.3.1) between the points of sign change of consecutive derivatives of $T_{N-1,r}$, we have the following relations between the zeroes $\{z_i\}_{i=1}^N$ and the knots $\{\zeta_i\}_{i=1}^{N-1-r}$ of $T_{N-1,r}$:

$$z_i < \zeta_i < z_{r+1+i}, \qquad i = 1, \dots, N-1-r. \tag{1.10}$$

Inequalities (2.1) are precisely the conditions of Proposition 1.1.7, which guarantee that the problem

$$S(z_i) = y_i, \qquad i = 1, \dots, N, \tag{1.11}$$

of interpolation by splines $S \in \mathbb{S}^r[\zeta_1, \dots, \zeta_{N-1-r}]$ of degree $r+1$ with the fixed set of knots $\{\zeta_i\}_{i=1}^{N-1-r}$ (see Definition 1.1.2) has a unique solution for each of the vectors $(y_1, \dots, y_N) \in \mathbb{R}^{N+1}$.

Given a function $f \in W_\infty^{r+1}[0,1]$, we introduce $I(f;x)$ as the polynomial from $\mathbb{S}^r[\zeta_1, \dots, \zeta_{N-1-r}]$ interpolating f at the points $\{z_i\}_{i=1}^N$:

$$I(f; z_i) = f(z_i), \qquad i = 1, \dots, N. \tag{1.12}$$

LEMMA 13.1.2. *The following estimate holds for any* $f \in W_\infty^{r+1}[0,1]$:

$$|I(f;x)| \leq |T_{N-1,r}(x)|, \qquad x \in [0,1]. \tag{1.13}$$

Proof. Let us assume that there exists a point $z_0 \in [0,1] \setminus \{z_1, \dots, z_N\}$ such that

$$|I(f; z_0)| > |T_{N-1,r}(z_0)|.$$

Choose the constant λ, $|\lambda| < 1$, in such a way that the function

$$\Phi(t) := T_{N-1,r}(t) - \lambda[f(t) - I(f;t)] \tag{1.14}$$

vanishes at the point z_0.

Let us show that the function $h := \Phi^{(r-1)}$ belongs to the class $H_{N-r}[0,1]$ (with $\{\tau_i := \zeta_i\}_{i=0}^{N-r}$) introduced in Definition 4.4.1. The nontrivial part in the verification of the inclusion $h \in H$ involves a verification of the property (4.4.26), (iii). Notice that

$$h''(t) = \Phi^{(r+1)}(t) = T_{N-1,r}^{(r+1)}(t) - \lambda f^{(r+1)}(t), \qquad t \in (\zeta_{j-1}, \zeta_j), \quad j = 1, \dots, N-r. \tag{1.15}$$

Using the properties

$$T_{N-1,r}^{(r+1)}(t) \equiv (-1)^j, \qquad \|\lambda f^{(r+1)}\|_{\mathbb{L}_\infty[0,1]} \leq |\lambda| < 1,$$

holding almost everywhere on the interval (ζ_{j-1}, ζ_j), we arrive at the desired relations

$$\text{sign } h''(x) = (-1)^j, \qquad x \in (\zeta_{j-1}, \zeta_j), \qquad j = 1, \dots, N. \tag{1.16}$$

The inclusion $h \in H_{N-r}[0,1]$ is established. By the definition of Φ,

$$\Phi(z_i) = 0, \qquad i = 0, \dots, N. \tag{1.17}$$

An application of the Rolle's theorem assures that the function $\Phi^{(r-1)}$ has at least $N - r + 2$ zeroes. However, Lemma 4.4.4 guarantees that $\Phi^{(r-1)}$ can have at most $N - r + 1$ zeroes.

The contradiction proves the result. $\square$

Let us define the mapping $\kappa_N : W_\infty^{r+1}[0,1] \to \mathbb{S}^r[\zeta_1, \dots, \zeta_{N-r-1}]$ by

$$\kappa_N(f) := I(f; \cdot), \qquad f \in W_\infty^{r+1}[0,1]. \tag{1.18}$$

Since $\mathbb{S}^r[\zeta_1, \dots, \zeta_{N-r-1}]$ is an N-dimensional subspace, we can apply the result of Lemma 13.1.2 to obtain the estimate for the N-width of the Sobolev class $W_\infty^{r+1}[0,1]$:

$$d_N \left(W_\infty^{r+1}; \mathbb{C}[0,1] \right) \leq \sup_{f \in W_\infty^{r+1}[0,1]} \| f - I(f; \cdot) \|_{\mathbb{C}[0,1]} = \| T_{N-1,r} \|_{\mathbb{C}[0,1]} = D_{N,r}. \tag{1.19}$$

Let us summarize our findings. The comparison of the estimates (1.9) from below and (1.19) from above enables to conclude that

$$d_N \left(W_\infty^{r+1}[0,1]; \mathbb{C}[0,1] \right) = D_{N,r}, \qquad N \geq r + 1. \tag{1.20}$$

The extremal N-dimensional subspace $\mathcal{E}_N$ in $\mathbb{C}[0,1]$ with the property

$$\mathrm{E} \left(W_\infty^{r+1}; \mathcal{E}_N \right)_{\mathbb{C}[0,1]} = d_N \left(W_\infty^{r+1}[0,1]; \mathbb{C}[0,1] \right)$$

(see Notation 13.1.1) was shown to coincide with $\mathbb{S}^r[\zeta_1, \dots, \zeta_{n-r-1}]$. The interpolation at the zeroes of the Chebyshev spline $T_{N-1,r}$ was proved to be an optimal approximating algorithm.

In a conclusion of this section we remark that in various settings, the even n-widths $d_{2n} \left(\widetilde{W}_\infty^r, \widetilde{\mathbb{L}}_q \right)$, $1 \leq q \leq \infty$, in *the periodic $\mathbb{L}_q$-spaces* were found by V. M. Tihomirov [87] and A. A. Ligun [55]. It turns out that the even n-widths coincide with the $\mathbb{L}_q$-norm of the corresponding Euler spline which deviates most from the extremal approximating spline subspace.

The n-widths $d_n \left(W_\infty^r, \mathbb{L}_q \right)$, $1 \leq q < \infty$, of non-periodic Sobolev classes were computed in terms of the $\mathbb{L}_q$-norms of certain perfect splines $\phi_{n,r}(q)$ by C. Micchelli and A. Pinkus [66] (see also [49], p. 265). For the proofs of all these results and the bibliography, the reader is referred to Ch. 6 of [49], Ch. 10 of [48], and the respective commentaries to those chapters ([49], p. 339, [48], p. 312). The solution of some other problems in the theory of N-widths of Sobolev classes could be found in A. Pinkus [73], [74], and V. M. Tihomirov, S. B. Babadjanov [86].

13.2.   Chebyshev ω-splines of the problem of n-widths $W^r H^\omega[0,1]$

In this section we describe the analog of the function T_n in $W^r H^\omega[0,1]$.

Let the coefficients $\{\alpha_i\}_{i=0}^{n+1}$ of the kernels

$$F_n(t) = \frac{1}{r!} \sum_{i=0}^{n+1} \alpha_i(\nu_i - t)_+^r, \quad K_n(t) = \frac{d}{dt} F_n(t) \tag{2.1}$$

be derived from the equations

$$\begin{cases} \displaystyle\sum_{i=0}^{n+1}(-1)^i\alpha_i = 1; \\[2mm] \displaystyle\sum_{i=0}^{n+1}\alpha_i\nu_i{}^j = 0, \qquad j = 0,\dots,r; \\[2mm] \displaystyle\sum_{i=0}^{n+1}\alpha_i(\nu_i - \vartheta_l)_+^r = 0, \qquad l = 1,\dots,n-r, \end{cases} \tag{2.2}$$

where the points $\{\nu_i\}_{i=0}^{n+1}$ and $\{\vartheta_i\}_{i=1}^{n-r}$ satisfy the inequalities (5.1.1). By Proposition 3.1.1, the support $\operatorname{supp} K_n(t)$ coincides with $[0,1]$. A (simpler) version of Theorem 6.0.1 for kernels of the form (2.1), (2.2) guarantees the existence of the function $C_n \in W^r H^\omega[0,1]$ with the complete alternance at the points $\{\nu_i\}_{i=0}^{n+1}$,

$$C_n(\nu_i) = (-1)^i \|C_n\|_{C[0,1]}, \qquad i = 0,\dots,n+1, \tag{2.3}$$

and such that

$$\sup_{h\in H^\omega[0,1]} \int_0^1 h(t)\frac{d}{dt}F_n(t)\,dt = \int_0^1 C_n^{(r)}(t)\frac{d}{dt}F_n(t)\,dt. \tag{2.4}$$

As in (1.2), the following property distinguishes the Chebyshev ω-spline C_n among other Chebyshev ω-splines – *the best element of approximation of C_n by the $n+1$-dimensional space of splines $\mathbb{S}^r[\vartheta_1,\dots,\vartheta_{n-r}]$ is the zero spline $\mathbf{O}$.*

PROPOSITION 13.2.1. *If C_n is the function with the properties (2.1)–(2.4), then*

$$\mathrm{E}\left(C_n; \mathbb{S}^r[\vartheta_1,\dots,\vartheta_{n-r}]\right) = \|C_n - \mathbf{O}\|_{C[0,1]} = \|C_n\|_{C[0,1]}. \tag{2.5}$$

Proof. To prove (2.5), we use Theorem 1.1.4, the criterion for elements of the best approximation of a function by a subspace in $\mathbb{C}[a,b]$. We need to verify all three conditions (1.1.4), (i)–(iii) for

$$f(x) = C_n(x), \quad \mathcal{F} = \mathbb{S}^r[\vartheta_1, \ldots, \vartheta_{n-r}], \quad \varphi_0 = \mathbf{0}, \quad g_0(t) := F_n^{(r)}(t). \qquad (2.6)$$

Indeed, by (2.2), $\overset{1}{\underset{0}{\bigvee}} g_0 = \sum_{i=0}^{n} |\alpha_i| = 1$, proving (1.1.4), (i). Also by (2.2), the kernel F_n satisfies the zero boundary conditions at the endpoints of $[0,1]$:

$$F_n^{(i)}(0) = F_n^{(i)}(1) = 0, \qquad i = 0, \ldots, r-1. \qquad (2.7)$$

Therefore, we can verify the condition (1.1.4), (ii):

$$\|C_n - \mathbf{0}\|_{\mathbb{C}[0,1]} = \|C_n\|_{\mathbb{C}[0,1]} = |\sum_{i=0}^{n+1} \alpha_i C_n(\nu_i)| =$$

$$= |\int_0^1 C_n(t) \sum_{i=0}^{n+1} \alpha_i \delta(\nu_i - t)\, dt| = |\int_0^1 C_n(t)\, dg_0(t)|. \qquad (2.8)$$

Finally, for any spline $\varphi \in S_{n+1}^r[\vartheta_1, \ldots, \vartheta_{n-r}]$,

$$\varphi^{(r)}(t) = \gamma_i, \quad t \in [\vartheta_i, \vartheta_{i+1}], \quad i = 0, \ldots, n-r,$$

we have the following identities:

$$|\int_0^1 \varphi(t)\, dg_0(t)| = |\sum_{i=0}^{n+1} \alpha_i \varphi(\nu_i)| = |\int_0^1 \varphi^{(r)}(t)\, \frac{d}{dt} F(t)\, dt| =$$

$$= |\sum_{i=0}^{n-r} \gamma_i \int_{\vartheta_i}^{\vartheta_{i+1}} \frac{d}{dt} F(t)\, dt| = \sum_{i=0}^{n-r} \gamma_i (F(\vartheta_{i+1}) - F(\vartheta_i)) = 0, \qquad (2.9)$$

because $F(\vartheta_i) = 0, \quad i = 0, \ldots, n-r+1$, by the equations (2.4). The last condition (1.1.4), (iii) is also verified. $\qquad\qquad\qquad\qquad\qquad\qquad\qquad\qquad\qquad\qquad\qquad\qquad\qquad\square$

Now we can state our hypothesis – an analog of Theorem 13.1.1 for functional classes $W^r H^\omega[0,1]$.

CONJECTURE 13.2.2. *Let r, $n \in \mathbb{N}$ and $n \geq r$. Then,*

$$d_{n+1}(W^r H^\omega[0,1], \mathbb{C}[0,1]) = \|\mathscr{Z}_{n,r}\|_{\mathbb{C}[0,1]}.$$

The set $S_{n,r} := \mathbb{S}^r[\vartheta_1, \ldots, \vartheta_{n-r}]$ is an optimal $(n+1)$-dimensional approximating subspace for the class $W^r H^\omega [0,1]$ in the uniform metrics:

$$\mathrm{E}\left(W^r H^\omega [0,1], S_{n,r}\right)_{\mathbb{C}[0,1]} = d_{n+1}\left(W^r H^\omega [0,1], \mathbb{C}[0,1]\right).$$

In Chapter 15 the Chebyshev functions (for $r = 1$) were shown to be extremal in the problem of n-width of the nonperiodical class $W^1 H^\omega [0,1]$, that is

$$d_{n+1}\left(W^1 H^\omega [0,1], \mathbb{C}[0,1]\right) = \|C_n\|_{\mathbb{C}[0,1]}, \qquad n \in \mathbb{N}, \tag{2.10}$$

if the modulus of continuity ω satisfies the restrictions

$$(2 - \sqrt{2})\left(\omega(a) + \omega(b)\right) \leq \omega(a+b), \qquad \text{for all } a, b \geq 0, \quad a + b \leq 1. \tag{2.11}$$

REMARK 13.2.1. The functional class $W^r H^\omega [0,1]$ contains the linear space P_r of polynomials of degree r whose dimension is $r + 1$. Therefore, the first r widths $\{d_i\left(W^r H^\omega [0,1], \mathbb{C}[0,1]\right)\}_{i=1}^r$ are *infinite*.

In Chapter 14 we determine the first finite width $d_{r+1}\left(W^r H^\omega [0,1], \mathbb{C}[0,1]\right)$ of the class $W^r H^\omega [0,1]$, which provides a partial confirmation of our hypothesis formulated in Conjecture 13.2.2.

Chapter 14

Function in $W^r H^\omega[-1, 1]$ Deviating Most from Polynomials $\displaystyle\sum_{i=1}^{r} a_i t^i$

The classical Chebyshev polynomial $\mathcal{T}_r$ of degree $r + 1$ is given by the formula

$$\mathcal{T}_r(x) = \frac{2^{-r}}{(r+1)!} \cos[(r+1)\arccos(x)], \qquad x \in [-1, 1]. \tag{0.1}$$

The polynomial $\mathcal{T}_r$ is a function of the most deviation from the linear space P_r of polynomials of degree r among all functions from the Sobolev class $W_\infty^r[-1, 1]$ (cf. [1]).

Let $\{\tau_i^* := \cos \frac{(-r-1+2i)\pi}{2r+2}\}_{i=0}^{r+1}$ be the set of the alternance points of the Chebyshev polynomial:

$$\mathcal{T}_r(\tau_i^*) = (-1)^{r+1+i} \|\mathcal{T}_r\|_{C[-1,1]} = (-1)^{r+1+i} \frac{2^{-r}}{(r+1)!}, \qquad i = 0, \dots, r+1, \tag{0.2}$$

and the coefficients of the kernels

$$F(t) := \frac{1}{r!} \sum_{i=0}^{r+1} \alpha_i (\tau_i - t)_+^r, \quad K(t) := -\frac{1}{(r-1)!} \sum_{i=0}^{r+1} \alpha_i (\tau_i - t)_+^{r-1}, \tag{0.3}$$

be determined from the equations

$$\sum_{i=0}^{r+1} \alpha_i \tau_i^j = 0, \quad j = 0, \dots, r; \quad \sum_{i=0}^{r+1} (-1)^{r+1+i} \alpha_i = 1. \tag{0.4}$$

Results of Section 3.3 guarantee that $\operatorname{sign}\alpha_i = (-1)^{i+r+1}$, $i = 0, \dots, r+1$, and that F is a simple kernel such that $\operatorname{sign} F(t) = 1$ on $(-1, 1)$. Thus,

$$\|\mathcal{T}_r\|_{C[-1,1]} = \sum_{i=0}^{r+1} \alpha_i \mathcal{T}_r(\tau_i) =$$

$$= \int_{-1}^{1} \mathcal{T}_r^{(r+1)}(t) F(t)\, dt = \int_{0}^{2} \Re\,(F; t)\, dt = \sup_{h \in H^\omega[-1,1]} \int_{-1}^{1} h(t) K(t)\, dt, \tag{0.5}$$

for $\omega(t) = t$.

In this chapter we will show that any function $g^* \in H^\omega[-1,1]$ (up to the change of sign) of the most deviation from the space of polynomials of degree r can be described by the properties (0.2) and (0.5).

14.1. Preliminary observations

DEFINITION 14.1. The functional $D[-1,1]$ class is defined as follows:
$$D[-1,1] := \{ f \in W^r H^\omega[-1,1] \mid f(\tau_i) = (-1)^{r+1+i} \|f\|_{\mathbb{C}[-1,1]}, \ i = 0, \ldots, r+1 \} \tag{1.1}$$
for some $\{\tau_i\}_{i=0}^{r+1} : \ -1 \le \tau_0 < \tau_1 < \tau_{r+1} \le 1$.

From Chebyshev's Theorem 1.1.2 it follows that
$$\sup_{h \in W^r H^\omega[-1,1]} \inf_{u \in P_r} \|h - u\|_{\mathbb{C}[-1,1]} = \sup_{g \in D[-1,1]} \|g\|_{\mathbb{C}[-1,1]}. \tag{1.2}$$
Note also that the derivative of any function f from $D[-1,1]$ has at least r zeroes at the interior points of alternance of f on $(-1,1)$. Then, by the Rolle theorem, each of the derivatives $f^{(k)}$ has at least one zero on $[-1,1]$. Therefore, we have the following constraints for the upper bounds and modulii of continuity of the derivatives of functions from $D[-1,1]$:
$$\|f^{(k)}\|_{\mathbb{C}[-1,1]} \le 2^{r-k}\omega(2), \qquad \omega(f^{(k)};t) \le \omega(t) + 2^{r-k-1}\omega(2)t, \tag{1.3}$$
for $k = 0, \ldots, r$. Consequently, by the Arzela–Ascoli theorem, $D[-1,1]$ is a compact in $\mathbb{C}[-1,1]$ with a function $g^*(t)$ of the greatest uniform norm.

14.2. Generating kernels

Let the collection $\tau = \{\tau_i\}_{i=0}^{r+1}$ be such that $-1 \le \tau_0 < \tau_1 < \cdots < \tau_{r+1} \le 1$. Let the coefficients of polynomial kernels
$$F_\tau(t) := \frac{1}{r!} \sum_{i=0}^{r+1} \alpha_i(\tau_i - t)_+^r, \quad K_\tau(t) := -\frac{1}{(r-1)!} \sum_{i=0}^{r+1} \alpha_i(\tau_i - t)_+^{r-1}, \tag{2.1}$$
be determined from the equations
$$\begin{cases} \sum\limits_{i=0}^{r+1} \alpha_i \tau_i^j = 0, & j = 0, \ldots, r; \\ \sum\limits_{i=0}^{r+1} (-1)^{r+1+i} \alpha_i = 1. \end{cases} \tag{2.2}$$
We mention some properties of the kernel $F(t)$ of type III introduced in Section 3.3. By Proposition 3.3.1,
$$\operatorname{sign} \alpha_i = (-1)^{r+1+i}, \qquad i = 0, \ldots, r+1. \tag{2.3}$$
The function $F_\tau^{(i)}$ has precisely $r - i$ simple zeroes on the interval $(-1,1)$. In particular, the kernel $K_\tau(t) := \dfrac{d}{dt} F_\tau(t)$ has one point of sign change on the interval $(-1,1)$, so the kernel $F(t)$ is simple in the sense of Definition 2.1.1.

14.3. Preliminary remarks

Let $f \in W^r H^\omega[-1, 1]$. From Taylor's formula

$$f(t) = \sum_{j=0}^{r-1} \frac{f^{(j)}(-1)}{j!} t^j + \frac{1}{(r-1)!} \int_{-1}^{1} f^{(r-1)}(y)(t-y)_+^{r-1} \, dy, \quad t \in [-1, 1], \quad (3.1)$$

and equations (2.2) for coefficients $\{\alpha_i\}_{i=0}^{r+1}$ we derive the formula

$$\sum_{i=0}^{r+1} \alpha_i f(\tau_i) = \int_{-1}^{1} f^{(r)}(t) K(t) \, dt. \qquad (3.2)$$

14.3.1. Chebyshev ω-polynomials

THEOREM 14.3.1. *Let $g^* \in M[-1, 1]$ be a function extremal in (1.2). There exists such a collection of points $\tau^* = \{\tau_i^*\}_{i=0}^{r+1} : -1 = \tau_0^* < \tau_1^* < \cdots < \tau_{r+1}^* = 1$, that*

$$(i) \quad \sup_{h \in H^\omega[-1,1]} \int_{-1}^{1} h(t) K_{\tau^*}(t) \, dt = \int_{-1}^{1} \frac{d^r}{dt^r} g^*(t) K_{\tau^*}(t) \, dt;$$

$$(3.3)$$

$$(ii) \quad g^*(\tau_i^*) = (-1)^{i+r+1} \|g^*\|_{\mathbb{C}[-1,1]}, \qquad i = 0, \ldots, r+1.$$

Proof. Let $\{\tau_i^*\}_{i=0}^{r+1}$ be any sequence of $r + 2$ consecutive alternance points of the function g^*. First of all, let us show that g^* has precisely $r + 2$ alternance points, and $\tau_0^* := -1$, $\tau_{r+1}^* := 1$.

Indeed, otherwise, if

$$l(t) = \frac{1}{2}(\tau_{r+1}^* + \tau_0^*)t + \frac{1}{2}(\tau_{r+1}^* - \tau_0^*) \quad \text{and} \quad f^*(t) = \left(\frac{2}{\tau_{r+1}^* - \tau_0^*}\right)^r g^*(l(t)),$$

then

$$f^* \in D[-1, 1] \quad \text{and} \quad \|f^*\|_{\mathbb{C}[-1,1]} > \|g^*\|_{\mathbb{C}[-1,1]}, \qquad (3.4)$$

contradicting the extremal property (1.2) of $g^*(t)$.

Now, let us consider the function $h^* \in H^\omega[-1, 1]$ extremal in the problem

$$\int_{-1}^{1} h(t) K_{\tau^*}(t) \, dt \to \sup, \qquad h \in H_0^\omega[-1, 1]. \qquad (3.5)$$

Let

$$\widehat{H}(t) = \frac{1}{r!} \int_{-1}^{1} h^*(y)(t-y)_+^{r-1} \, dy, \qquad -1 \le t \le 1,$$

$$(3.6)$$

$$H(t) = \widehat{H}(t) - q(t), \qquad -1 \le t \le 1,$$

where $q(t)$ is a polynomial of the best approximation for $\widehat{H}(t)$ on the interval $[-1,1]$. Then, by (3.2) and the extremality of h^* in the problem (3.5),

$$\|H\|_{C[-1,1]} \geq \sum_{i=0}^{r+1} \alpha_i H(\tau_i^*) =$$

$$= \int_{-1}^{1} \frac{d^r}{dt^r} H(t) K_{\tau^*}(t)\, dt = \sup_{h \in H^\omega[-1,1]} \int_{-1}^{1} h^*(t) K_{\tau^*}(t)\, dt. \quad (3.7)$$

On the other hand, by our choice of points $\{\tau_i^*\}_{i=0}^{r+1}$,

$$\|g^*\|_{C[-1,1]} = \sum_{i=0}^{r+1} \alpha_i g^*(\tau_i^*) = \int_{-1}^{1} \frac{d^r}{dt^r} g^*(t) K_{\tau^*}(t)\, dt \leq$$

$$\leq \sup_{h \in H^\omega[-1,1]} \int_{-1}^{1} h(t) K_{\tau^*}(t)\, dt. \quad (3.8)$$

Combining (3.7) and (3.8), we infer that

$$\|H\|_{C[-1,1]} \geq \|g^*\|_{C[-1,1]}. \quad (3.9)$$

However, by the definition (3.6), $H \in D[-1,1]$, so the property of extremality of g^* in (1.2) implies that the equality persists in (3.9), as well as everywhere in (3.7) and (3.8).

But the equality in (3.7) is possible, only if

$$H(\tau_i^*) = (-1)^{i+r+1}\|H\|_{C[-1,1]} = (-1)^{i+r+1}\|g^*\|_{C[-1,1]}, \quad i = 0, \ldots, r+1, \quad (3.10)$$

while the equality in (3.8) leads us to the conclusion that

$$\sup_{h \in H^\omega[-1,1]} \int_{-1}^{1} h(t) K_{\tau^*}(t)\, dt = \int_{-1}^{1} \frac{d^r}{dt^r} g^*(t) K_{\tau^*}(t)\, dt. \quad (3.11)$$

$\square$

COROLLARY 14.3.2. *The function $g^*(t) \in M[-1,1]$ with the property (3.3) is unique.*

Proof. By Korneichuk's Lemma 2.1.1, any function $T^{(r)}(t) \in H^\omega[-1,1]$ with the property (3.3), (i), is defined up to an additive constant. Therefore, the function T is determined up to a polynomial of degree r. However, the property (3.3), (ii) implies that $T \equiv g^*$. $\square$

14.4. Concluding remarks

14.4.1. The norm of the Chebyshev function in $W^r H^\omega[-1,1]$

Let $C_r(x) = g^*(x)$, $-1 \le x \le 1$, where the function g^* with the property (1.2) is described in Theorem 14.3.1 and its corollaries.

It follows from the formula (3.2) and the expression (3.12) that

$$\|C_r\|_{C[-1,1]} = \int_{-1}^{1} C_r^{(r)}(x) K_{\tau^*}(x)\, dx = \int_{0}^{2} \Re\,(F_{\tau^*}; x)\, \omega'(x)\, dx, \tag{4.1}$$

where $\Re(F_{\tau^*}; \cdot)$ is the rearrangement of the simple kernel F_{τ^*}.

14.4.2. Solution of one extremal problem

DEFINITION 14.4.1. The functional class $V_l[a,b]$ for $l \in \mathbb{N}$ is defined as follows:

$$V_l[a,b] := \{ f \in \mathbb{C}^{l-1}[a,b] \ : \ \bigvee_a^b f^{(l)} \le 1, \quad f^{(i)}(a) = f^{(i)}(b) = 0, \quad i = 0,\dots,l-1 \},$$

where $\bigvee_a^b g$ is a variation of the function g.

From the duality theorem it follows that

$$\sup_{g \in W^r H^\omega[-1,1]} \ \inf_{p \in P_r} \|g - p\|_{C[-1,1]} = \sup_{G \in V_{r-1}[-1,1]} \ \sup_{h \in H^\omega[-1,1]} \int_{-1}^{1} h(t) G(t)\, dt. \tag{4.2}$$

We showed that the pair $(C^{(r)}, K_{\tau^*})$ is extremal in (4.2) among all pairs $(h, G) \in H^\omega[-1,1] \times V_{r-1}[-1,1]$.

The inclusion $P_r \in W^r H^\omega[-1,1]$ implies that the polynomial space P_r is the optimal $(r+1)$-dimensional approximative space for $W^r H^\omega[-1,1]$, i.e.

$$d_{r+1}(W^r H^\omega[-1,1], \mathbb{C}[-1,1]) = \mathrm{E}(W^r H^\omega[-1,1]; P_r). \tag{4.3}$$

We conclude the chapter with the following *hypothesis* on the uniqueness of the function g^* and its consequences.

CONJECTURE 14.4.1. *There exists a unique function of the maximal norm in $D[-1,1]$.*

If this conjecture is true, we can give explicit formulas for the rth derivative of the function $g^*(x)$.

COROLLARY 14.4.2. *If the function $g^* \in D[-1,1]$ of the maximal norm in $D[-1,1]$ is unique, then*

$$\frac{d^r}{dt^r} g^*(t) = \begin{cases} \dfrac{1}{2}\omega(2t), & 0 \le t \le 1; \\[2mm] -\dfrac{1}{2}\omega(-2t), & -1 \le t \le 0. \end{cases} \tag{4.4}$$

Proof. Let $\hat{g}(t) = (-1)^{r+1} g^*(-t)$, $-1 \leq t \leq 1$. By the definition of the class $D[-1,1]$ and Theorem 14.3.1, $\hat{g} \in D[-1,1]$ and $\|\hat{g}\|_{C[-1,1]} = \|g^*\|_{C[-1,1]}$, and the sets of alternance points of g^* and $\hat{g}$ are symmetric with respect to the origin. Thus, by our assumption of Conjecture 14.4.1, $\hat{g}(t) = g^*(t)$, or, equivalently,

$$g^*(t) = (-1)^{r+1} g^*(-t), \qquad t \in [-1,1]. \tag{4.5}$$

Consequently, the set of alternance points of g^* is symmetric with respect to the origin:

$$\tau_i^* = \tau_{r+1-i}^*, \qquad i = 0, \ldots, r+1. \tag{4.6}$$

Then, the derivative $K_{\tau^*}(t)$ of the simple kernel $F_{\tau^*}(t)$ is odd:

$$K(-t) = -K(t), \qquad t \in [-1,1]. \tag{4.7}$$

The property (4.7) and Corollary 2.1.2 of Korneichuk's lemma lead us to the formula (4.4). □

Chapter 15

N-widths of the class $W^1 H^\omega[-1, 1]$

In Chapter 15 we compute the N-widths of the functional class $W^1 H^\omega[-1,1]$, where $\omega(t)$ is a concave modulus of continuity on the interval $[0,2]$, satisfying the restrictions

$$(2 - \sqrt{2})\,(\omega(a) + \omega(b)) \le \omega(a+b), \qquad \text{for all} \quad a, b \ge 0, \quad a + b \le 2. \qquad (\bigstar)$$

In the special case of Hölder's modulii of continuity $\omega_\alpha(t) = t^\alpha$ the condition $(\bigstar)$ is satisfied if $1 \ge \alpha \ge \log_2(4 - 2\sqrt{2}) = 0.2284...$

Before proceeding with the computation of N-widths of $W^1 H^\omega[-1,1]$, we mention the following result on the exact values of $\{d_n\left(H^\omega[-1,1], \mathbb{C}[-1,1]\right)\}_{n \in \mathbb{N}}$ due to Yu. Grigoryan [29].

THEOREM 15.0.1. *Let ω be a concave modulus of continuity on $[0,2]$. Then, for any $n \in \mathbb{N}$,*

$$d_n\left(H^\omega[-1,1],\ \mathbb{C}[-1,1]\right) = d_n\left(H^\omega[-1,1],\ \mathbb{L}_\infty[-1,1]\right) = \frac{1}{2}\omega(\frac{2}{n}).$$

For the estimate from above one can use *the Steklov approximations* of piece-wise constant splines from the class $\mathbb{S}^0[\tau_1, \tau_2, \ldots, \tau_{n-1}]$, where $\{\tau_i = -1 + \frac{2i}{n}\}_{i=1-n}^{n-1}$ (see Appendix in [49], [50] for a discussion various properties of the Steklov functions).

The estimate from below is analogous to the one for the class $W^1 H^\omega[-1,1]$ in Section 15.2.

15.1. Formulation of the main results

Let the collections of points $\bar{\nu} = \{\nu_i\}_{i=0}^{n+1}$ and $\bar{\vartheta} = \{\vartheta_i\}_{i=0}^{n}$ on the interval $[-1,1]$ be such that

(I) $\quad -1 = \nu_0 = \vartheta_0 < \nu_1 < \vartheta_1 < \nu_2 < \vartheta_2 < \cdots < \nu_{n-1} < \vartheta_{n-1} < \nu_n = \vartheta_n = 1;$

(II) $\quad \nu_i = \nu_{n+1-i}, \qquad i = 1, \ldots, n+1; \qquad \vartheta_j = \vartheta_{n-j}, \quad j = 0, \ldots, n.$

$$\tag{1.1}$$

The inequalities (1.1), (I) guarantee the existence and uniqueness of such a piece-wise linear continuous kernel

$$F(x) = F_{\bar{\nu}, \bar{\vartheta}}(x) = \sum_{i=1}^{n+1} \alpha_i (\nu_i - x)_+ \tag{1.2}$$

with the knots $\{\nu_i\}_{i=1}^{n+1}$ and coefficients $\{\alpha_i\}_{i=0}^{n+1}$, $\sum_{i=0}^{n+1}|\alpha_i| = 1$, that

$$F(\vartheta_i) = 0, \qquad i = 0, \ldots, n+1. \tag{1.3}$$

The corresponding matrix of the system of linear equations (1.3) for $\{\alpha_i\}_{i=1}^{n+1}$ is *upper triangular*. Therefore, we can give explicit formulas for the computation of the coefficients $\{\alpha_i = \alpha_i(\bar\nu, \bar\vartheta)\}_{i=1}^{n+1}$:

$$\beta_{n+1} := 1, \qquad \beta_i := [\vartheta_{i-1} - \nu_i]^{-1} \sum_{j=i+1}^{n+1} \beta_j(\nu_j - \vartheta_i), \qquad i = 1, \ldots, n.$$

$$\alpha_i := \beta_i \cdot \left[\sum_{j=1}^{n+1}|\beta_j|\right]^{-1}. \tag{1.4}$$

THEOREM 15.1.1. *Let ω be a concave modulus of continuity on $[0,2]$, and $n \in \mathbb{N}$. There exist collections of points $\bar\vartheta = \{\vartheta_i = \vartheta_i(n,\omega)\}_{i=0}^n$ and $\bar\nu = \{\nu_i = \nu_i(n,\omega)\}_{i=0}^{n+1}$ satisfying (1.1), the kernel F defined by (1.2), (1.3), and the function $x_{n,\omega} \in \mathbb{C}^1[-1,1]$ endowed with the properties*

$$(I) \qquad \sup_{h \in H^\omega[\vartheta_{i-1},\vartheta_i]} \int_{\vartheta_{i-1}}^{\vartheta_i} h(t)\frac{dF}{dt}(t)\,dt = \int_{\vartheta_{i-1}}^{\vartheta_i} \frac{dx_{n,\omega}}{dt}(t)\frac{dF}{dt}(t)\,dt, \qquad i = 1, \ldots, n.$$

$$(II) \qquad x_{n,\omega}(\nu_i) = (-1)^i\|x_{n,\omega}\|_{\mathbb{C}[-1,1]}, \qquad i = 0, \ldots, n+1. \tag{1.5}$$

Notice that the property (1.5) guarantees that

$$\dot{x}_{n,\omega}\big|_{[\vartheta_{i-1},\vartheta_i]} \in H^\omega[\vartheta_{i-1},\vartheta_i], \qquad i = 1, \ldots, n. \tag{1.6}$$

The following result provides a sufficient condition for the global inclusion $x_{n,\omega} \in W^1H^\omega[-1,1]$, gives the exact value of $d_n(W^1H^\omega[-1,1], \mathbb{C}[-1,1])$ for all $n \in \mathbb{N}$ and ω under the imposed restriction ($\bigstar$), and describes the optimal n-dimensional approximating subspaces $\{\mathcal{A}_n\}_{n\in\mathbb{N}}$ with the properties

$$d_n\left(W^1H^\omega[-1,1],\, \mathbb{C}[-1,1]\right) = \mathrm{E}\left(W^1H^\omega[-1,1],\, \mathcal{A}_n\right)$$

THEOREM 15.1.2. *Let a concave modulus of continuity ω satisfy the restriction ($\bigstar$), and the functions $\{x_{n,\omega}\}_{n\in\mathbb{N}}$ be as in Theorem 15.1.1. Then,*

(I) $x_{n,\omega} \in W^1H^\omega[-1,1]$;

(II) $d_{n+1}\left(W^1H^\omega[-1,1],\, \mathbb{C}[-1,1]\right) = \|x_{n,\omega}\|_{\mathbb{C}[-1,1]}, \qquad n \in \mathbb{N}.$

  $d_1\left(W^1H^\omega[-1,1],\, \mathbb{C}[-1,1]\right) = \infty;$

(III) $d_{n+1}\left(W^1H^\omega[-1,1],\, \mathbb{C}[-1,1]\right) = \mathrm{E}\left(W^1H^\omega[-1,1],\, \mathbb{S}^1[\vartheta_1,\ldots,\vartheta_{n-1}]\right),$

where $\{\vartheta_i = \vartheta_i(n,\omega)\}_{i=1}^{n-1}$ *are the knots (points of sign change of* $x_{n,\omega}''$*) of the function* $x_{n,\omega}(t)$.

REMARK 15.1.1. By Theorem 15.1.2, the space $\mathbb{S}^1[\vartheta_1,\ldots,\vartheta_{n-1}]$ of piecewise linear splines with the knots $\{\vartheta_i\}_{i=1}^{n-1}$ is an extremal approximating subspace of $\mathbb{C}[-1,1]$ of dimension $n+1$. However, in contrast with the case of linear modulus of continuity $\omega(t) = t$, the optimal method of approximation is nonlinear (see Lemma 15.3.3 in this context).

REMARK 15.1.2. The relation $d_1\left(W^1 H^\omega[-1,1], \mathbb{C}[-1,1]\right) = \infty$, and the more general properties

$$d_k\left(W^r H^\omega[-1,1], \mathbb{C}[-1,1]\right) = \infty, \qquad k = 1,\ldots,r, \tag{1.8}$$

follow from the existence of the $(r+1)$-dimensional linear space *span* $\{1, t, \ldots, t^r\}$ in $W^r H^\omega[-1,1]$.

REMARK 15.1.3. In Chapter 14 we found $d_{r+1}\left(W^r H^\omega[-1,1], \mathbb{C}[-1,1]\right)$ for all $r \in \mathbb{N}$. In particular, if $r = 1$, then

$$d_2\left(W^1 H^\omega[-1,1], \mathbb{C}[-1,1]\right) = \|x_{2,\omega}\|_{\mathbb{C}[-1,1]} = \frac{1}{8}\int_0^2 \omega(x)\,dx. \tag{1.9}$$

If ω satisfies ($\bigstar$), then the function $\dot{x}_{3,\omega}(t)$ is even and given on the interval $[0,1]$ by the formula

$$\frac{d}{dt}x_{3,\omega}(t) = \begin{cases} \dfrac{1}{\sqrt{2}+1}\omega\left(1 - (\sqrt{2}+1)x\right), & 0 \le x \le \dfrac{1}{\sqrt{2}+1}; \\ -\dfrac{\sqrt{2}}{\sqrt{2}+1}\omega\left(\dfrac{\sqrt{2}+1}{\sqrt{2}}x - \dfrac{1}{\sqrt{2}}\right), & \dfrac{1}{\sqrt{2}+1} \le x \le 1. \end{cases} \tag{1.10}$$

Therefore,

$$d_3\left(W^1 H^\omega[-1,1], \mathbb{C}[-1,1]\right) = \|x_{3,\omega}\|_{\mathbb{C}[-1,1]} = \frac{1}{\sqrt{6+4\sqrt{2}}}\int_0^2 \omega(t)\,dt. \tag{1.11}$$

REMARK 15.1.4. Let $\{\alpha_i = \alpha_i(n,\omega)\}_{i=1}^{n+1}$ be the coefficients of the kernel F from the statement of Theorem 15.1.2. Let us introduce the constants $\{l_i\}_{i=0}^n$ and $\{m_i\}_{i=0}^n$:

$$l_i := \left|\sum_{j=i}^{n+1}\alpha_j\right|, \qquad i = 1,\ldots,n+1; \tag{1.12}$$

$$m_i := \frac{l_i}{l_i + l_{i+1}}, \qquad i = 1,\ldots,n. \tag{1.13}$$

Notice that by the defnition of the kernel F,

$$\frac{dF}{dx}(x) = (-1)^i \chi l_i, \qquad \nu_{i-1} < x < \nu_i, \quad i = 1, \ldots, n+1, \tag{1.14}$$

where $\chi = \operatorname{sign} \alpha_1$. The Korneichuk's Lemma 2.1.1 and the formula (2.2.34) enable us to find the formula for $\dot{x}_{n,\omega}(t)$ on each of the intervals $\{[\vartheta_{i-1}, \vartheta_i]\}_{i=1}^n$:

$$\dot{x}_{n,\omega}(t - \nu_i) = \begin{cases} (-1)^i m_i \cdot \omega(m_i^{-1} t), & t \in [\vartheta_{i-1} - \nu_i, 0]; \\ (-1)^{i+1}(1 - m_i) \cdot \omega((1 - m_i)^{-1} t), & t \in [0, \vartheta_i - \nu_i]. \end{cases} \tag{1.15}$$

The formulas (1.15) enable us to find the formula for the N-widths of the class $W^1 H^\omega[-1,1]$ in terms of the constants $\{m_i\}_{i=0}^n$:

$$d_{n+1}\left(W^1 H^\omega[-1,1], \mathbb{C}[-1,1]\right) = \|x_{n,\omega}\|_{\mathbb{C}[-1,1]} =$$

$$= \frac{1}{2} m_1^2 \int_0^{\nu_1 - \nu_0} \omega(t)\, dt = \frac{1}{2}(1 - m_n)^2 \int_0^{\nu_{n+1} - \nu_n} \omega(t)\, dt =$$

$$= \frac{1}{2}(1 - m_k)^2 \int_0^{\nu_k - \nu_{k-1}} \omega(t)\, dt + \frac{1}{2} m_{k+1}^2 \int_0^{\nu_{k+1} - \nu_k} \omega(t)\, dt, \qquad k = 1, \ldots, n. \tag{1.16}$$

The proof of Theorem 15.1.2 divides into two parts – the estimates of N-width from below and from above.

15.2. Estimate of N-widths from below

Let $\{\nu_i = \nu_i(n, \omega)\}_{i=0}^{n+1}$ be the alternance points of the function $x_{n,\omega}(t)$. We introduce the collection of $n + 1$ continuous functions

$$\phi_i(t) = \begin{cases} \dot{x}_{n,\omega}(t), & t \in [z_i, z_{i+1}], \\ 0, & t \in [-1,1] \setminus [z_i, z_{i+1}], \end{cases} \qquad i = 0, \ldots, n, \tag{2.1}$$

and define the set of functions Γ_n as follows:

$$\Gamma_n := \left\{ \sum_{i=0}^n c_i \phi_i(t) \mid c_i = 0, \pm 1 \right\}. \tag{2.2}$$

REMARK 15.2.1. It will be shown below in Lemma 15.2.1 that the inequality ($\bigstar$) is a sufficient condition for the inclusion $\Gamma_n \subset H^\omega[-1,1]$.

Put

$$v_n(f) := \{f(z_0), f(z_1), \ldots, f(z_{n+1})\}, \qquad f \in \mathbb{C}[-1,1], \tag{2.3}$$

$$V_n := \{v_n(f) \mid f \in W^1 H^\omega[-1,1]\}, \tag{2.4}$$

Let L_{n+1} be a fixed $(n+1)$-dimensional subspace in $\mathbb{C}[-1,1]$. Then,

$$W_n := \{v_n(g) \mid g \in L_{n+1}\} \tag{2.5}$$

is a linear subspace in $\mathbb{R}^{n+2}$ whose dimension does not exceed $n+1$. The trivial inequality

$$\inf_{g \in L_{n+1}} \|f - g\|_{\mathbb{C}[-1,1]} \geq \inf_{g \in L_{n+1}} \|v_n(f) - v_n(g)\|_{l_\infty^{n+2}}, \tag{2.6}$$

implies that

$$d_{n+1}\left(W^1 H^\omega[-1,1], \mathbb{C}[-1,1]\right) :=$$

$$:= \inf_{L_{n+1}} \sup_{f \in W^1 H^\omega[-1,1]} \inf_{g \in L_{n+1}} \|f - g\|_{\mathbb{C}[-1,1]} \geq d_{n+1}\left(V_n, l_\infty^{n+2}\right). \tag{2.7}$$

Let us show that V_n contains the set

$$X_{n+2} = \left\{\bar{x} = (x_0, \ldots, x_{n+1}) \mid x_i = \pm \|x_{n,\omega}\|_{\mathbb{C}[-1,1]}, \quad i = 0, \ldots, n+1\right\}. \tag{2.8}$$

Indeed, let ζ be the leftmost zero of $x_{n,\omega}$ on $[-1,1]$, i.e. $\zeta \in [\nu_0, \nu_1] : x_{n,\omega}(\zeta) = 0$. Then, the function $f_{\bar{x}}$ with the properties $v_n(f_{\bar{x}}) = \bar{x} = (x_0, \ldots, x_{n+1}) \in X_{n+2}$ and $\dfrac{d}{dx} f_{\bar{x}} \in \Gamma_n$ is expressed by the formula

$$f_{\bar{x}}(t) := \int_\zeta^t \sum_{i=0}^n \operatorname{sign}(x_{i+1} - x_i) \cdot |\phi_i(y)| \, dy. \tag{2.9}$$

Now an application of Proposition 1.1.7 coupled with our observation (1.1.9) and the inequality (2.7) leads us to the desired estimate

$$d_{n+1}\left(W^1 H^\omega[-1,1], \mathbb{C}[-1,1]\right) \geq d_{n+1}\left(V_n, l_\infty^{n+2}\right) \geq$$

$$\geq d_{n+1}\left(X_{n+2}, l_\infty^{n+2}\right) = \|x_{n,\omega}\|_{\mathbb{C}[-1,1]}. \tag{2.10}$$

The inference of the estimate from below will be complete, once we verify the key inclusion $\Gamma_n \subset H^\omega[-1,1]$.

LEMMA 15.2.1. *Let Γ_n be defined by (2.2). If the concave modulus of continuity ω satisfies the inequality ($\bigstar$), then $\Gamma_n \subset H^\omega[-1,1]$.*

Proof. The proof consists of the verification of the following two propositions.

PROPOSITION 15.2.2. *Let the constants $\{m_k\}_{k=1}^n$ be introduced by (1.12). Then,*

$$\frac{\max\{m_k, 1 - m_k\}}{\min\{m_k, 1 - m_k\}} \leq \sqrt{2}, \ k = 1, \ldots, n,$$

or, equivalently,

$$\sqrt{2} - 1 \leq m_k \leq 2 - \sqrt{2}, \qquad k = 1, \ldots, n. \tag{2.11}$$

Proof. The function $x_{n,\omega}(t)$ is either even or odd, so it is enough to consider the restriction $x_{n,\omega}|_{[0,1]}$. Next, only the case $n = 4j$ will be treated in details. The proofs in the remaining three cases of n's could be elaborated in a similar fashion.

Fix $n = 4j$, $m \in \mathbb{N}$. We introduce a more convenient enumeration of the constants $\{m_i\}_{i=1}^{2j-1}$ and the points of alternance and knots of the function $x_{n,\omega}$:

$$\begin{aligned} \tau_i &:= \vartheta_{i+2j}, \quad z_i := \nu_{i+2j+1}, \qquad i = 0, \ldots, 2j; \\ k_i &:= m_{i+1+2j}, \qquad i = 0, \ldots, 2j - 1. \end{aligned} \tag{2.12}$$

In particular,

$$0 =: \tau_0 < z_0 < \tau_1 < z_1 < \tau_2 < \cdots < \tau_{2j-1} < z_{2j-1} < \tau_{2j} = z_{2j} := 1.$$

By the definition (1.15) of the derivative of the function $x_{4j,\omega}(t)$,

$$\dot{x}_{n,\omega}(t - z_i) = \begin{cases} (-1)^i k_i \cdot \omega(k_i^{-1} t), & t \in [\tau_i - z_i, 0], \\ (-1)^{i+1}(1 - k_i) \cdot \omega((1 - k_i)^{-1} t), & t \in [0, \tau_{i+1} - z_i], \end{cases} \tag{2.13}$$

where $i = 0, \ldots, 2j - 1$. We also introduce the notation

$$\delta_{a,b} := \int_a^b |\dot{x}_{4j,\omega}(t)| \, dt. \tag{2.14}$$

We use the following properties of the function $x_{n,\omega}(t)$:

$(i) \quad 2\delta_{\tau_0, z_0} = \delta_{z_0, \tau_1} + \delta_{\tau_1, z_1};$

$(ii) \quad \delta_{z_{i-1}, \tau_i} + \delta_{\tau_i, z_i} = \delta_{z_i, \tau_{i+1}} + \delta_{\tau_{i+1}, z_{i+1}}, \qquad i = 1, \ldots, 2j - 2;$

$(iii) \quad \delta_{z_{2j-1}, \tau_{2j}} = \delta_{\tau_{2j-1}, z_{2j-1}} + \delta_{z_{2j-2}, \tau_{2j-1}};$

$(iv) \quad \delta_{\tau_i, z_i} = k_i^2 \displaystyle\int_0^{\tau_{i+1} - \tau_i} \omega(t) \, dt; \qquad i = 0, \ldots 2j - 1;$

$(v) \quad \delta_{z_i, \tau_{i+1}} = (1 - k_i)^2 \displaystyle\int_0^{\tau_{i+1} - \tau_i} \omega(t) \, dt; \qquad i = 0, \ldots, 2j - 1;$

$(vi) \quad (1 - k_i)\, \omega(\tau_{i+1} - \tau_i) = k_{i+1} \omega(\tau_{i+2} - \tau_{i+1}), \qquad i = 0, \ldots, 2j - 2.$

$$\tag{2.15}$$

The properties (2.15), (i)–(iii) follow from the facts that the function $\dot{x}_{4j,\omega}(t)$ is even and $\{z_i\}_{i=0}^{2j}$ are the alternance points of $x_{4j,\omega}$:

$$\begin{aligned} \delta_{\tau_0, z_0} &= \|x_{4j,\omega}\|_{C[-1,1]}, \quad \delta_{z_{2j-1}, \tau_{2j}} = 2\|x_{4j,\omega}\|_{C[-1,1]}, \\ \delta_{z_i, \tau_{i+1}} + \delta_{\tau_{i+1}, z_{i+1}} &= 2\|x_{4j,\omega}\|_{C[-1,1]}, \qquad i = 0, \ldots, 2j - 1. \end{aligned}$$

The properties (2.15), (iv), (v) follow from the definition (2.13) of $x_{4j,\omega}(t)$, and, finally, the relation (2.15), (vi) is a consequence of the continuity of the derivative $\dot{x}_{n,\omega}(t)$ at each of the points $\{\tau_i\}_{i=1}^{2j-1}$.

Our first goal is to establish the inequality

$$k_0 \geq \frac{1}{2}. \tag{2.16}$$

If the inequality (2.16) holds, then from (2.15), (iv), (v) it follows that

$$\delta_{\tau_0, z_0} = k_0^2 \int_0^{\tau_1} \omega(t)\, dt \geq (1 - k_0)^2 \int_0^{\tau_1} \omega(t)\, dt = \delta_{z_0, \tau_1}. \tag{2.17}$$

Therefore, properties (2.15), (i) and (2.17) imply that $\delta_{z_0, \tau_0} \leq \delta_{\tau_1, z_1}$, and

$$\delta_{\tau_1, z_1} \geq \delta_{z_0, \tau_0} \geq \delta_{z_0, \tau_1}. \tag{2.18}$$

Now we can utilize the property of concave modulii of continuity formulated in Proposition 1.2.5. Indeed, by the equation (2.15), (vi) for $i = 0$,

$$(1 - k_0)\omega(\tau_1 - \tau_0) = k_1 \omega(\tau_2 - \tau_1), \tag{2.19}$$

while by (2.18) and (2.15)-(iv), (v),

$$(1 - k_0)^2 \int_0^{\tau_1 - \tau_0} \omega(t)\, dt = \delta_{z_0, \tau_1} \leq \delta_{\tau_1, z_1} = k_1^2 \int_0^{\tau_2 - \tau_1} \omega(t)\, dt. \tag{2.20}$$

Then, the relations (2.19), (2.20) enable us to apply Proposition 1.2.5 and (2.16) to conclude that

$$\frac{1}{2} \geq 1 - k_0 \geq k_1, \tag{2.21}$$

Consequently, $1 - k_1 \geq \frac{1}{2}$, so by (2.15), (iv), (v),

$$\delta_{z_1, \tau_2} := (1 - k_1)^2 \int_0^{\tau_2 - \tau_1} \omega(t)\, dt \geq k_1^2 \int_0^{\tau_2 - \tau_1} \omega(t)\, dt =: \delta_{\tau_1, z_1}. \tag{2.22}$$

Combining inequalities (2.18) and (2.22) we infer that

$$\delta_{z_0, \tau_1} \leq \delta_{\tau_1, z_1} \leq \delta_{z_1, \tau_2}. \tag{2.23}$$

Then, the equation (2.15), (ii) for $i = 1$,

$$\delta_{z_0, \tau_1} + \delta_{\tau_1, z_1} = \delta_{z_1, \tau_2} + \delta_{\tau_2, z_2} \tag{2.24}$$

and inequalities (2.23) imply that

$$(1 - k_1)^2 \int_0^{\tau_2 - \tau_1} \omega(t)\, dt =: \delta_{z_1, \tau_2} \geq \delta_{\tau_2, z_2} := k_2^2 \int_0^{\tau_3 - \tau_2} \omega(t)\, dt. \tag{2.25}$$

The property (2.25), the equality $(1 - k_1)\omega(\tau_2 - \tau_1) = k_2\omega(\tau_3 - \tau_2)$ of (2.15), (vi) and the result of Proposition 1.2.5 lead us to the conclusion that $1 - k_1 \leq k_2$, or, equivalently,

$$k_1 \geq 1 - k_2. \tag{2.26}$$

A repeated application of the above analysis demonstrates that

$$(i) \ \frac{1}{2} \geq 1 - k_0 \geq k_1 \geq 1 - k_2 \geq k_3 \geq \cdots \geq 1 - k_{2j-2} \geq k_{2j-1},$$

$$(ii) \ \begin{cases} \delta_{z_{2i},\tau_{2i+1}} \leq \delta_{\tau_{2i+1},z_{2i+1}} \leq \delta_{z_{2i+1},\tau_{2i+2}}, \\ \delta_{z_{2i+1},\tau_{2i+2}} \geq \delta_{\tau_{2i+2},z_{2i+2}} \geq \delta_{z_{2i+2},\tau_{2i+3}}, \end{cases} \quad i = 0, \ldots, m-1. \tag{2.27}$$

Consequently, by (2.27)-(i),

$$1 \geq \frac{1 - k_0}{k_0} \geq \frac{k_1}{1 - k_1} \geq \frac{1 - k_2}{k_2} \geq \cdots \geq \frac{k_{2j-1}}{1 - k_{2j-1}}, \tag{2.28}$$

under the assumption $k_0 \geq \dfrac{1}{2}$.

If we assumed that $k_0 < \dfrac{1}{2}$ instead of (2.15), then the repetition of our analysis, based on formulas (2.15), would produce the strict inequalities

$$\frac{1}{2} < 1 - k_0 < k_1 < 1 - k_2 < \cdots < k_{2j-1}. \tag{2.29}$$

However, the property (2.15), (iii) leads us to the estimate

$$\delta_{z_{2j-1},\tau_{2j}} \geq \delta_{\tau_{2j-1},z_{2j-1}}, \tag{2.30}$$

which, by (2.15), (iv), (v) for $i = 2j - 1$, is equivalent to the inequality $k_{2j-1} \leq 1 - k_{2j-1}$, so that

$$k_{2j-1} \leq \frac{1}{2}. \tag{2.31}$$

The inequality (2.31) precludes the chain (2.29) and its cause, the inequality $k_0 < \dfrac{1}{2}$, from arising. This verifies inequalities (2.16), (2.27) and (2.28). Furthermore, by (2.15), (iii),

$$\delta_{z_{2j-1},\tau_{2j}} = \delta_{\tau_{2j-1},z_{2j-1}} + \delta_{z_{2j-2},\tau_{2j-1}}, \tag{2.32}$$

and by (2.27), (ii) for $i = m - 1$,

$$\delta_{\tau_{2j-1},z_{2j-1}} \geq \delta_{z_{2j-2},\tau_{2j-1}}. \tag{2.33}$$

Combining (2.32), (2.33), we obtain the inequality

$$\frac{\delta_{z_{2j-1},\tau_{2j}}}{\delta_{z_{2j-1},\tau_{2j-1}}} \geq 2, \tag{2.34}$$

which by (2.15), (iv), (v) is equivalent to

$$\frac{k_{2j-1}}{1-k_{2j-1}} \geq \frac{1}{\sqrt{2}}. \tag{2.35}$$

Finally, the combination of inequalities (2.28) and (2.35) enables us to conclude that

$$\frac{1-k_0}{k_0} \geq \frac{k_1}{1-k_1} \geq \frac{1-k_2}{k_2} \geq \cdots \geq \frac{k_{2j-1}}{1-k_{2j-1}} \geq \frac{1}{\sqrt{2}}, \tag{2.36}$$

as desired. The proof in the case $l = 4j$ is complete.   □

PROPOSITION 15.2.3. *Let $\phi(x) \in \Gamma_n$. Then,*

$$|\phi(x_2) - \phi(x_1)| \leq \omega(|x_2 - x_1|), \qquad \forall\, (x_1, x_2) \in [-1, 1] \times [-1, 1]. \tag{2.37}$$

The points x_1, $x_2 \in [0, 1]$ lie in one of the intervals $[\vartheta_i, \nu_{i+1}]$ or $[\nu_i, \vartheta_i]$, for $i = 0, \ldots, n$. In the proof of the inequality (2.37) we consider the "worst" possible case, where

$$x_1 \in [\vartheta_i, \nu_{i+1}], \quad x_2 \in [\nu_j, \vartheta_j], \qquad 0 \leq i < j \leq n-1, \tag{2.38}$$

and by (1.16),

$$\phi(t) = \begin{cases} (-1)^i f_{m_i, \vartheta_i, \vartheta_{i+1}}(t) = m_i \omega\left(m_i^{-1}(\nu_{i+1} - t)\right), & t \in [\vartheta_i, \nu_{i+1}]; \\ (-1)^j f_{m_j, \vartheta_{j-1}, \vartheta_j}(t) = (1 - m_j)\omega\left((1 - m_j)^{-1}(t - \nu_j)\right), & t \in [\nu_j, \vartheta_j]; \end{cases} \tag{2.39}$$

where the functions $f_{v,A,B}$ for $0 < v < 1$ and $A < B$ are introduced in (2.1.14) of Definition 2.1.4. Let us introduce the points $x_1' \in [\nu_{i+1}, \vartheta_{i+1}]$ and $x_2' \in [\vartheta_{j-1}, \nu_j]$:

$$x_1' := \nu_{i+1} + \frac{m_i}{1 - m_i}(\nu_{i+1} - x_1), \qquad x_2' := \nu_j - \frac{1 - m_j}{m_j}(x_2 - \nu_j). \tag{2.40}$$

REMARK 15.2.2. The points x_1' and x_2' are chosen in (2.40) in such a way that

$$a := x_1' - x_1 = \frac{1}{m_i}(\nu_{i+1} - x_1), \qquad b := x_2 - x_2' = \frac{1}{1 - m_j}(x_2 - \nu_j). \tag{2.41}$$

Therefore, by the formulae (2.39) and (2.41), Proposition 15.2.2, and our assumption (★),

$$|\phi(x_2) - \phi(x_1)| = m_i\, \omega(a) + (1 - m_j)\, \omega(b) \leq (2 - \sqrt{2})\,(\omega(a) + \omega(b)) \leq$$
$$\leq \omega(a + b) \leq \omega(x_2 - x_1), \tag{2.42}$$

which concludes the proof of Lemma 15.2.1.   □

COROLLARY 15.2.4. *Let $\{\vartheta_i = \vartheta_i(n, \omega)\}_{i=0}^n$ be the knots of the function $x_{n,\omega}$. Then, the function $x_{n,\omega}$ belongs to $W^1 H^\omega[-1, 1]$, and*

$$\sum_{l=i}^{j-1}(-1)^{i+l}\omega(\vartheta_{l+1} - \vartheta_l) \leq \omega(\vartheta_j - \vartheta_i), \qquad \forall\, (i, j) \in \mathcal{P}(n), \tag{2.43}$$

where $\mathcal{P}(n) := \{(i, j) \,|\, 1 \leq i < j \leq n, \; j - i = 2k + 1, \; k \in \mathbb{N}\}$.

Proof. First of all, $\dot{x}_{n,\omega}(\cdot) = \sum\limits_{i=1}^{n} \phi_i(\cdot)$, so the definition (2.2) of the set Γ_n implies that $\dot{x}_{n,\omega} \in \Gamma_n$. Then, Lemma 15.2.1 guarantees the inclusion $x_{n,\omega} \in W^1 H^\omega[-1,1]$. Then, by Corollary 2.2.8, the collection of inequalities (2.43) is precisely the criterion for the function $\dot{x}_{n,\omega}$ to belong to $H^\omega[-1,1]$. $\qquad\qquad\square$

15.3. Estimate of *N*-widths from above. Optimal subspaces

Let $\mathbb{S}^1[\vartheta_1,\ldots,\vartheta_{n-1}]$ be the space of linear splines with the knots $\{\vartheta_i = \vartheta_i(n,\omega)\}_{i=1}^{n-1}$ coinciding with the points of sign change of the function $\ddot{x}_{n,\omega}(t)$. Notice that the dimension of the space $\mathbb{S}^1[\vartheta_1,\ldots,\vartheta_{n-1}]$ is equal to $n+1$.

DEFINITION 15.3.1. *The subset M of functions of the finite variation* is defined as follows:

$$M := \left\{ g \ \middle| \ \bigvee_{-1}^{1} g \le 1, \quad \int\limits_{-1}^{1} u(t)\, dg(t) = 0, \quad \forall u \in \mathbb{S}^1[\vartheta_1,\ldots,\vartheta_{n-1}] \right\}. \qquad (3.1)$$

LEMMA 15.3.1. *Let the set M be defined by* (3.1). *Then,*

$$g \in M \quad\Longleftrightarrow\quad \left\{ \bigvee_{-1}^{1} g \le 1, \quad \int\limits_{\vartheta_{i-1}}^{\vartheta_i} g(t)\, dt = 0, \qquad i = 1,\ldots,n \right\}. \qquad (3.2)$$

Proof. From the inclusion $span\,\{1,\, t\} \subset \mathbb{S}^1[\vartheta_1,\ldots,\vartheta_{n-1}]$ we derive the implication

$$g \in M \implies \int\limits_{-1}^{1} dg = 0, \quad \int\limits_{-1}^{1} t\, dg(t) = 0. \qquad (3.3)$$

Therefore, for any functions $g \in M$ and $w \in \mathbb{C}^1[-1,1]$, we can integrate by parts as follows:

$$\int\limits_{-1}^{1} w(t)\, dg(t) = \int\limits_{-1}^{1} w'(t)\, g(t)\, dt, \qquad (3.4)$$

By (3.4) and the definition (3.1) of the class M,

$$0 = \int\limits_{-1}^{1} u(t)\, dg(t) = \int\limits_{-1}^{1} u'(t)\, g(t)\, dt = \sum_{i=1}^{n} \int\limits_{\vartheta_{i-1}}^{\vartheta_i} u'(t)\, g(t)\, dt, \qquad (3.5)$$

for all $u \in \mathbb{S}^1[\vartheta_1,\ldots,\vartheta_{n-1}]$. Choosing the splines $\{u_i\}_{i=1}^n$ from $\mathbb{S}^1[\vartheta_1,\ldots,\vartheta_{n-1}]$ in (3.4) in sich a way that

$$u_i'(t) = \delta_{ij}, \quad \vartheta_{j-1} \le t \le \vartheta_j; \qquad (i,\, j) \in \{1,\ldots,n\},$$

and employing the property (3.5), we arrive at the equations

$$\int_{\vartheta_{i-1}}^{\vartheta_i} g(t)\, dt = 0, \quad i = 1, \ldots, n,$$

for any function $g \in M$. Reversing the argument, we can prove the other implication in (3.2). $\square$

From the duality Theorem 1.1.4, the integration by parts formula (3.4), and the possibility of interchanging the supremums we infer the series of the equalities

$$\sup_{x \in W^1 H^\omega[-1,1]} \inf_{u \in \mathbb{S}^1[\vartheta_1, \ldots, \vartheta_{n-1}]} \|x - u\|_{C[-1,1]} = \sup_{x \in W^1 H^\omega[-1,1]} \sup_{g \in M} \int_{-1}^{1} x(t)\, dg(t) =$$

$$= \sup_{g \in M} \sup_{x \in W^1 H^\omega[-1,1]} \int_{-1}^{1} x(t)\, d\,g(t) = \sup_{g \in M} \sup_{y \in H^\omega[-1,1]} \int_{-1}^{1} y(t) g(t)\, dt. \quad (3.6)$$

DEFINITION 15.3.2. For a fixed $g \in M$, let

$$L_i := \operatorname*{ess\,inf}_{t \in [\vartheta_{i-1}, \vartheta_i]} g(t), \qquad M_i := \operatorname*{ess\,sup}_{t \in [\vartheta_{i-1}, \vartheta_i]} g(t). \quad (3.7)$$

By the equations in (3.2), $M_i \geq 0$ and $L_i \leq 0$ for all $i = 1, \ldots, n$. Let $\{a_i\}_{i \in \mathbb{N}}$ and $\{b_i\}_{i \in \mathbb{N}}$ be two convergent sequences on $[\vartheta_0, \vartheta_1]$ such that

$$A := \lim_{i \to \infty} a_i : \quad \lim_{i \to \infty} g(a_i) = L_1; \quad B := \lim_{i \to \infty} b_i : \quad \lim_{i \to \infty} g(b_i) = M_1. \quad (3.8)$$

We distinguish two cases in the definition of the function $G(t)$.

Case 1. If $A < B$, then

$$G(t) := \begin{cases} L_i \mathcal{X}([\vartheta_{i-1}, \zeta_i]; t) + M_i \mathcal{X}([\zeta_i, \vartheta_i]; t), & t \in [\vartheta_{i-1}, \vartheta_i], \quad i = 1, 3, \ldots \\ M_i \mathcal{X}[\vartheta_{i-1}, \zeta_i]; t) + L_i \mathcal{X}[\zeta_i, \vartheta_i]; t), & t \in [\vartheta_{i-1}, \vartheta_i], \quad i = 2, 4, \ldots \end{cases}$$
$$(3.9)$$

where $\{\zeta_i\}_{i=1}^n$ satisfy the equations

$$\begin{aligned} L_i(\zeta_i - \vartheta_{i-1}) + M_i(\vartheta_i - \zeta_i) &= 0, & i = 1, 3, \ldots \\ M_i(\zeta_i - \vartheta_{i-1}) + L_i(\vartheta_i - \zeta_i) &= 0, & i = 2, 4, \ldots \end{aligned} \quad (3.10)$$

Case 2. If $B < A$, then

$$G(t) = \begin{cases} M_i \mathcal{X}([\vartheta_{i-1}, \zeta_i]; t) + L_i \mathcal{X}([\zeta_i, \vartheta_i]; t), & t \in [\vartheta_{i-1}, \vartheta_i), \quad i = 1, 3, \ldots \\ L_i \mathcal{X}([\vartheta_{i-1}, \zeta_i]; t) + M_i \mathcal{X}([\zeta_i, \vartheta_i]; t), & t \in (\vartheta_{i-1}, \vartheta_i), \quad i = 2, 4, \ldots \end{cases}$$
$$(3.11)$$

where $\{\zeta_i\}_{i=1}^n$ are determined from the equations

$$
\begin{aligned}
M_i(\zeta_i - \vartheta_{i-1}) + L_i(\vartheta_i - \zeta_i) = 0, && i = 1, 3, \ldots; \\
L_i(\zeta_i - \vartheta_{i-1} + M_i(\vartheta_i - \zeta_i) = 0, && i = 2, 4, \ldots
\end{aligned}
\tag{3.12}
$$

The equations (3.10) and (3.12) guarantee that the equations in (3.5) hold. The definitions (3.9) and (3.11) assure that $\overset{1}{\underset{-1}{\bigvee}} G \leq \overset{1}{\underset{-1}{\bigvee}} g$. Therefore, by the criterion (3.2), $G \in M$. Next, by the definition of the constants $\{L_i\}_{i=1}^n$ and $\{M_i\}_{i=1}^n$, we have the inclusions

$$
g\big|_{[\vartheta_{i-1},\vartheta_i]} \in \Theta_{-L_i,M_i} \quad \text{or} \quad g\big|_{[\vartheta_{i-1},\vartheta_i]} \in \Theta_{-M_i,L_i}[\vartheta_{i-1},\vartheta_i], \qquad i = 1,\ldots,n,
\tag{3.13}
$$

where classes $\Theta_{m,M}[a,b]$ are introduced in (2.1.13). Thus, we can apply Lemma 2.1.4 to conclude that

$$
\sum_{i=1}^n \sup_{y \in H^\omega[\vartheta_{i-1},\vartheta_i]} \int_{\vartheta_{i-1}}^{\vartheta_i} y(t) g(t)\, dt \leq \sum_{i=1}^n \sup_{y \in H^\omega[\vartheta_{i-1},\vartheta_i]} \int_{\vartheta_{i-1}}^{\vartheta_i} y(t) G(t)\, dt.
\tag{3.14}
$$

Also by Proposition 2.1.4, the upper bounds in (3.14) are attained on the functions $\pm x_{j_1,\ldots,j_n}(t)$ defined by the equations

$$
x_{j_1,\ldots,j_n}(t) = (-1)^i f_{j_i,\vartheta_{i-1},\vartheta_i}(t) + c_i, \qquad \vartheta_{i-1} \leq t \leq \vartheta_i, \quad i = 1,\ldots,n,
\tag{3.15}
$$

where $j_i = \dfrac{|L_i|}{L_i + M_i}$ or $j_i = \dfrac{M_i}{|L_i| + M_i}$ in Cases 1 and 2 of Definition 15.3.2, respectively, and $\{c_i\}_{i=1}^n$ are chosen to guarantee the continuity of $x_{j_1,\ldots,j_n}(t)$ on $[-1,1]$.

DEFINITION 15.3.3. The functional set $\mathcal{N}$ is defined as follows:

$$
\mathcal{N} := \{ x_{j_1,\ldots,j_n} \mid 0 < j_i < 1 \},
$$

where functions $x_{j_1,\ldots,j_n}(t)$ are introduced in (3.15).

PROPOSITION 15.3.2. *The following inclusion holds: $\mathcal{N} \subset H^\omega[-1,1]$.*

Proof. Fix $x = x_{j_1,\ldots,j_n} \in \mathcal{N}$ and the corresponding generating kernel $G = G_{j_1,\ldots,j_n}$. Notice that each of the restrictions $x\big|_{[\vartheta_{i-1},\vartheta_i]}$ is extremal in the problem

$$
\int_{\vartheta_{i-1}}^{\vartheta_i} G(t) h(t)\, dt \to \sup, \qquad h \in H^\omega[\vartheta_{i-1},\vartheta_i], \qquad i = 1,\ldots,n.
\tag{3.16}
$$

Consequently, by Corollary 2.2.8, the function x belongs to $H^\omega[-1,1]$, if and only if the following inequalities are satisfied:

$$\sum_{l=i}^{j-1}(-1)^{i+l}\omega(\vartheta_{l+1}-\vartheta_l) \leq \omega(\vartheta_j-\vartheta_i), \qquad \forall\,(i,j)\in\mathcal{P}(n), \qquad (3.17)$$

where $\{\vartheta_i\}_{i=1}^{n_1}$ are the points of sign change of $\ddot{x}_{n,\omega}(t)$.

REMARK 15.3.1. Notice that the conditions (3.17) are independent of the choice of the function $x\in\mathcal{N}$. In particular, if *one function* from $\mathcal{N}$ belongs to $H^\omega[a,b]$, then the whole set $\mathcal{N}$ lies in $H^\omega[-1,1]$.

By the definition (1.15), $\dot{x}_{n,\omega}\in\mathcal{N}$ (with $j_i := m_i$, $i=1,\ldots,n$). On the other hand, Corollary 15.2.4 ascertains that the the condition ($\bigstar$) implies the inclusion $\dot{x}_{n,\omega} = x_{m_1,\ldots,m_n} \in H^\omega[-1,1]$. Now the referrence to Remark 15.3.1 proves the result. $\square$

The inequalities (3.6), the conditions of extremality in (3.14), and the result of Proposition 15.3.2 show that

$$\sup_{x\in W^1 H^\omega[-1,1]} \mathrm{E}\left(x,\mathbb{S}^1[\vartheta_1,\ldots,\vartheta_{n-1}]\right) = \sup_{y\,|\,\dot{y}\in\mathcal{N}} \mathrm{E}\left(y,\mathbb{S}^1[\vartheta_1,\ldots,\vartheta_{n-1}]\right). \quad (3.18)$$

The result of the following lemma completes the estimate from above of the maximal deviation of the functional class $W^1 H^\omega[-1,1]$ from the space $\mathbb{S}^1[\vartheta_1,\ldots,\vartheta_{n-1}]$ of linear splines.

LEMMA 15.3.4. *Let y be such that $\dot{y}\in\mathcal{N}$. Then,*

$$\mathrm{E}\left(y,\mathbb{S}^1[\vartheta_1,\ldots,\vartheta_{n-1}],\mathbb{C}[-1,1]\right) \leq$$
$$\leq \mathrm{E}\left(y,\mathbb{S}^1[\vartheta_1,\ldots,\vartheta_{n-1}],\mathbb{C}[-1,1]\right) = \|x_{n,\omega}\|_{\mathbb{C}[-1,1]}. \quad (3.19)$$

Proof. Given a function $y\in\mathbb{C}^1[-1,1] : \dot{y}\in\mathcal{N}$, we find an explicit formula for an approximating spline $l_y\in\mathbb{S}^1[\vartheta_1,\ldots,\vartheta_{n-1}]$ with the property

$$\|y-l_y\|_{\mathbb{C}[-1,1]} \leq \|x_{n,\omega}\|_{\mathbb{C}[-1,1]}. \quad (3.20)$$

Let $\{m_i\}_{i=1}^n$, be the set of constants from the definitions (1.13), (1.15) of the function $x_{n,\omega}(t)$.

As extremal functions of problems (3.16), each of the restrictions $\dot{y}\big|_{[\vartheta_{i-1},\vartheta_i]}$ is a strictly monotone function on $[\vartheta_{i-1},\vartheta_i]$ for $i=1,\ldots,n$ (see the Korneichuk's formula (2.1.7) and Figure 2.1.1). Therefore, for each $i=1,\ldots,n$, there exists such a point $\xi_i\in[\vartheta_{i-1},\vartheta_i]$ that

$$\frac{\delta_{\xi_i,\vartheta_i}(\dot{y})}{\delta_{\xi_i,\vartheta_{i-1}}(\dot{y})} = \frac{m_i^2}{(1-m_i)^2}, \qquad i=1,\ldots,n, \qquad (3.21)$$

where

$$\delta_{a,b}(f) := \int_a^b (f(t)-f(a))\,dt.$$

Proposition 2.2.12, applied to the function $h(t) = f(t) - f(\xi_i)$, (more precisely, its variant for the interval $[\vartheta_{i-1}, \vartheta_i]$), provides the estimates

$$
|\delta_{\xi_i, \vartheta_i}(\dot y)| \le m_i^2 \int_0^{\vartheta_i - \vartheta_{i-1}} \omega(t)\, dt, \qquad |\delta_{\xi_i, \vartheta_{i-1}}(\dot y)| \le (1 - m_i)^2 \int_0^{\vartheta_i - \vartheta_{i-1}} \omega(t)\, dt.
$$
(3.22)

Let us define l_y by the equations

$$
\begin{aligned}
\dot l_y(t) &= \dot y(\xi_i), \qquad \vartheta_{i-1} < t < \vartheta_i, \qquad 1 \le i \le n; \\
l_y(1) - y(1) &= y(\xi_n) - l_y(\xi_n).
\end{aligned}
$$
(3.23)

The inequalities (3.22) and the formula (1.16) for the norm of the function $x_{n,\omega}(t)$ lead us to the final estimate

$$
\mathrm{E}(y, \mathbb{S}^1[\vartheta_1, \ldots, \vartheta_n], \mathbb{C}[-1,1]) \le \|y - l_y\|_{\mathbb{C}[-1,1]} \le
$$

$$
\frac{1}{2} \max \left\{ \|\dot y - \dot l_y\|_{\mathrm{L}_1[\vartheta_0, \xi_1]}, \ \max_{1 \le i \le n-1} \|\dot y - \dot l_y\|_{\mathrm{L}_1[\xi_i, \xi_{i+1}]}, \ \|\dot y - \dot l_y\|_{|\mathrm{L}_1[\xi_n, \vartheta_n]} \right\} \le
$$

$$
\le \frac{1}{2} \max \left\{
\begin{array}{l}
m_1^2 \displaystyle\int_0^{\vartheta_1 - \vartheta_0} \omega(\xi)\, d\xi \\[2em]
\displaystyle\max_{1 \le i \le n-1} \left\{ (1 - m_i)^2 \int_0^{\vartheta_i - \vartheta_{i-1}} \omega(\xi)\, d\xi + m_{i+1}^2 \int_0^{\vartheta_{i+1} - \vartheta_i} \omega(\xi)\, d\xi \right\} \\[2em]
(1 - m_n)^2 \displaystyle\int_0^{\vartheta_n - \vartheta_{n-1}} \omega(\xi)\, d\xi
\end{array}
\right\} =
$$

$$
= \|x_{n,\omega}\|_{\mathbb{C}[-1,1]} = \mathrm{E}\left(x_{n,\omega}, \mathbb{S}^1[\vartheta_1, \ldots, \vartheta_{n-1}]\right). \quad (3.24)
$$

REMARK 15.3.2. The verification of the last equality in (3.24) proceeds along the lines of the argument in the proof of Proposition 13.2.1.

Therefore, taking into account the fact that $\mathbb{S}^1[\vartheta_1, \ldots, \vartheta_{n-1}]$, we arrive at the estimate

$$
d_{n+1}\left(W^1 H^\omega[-1,1], \mathbb{C}[-1,1]\right) \le \|x_{n,\omega}\|_{\mathbb{C}[-1,1]}. \quad (3.25)
$$

Finally, the juxtaposition of inequalities (3.24) shows that both the estimates from above from Section 15.3 and from below from Section 15.2 are sharp.

Chapter 16

Lower Bounds for the N-Widths of the Class $W^r H^\omega[n]$

16.1. Definition of the class $W^r H^\omega[n]$

Let the collections of points $\{\nu_i = \nu_i(\omega, r, n)\}_{i=0}^{n+1}$, $\{\vartheta_i = \vartheta(\omega, r, n)\}_{i=0}^{n-r+1}$ satisfy inequalities

$$
\begin{aligned}
-1 &=: \nu_0 < \nu_1 < \cdots < \nu_n < \nu_{n+1} := 1; \\
-1 &=: \vartheta_0 < \vartheta_1 < \cdots < \vartheta_{n-r} < \vartheta_{n-r+1} := 1; \\
\nu_i &< \vartheta_i < \nu_{i+r}, \qquad i = 1, \ldots, n.
\end{aligned}
\tag{1.1}
$$

Let the coefficients $\{\alpha_i\}_{i=0}^{n+1}$ of the kernels

$$
F(t) = \frac{1}{r!} \sum_{i=0}^{n+1} \alpha_i(\nu_i - t)_+^r, \qquad K(t) = -\frac{1}{(r-1)!} \sum_{i=0}^{n+1} \alpha_i(\nu_i - t)_+^{r-1}
\tag{1.2}
$$

be determined from the system of linear equations

$$
\begin{cases}
\displaystyle\sum_{i=0}^{n+1} (-1)^i \alpha_i = 1; \\[2mm]
\displaystyle\sum_{i=0}^{n+1} \alpha_i(\nu_i + 1)^j = 0, \qquad j = 0, \ldots, r; \\[2mm]
\displaystyle\sum_{i=0}^{n+1} \alpha_i(\nu_i - \vartheta_l)_+^r = 0, \qquad l = 1, \ldots, n - r.
\end{cases}
\tag{1.3}
$$

By Proposition 3.3.1, the kernel F has precisely $n - r$ simple zeroes $\{\vartheta_i\}_{i=1}^{n-r}$ on the interval $(-1, 1)$, while the kernel $K(t) = \dfrac{d}{dt} F(t)$ vanishes at exactly one point $\{\eta_i\}_{i=1}^{n-1}$ on each of the intervals $\{\vartheta_{i-1}, \vartheta_i\}_{i=1}^{n-r+1}$. In particular, for each of the indices $i = 1, \ldots, n - r + 1$, the restriction

$$
F_i(t) := F\big|_{[\vartheta_{i-1}, \vartheta_i]}
$$

is a simple kernel on the interval $[\vartheta_{i-1}, \vartheta_i]$ in the sense of Definition 2.1.1.

THEOREM 16.1.1. *Let ω be a concave modulus of continuity on the interval $[0,2]$. For all r, $n \in \mathbb{N}$: $n \geq r$, there exist collections of points $\{\nu_i = \nu_i(\omega,r,n)\}_{i=0}^{n+1}$ and $\{\vartheta_i = \vartheta(\omega,r,n)\}_{i=0}^{n-r+1}$ as in (1.1) and the function $\mathcal{Z}(t) = \mathcal{Z}_{\omega,r,n}(t)$ endowed with the following properties.*

I.   For all $i = 1, \ldots, n - r + 1$,

$$\int_{\vartheta_{i-1}}^{\vartheta_i} \mathcal{Z}(t) F_i'(t)\, dt = \sup_{h \in H^\omega[\vartheta_{i-1},\vartheta_i]} \int_{\vartheta_{i-1}}^{\vartheta_i} h(t) F_i'(t)\, dt = \int_0^{\vartheta_i - \vartheta_{i-1}} \Re(F_i;t)\omega'(t)\, dt. \quad (1.4)$$

II.   The collection $\{\nu_i\}_{i=0}^{n+1}$ constitutes the set of alternance points of the function $\mathcal{Z}$:

$$\mathcal{Z}(\nu_i) = (-1)^i \|\mathcal{Z}\|_{\mathbb{C}[-1,1]}, \qquad i = 0, \ldots, n+1. \quad (1.5)$$

Next, we introduce the functional classes $W^r H^\omega[n]$ for any $r \in \mathbb{N}$.

DEFINITION 16.1.1. *The functional classes $W^r H^\omega[n]$, $n \in \mathbb{N}$: $n \geq r$, are defined as follows:*

$$W^r H^\omega[n] := \left\{ f \in \mathbb{C}^r[-1,1] \mid f^{(r)}\big|_{[\vartheta_{i-1},\vartheta_i]} \in H^\omega[\vartheta_{i-1},\vartheta_i],\ i = 1, \ldots, n-r+1 \right\}, \quad (1.6)$$

where points $\{\vartheta_i = \vartheta_i(\omega,r,n)\}_{i=1}^{n-r}$ are the knots of the function $\mathcal{Z}_{\omega,n,r}$ and $\vartheta_0 := -1$, $\vartheta_{n-r+1} := 1$.

The following theorem gives estimates from below for the n-widths of the classes $W^r H^\omega[n]$.

THEOREM 16.1.2. *Let the functional classes $W^r H^\omega[n]$ be introduced in Definition 16.1.1 for $n \geq r$. Then, the n-widths of the functional classes $W^r H^\omega[n]$ can be estimated as follows:*

$$d_{n+1}\left(W^r H^\omega[n], \mathbb{C}[-1,1]\right) \geq \|\mathcal{Z}_{\omega,n,r}\|_{\mathbb{C}[-1,1]}, \qquad n \geq r + 1. \quad (1.7)$$

In Section 16.2, we give two proofs of Theorem 16.2 – one of them employs Theorem 1.1.6 on the n-width of the ball, while the other does not.

16.2.   Linear spaces $\mathcal{R}_{n+2}$ and M_{n+2} and their properties

The function $\mathcal{Z}_n = \mathcal{Z}_{n,r,\omega}$ from Theorem 16.1.1, has $n+2$ alternance points $\{\nu_i\}_{i=0}^{n+1}$ such that $-1 =: \nu_0 < \nu_1 < \cdots < \nu_{n+1} := 1$,

$$\mathcal{Z}_n(\nu_i) = (-1)^i \|\mathcal{Z}_n\|_{\mathbb{C}[-1,1]}, \qquad i = 0, \ldots, n+1. \quad (2.1)$$

In addition, the function $\mathcal{Z}_n$ has $n - r$ knots $\{\vartheta_j\}_{j=1}^{n-r}$, i.e. points of sign change of the function $\mathcal{Z}_n^{(r+1)}$ on the interval $[-1,1]$. By the Rolle theorem, each of the derivatives $\mathcal{Z}_n^{(k)}$, $k = 1, \ldots, r$, has precisely $n+1-k$ points of sign change (zeroes) $\{\eta_i^k\}_{i=1}^{n+1-k}$ on the interval $[-1, 1]$. Let

$$\xi_0 := -1; \quad \xi_i := \eta_i^r, \quad i = 1, \ldots, n+1-r; \quad \xi_{n+2-r} := 1, \tag{2.2}$$

i.e., $\{\xi_i\}_{i=1}^{n+1-r}$ is the set of zeroes of the function $\mathcal{Z}_n^{(r)}$ on the interval $[-1, 1]$.

Next, we define the continuous functions $\{\phi_i\}_{i=1}^{n+2-r}$ on the interval $[-1, 1]$ by the formulas

$$\phi_i(t) := \begin{cases} \mathcal{Z}_n^{(r)}\big|_{[\xi_{i-1}, \xi_i]}(t), & \xi_{i-1} \le t \le \xi_i; \\ 0, & t \in [-1, 1] \setminus [\xi_{i-1}, \xi_i]. \end{cases} \tag{2.3}$$

For $i = 1, \ldots, n+2-r$, put

$$\Phi_i(x) = \frac{1}{(r-1)!} \int_{-1}^{1} \phi_i(t)(x-t)_+^{r-1}\, dt, \qquad -1 \le x \le 1, \tag{2.4}$$

i.e. Φ_i is the rth integral of ϕ_i such that $\Phi_i^{(k)}(-1) = 0$, $k = 0, \ldots, r-1$.

We introduce the set M_{n+2} as the $(n+2)$-dimensional linear space of linear combinations of the functions $\{\Phi_i\}_{i=1}^{n+2-r}$ and $\{t^i\}_{i=0}^{r-1}$:

$$M_{n+2} := \ \operatorname{span}\ \{\Phi_1(t), \ldots, \Phi_{n+2-r}(t),\ 1,\ t,\ \ldots, t^{r-1}\}. \tag{2.5}$$

Notice that $\mathcal{Z}_n \in M_{n+2}$, because

$$\mathcal{Z}_n(t) = \sum_{i=1}^{n+2-r} \Phi_i(t) + \sum_{i=0}^{r-1} \tilde{a}_i t^i, \tag{2.6}$$

for some coefficients $\tilde{a}_i \in \mathbb{R}$, $\quad i = 0, \ldots, r-1$.

For any function $f \in \mathbb{C}[-1, 1]$, put

$$\eta(f) := \{f(\nu_0), f(\nu_1), \ldots, f(\nu_{n+1})\}, \tag{2.7}$$

where $\{\nu_i\}_{i=0}^{n+1}$, $-1 = \nu_0 < \nu_1 < \ldots \nu_{n+1} = 1$ are the alternance points of the function $\mathcal{Z}_n$.

Put

$$\mathcal{R}_{n+2} := \{\eta(f) \mid f \in M_{n+2}\}, \tag{2.8}$$

where the class M_{n+2} is defined in (2.5). Clearly, the dimension of the linear subspace $\mathcal{R}_{n+2}$ does not exceed $n + 2$.

Recall Definition 16.1.1 of the functional classes $W^r H^\omega[n]$, $n \in \mathbb{N}: \ n \ge r$:

$$W^r H^\omega[n] := \Big\{ f \in \mathbb{C}^r[-1,1] \,\big|\, f^{(r)}\big|_{[\vartheta_{i-1}, \vartheta_i]} \in H^\omega[\vartheta_{i-1}, \vartheta_i], \ i = 1, \ldots, n-r+1 \Big\},$$

where the points $\{\vartheta_i = \vartheta_i(n, r, \omega)\}_{i=1}^{n-r}$ are the knots of the function $\mathcal{Z}_n$,   and $\vartheta_0 = -1$, $\vartheta_{n-r+1} = 1$.

PROPOSITION 16.2.1. *Let the function* $g \in M_{n+2}$, $g(t) = \displaystyle\sum_{i=1}^{n+2-r} c_i \Phi_i(t) + \sum_{i=0}^{r-1} a_i t^i$,

be such that $|c_i| \le 1$ *for all* $i = 1, \ldots, n+2-r$.
  Then, g belongs to $W^r H^\omega[n]$.

Proof. We need to prove that the following inequality

$$| g^{(r)}(t'') - g^{(r)}(t') | \le \omega(t'' - t') \tag{2.9}$$

is satisfied for any pair of points $t', t'' \in [\vartheta_i, \vartheta_{i+1}]$, $i = 0, \ldots, n-r$, $t' < t''$, Recall that the function $\mathcal{Z}_n$ has $n - r + 1$ zeroes $\xi_i \in [\vartheta_{i-1}, \vartheta_i]$, $i = 1, \ldots, n - r + 1$.

If $t', t'' \in [\vartheta_{i-1}, \xi_i]$, for some $i = 1, \ldots, n - r + 1$, then

$$|g^{(r)}(t'') - g^{(r)}(t')| := |c_i| \cdot |\phi_i(t'') - \phi_i(t')| \le \omega(t'' - t'), \tag{2.10}$$

since $\phi_i\big|_{[\vartheta_{i-1}, \xi_i]} := \mathcal{Z}_n^{(r)}\big|_{[\vartheta_{i-1}, \xi_i]} \in H^\omega[\vartheta_{i-1}, \xi_i]$.

The case of points t', t'' on the intervals $[\xi_i, \vartheta_i]$ for $i = 1, \ldots, n - r + 1$, is treated analogously.

It remains only to consider the case when $t' \in [\vartheta_{i-1}, \xi_i]$ and $t'' \in [\xi_i, \vartheta_i]$, for some $i = 1, \ldots, n - r + 1$. In this case,

$$|g^{(r)}(t'') - g^{(r)}(t')| := |c_{i+1}\phi_{i+1}(t'') - c_i\phi_i(t')| \le$$
$$\le \max\{|c_i|, |c_{i+1}|\} \left(|\phi_{i+1}(t'')| + |\phi_i(t')| \right) \le$$
$$\le |\phi_{i+1}(t'')| + |\phi_i(t')| := \left| \mathcal{Z}_n^{(r)}(t'') - \mathcal{Z}_n^{(r)}(t') \right| \le \omega(t'' - t'), \tag{2.11}$$

since $\mathcal{Z}_n^{(r)}\big|_{[\vartheta_{i-1}, \vartheta_i]} \in W^r H^\omega[\vartheta_{i-1}, \vartheta_i]$, $i = 1, \ldots, n-r+1$. The proof of Proposition 16.2.1 is complete. □

Let the constants $\{A_i = A_i(r, \omega)\}_{i=r+1}^\infty$ be defined as follows:

$$A_{n+1} := \|\mathcal{Z}_{n,r,\omega}\|_{C[-1,1]}, \qquad n \ge r. \tag{2.12}$$

The formulas for the constants $\{A_i = A_i(\omega, r)\}_{i=r+1}^\infty$ are given in (1.15).

PROPOSITION 16.2.2. *Let the vector $\eta(g) \in \mathcal{R}_{n+2}$ for $g(t) = \displaystyle\sum_{i=1}^{n+2-r} c_i \Phi_i(t) + \displaystyle\sum_{i=0}^{r-1} a_i t^i$ be such that*

$$\eta(g) \in [-A_{n+1}, A_{n+1}]^{n+2},$$

where the constants $\{A_i\}_{i \ge r+1}$ are defined in (2.12). Then,

$$|c_i| \le 1, \qquad i = 1, \ldots, n + 2 - r.$$

Proof. Suppose that there exists a function $g^*(t) = \sum\limits_{i=1}^{n+2-r} c_i^* \Phi_i(t) + \sum\limits_{i=0}^{r-1} a_i^* t^i$ from M_{n+2}, such that

$$|g^*(\nu_i)| \le A_{n+1}, \qquad i = 0, \dots, n+1, \tag{2.13}$$

and

$$\lambda := \max_{1 \le i \le n+2-r} |c_i^*| > 1. \tag{2.14}$$

Let the index j, $1 \le j \le n+2-r$, be such that $|c_j^*| = \lambda$. We consider the function

$$\hat{g}(t) = \frac{g^*(t)}{\lambda \operatorname{sign} c_j^*}, \qquad -1 \le t \le 1, \tag{2.15}$$

i.e.,

$$\hat{g}(t) = \sum_{i=1}^{n+2-r} \hat{c}_i \Phi_i(t) + \sum_{i=0}^{r-1} \hat{a}_i t^i, \qquad -1 \le t \le 1, \tag{2.16}$$

with

$$\hat{c}_i = \frac{c_i^*}{c_j^*}, \quad i = 1, \dots, n+2-r; \quad \hat{a}_i = \frac{a_i^*}{c_j^*}, \quad i = 0, \dots, r. \tag{2.17}$$

Notice that by the definition of $\lambda = |c_j^*|$ in (2.14),

$$|\hat{c}_i| \le 1, \qquad i = 1, \dots, j-1, j+1, \dots, n+2-r; \qquad \hat{c}_j = 1. \tag{2.18}$$

Also, since $\lambda > 1$, by our assumption (2.13),

$$|\hat{g}(\nu_i)| := \frac{|g^*(\nu_i)|}{\lambda} < A_{n+1}, \qquad i = 0, \dots, n+1. \tag{2.19}$$

Let us consider the difference

$$X(t) = \mathcal{Z}_n(t) - \hat{g}(t), \qquad -1 \le t \le 1. \tag{2.20}$$

On one hand, by the relations (2.6) and (2.18),

$$\frac{d^r}{dt^r} X(t) = \sum_{\substack{i=1 \\ i \ne j}}^{n+2-r} \alpha_i \phi_i(t), \qquad -1 \le t \le t, \tag{2.21}$$

where $\alpha_i = 1 - \hat{c}_i$, $\quad i = 1, \dots, j-1, j+1, \dots, n+2-r$.
  Thus, for all $t \in [\xi_{i-1}, \xi_i]$ and $i = 1, \dots, j-1, j+1, \dots, n+2-r$,

$$\operatorname{sign} X^{(r)}(t) = \operatorname{sign} \alpha_i \cdot \operatorname{sign} \Phi_i(t) = (-1)^{i+r+1} \operatorname{sign} \alpha_i, \tag{2.22}$$

and

$$X^{(r)}(t) \equiv 0, \qquad t \in [\xi_{j-1}, \xi_j]. \tag{2.23}$$

Thus, from (2.22), (2.23) it follows that the function $X^{(r)}(t)$ changes its sign *at most $n - r$ times* on the interval $[-1, 1]$.

On the other hand, by the relations (2.1), (2.19),

$$\operatorname{sign} X(\nu_i) = \operatorname{sign}\left((-1)^i\|\mathcal{Z}_n\|_{C[-1,1]} - \hat{g}(\nu_i)\right) = (-1)^i, \qquad i = 0, \ldots, n+1. \tag{2.24}$$

Thus, X has at least $n + 1$ distinct zeroes $\{\gamma_i \in (\nu_i, \nu_{i+1})\}_{i=0}^n$, such that

$$\gamma_0 < \gamma_1 < \cdots < \gamma_n. \tag{2.25}$$

Then, by the Rolle theorem, the rth derivative $X^{(r)}$ has to change its sign *at least $n + 1 - r$ times* on the interval $[-1, 1]$. This contradiction with the previous observation on the number of sign changes of the function $X^{(r)}$ proves the result.
□

COROLLARY 16.2.3. *Let the sets M_{n+2} and $\mathcal{R}_{n+2}$ be defined in (2.5), (2.8), respectively. Then, the set M_{n+2} interpolates $\mathbb{R}^{n+2}$, i.e.,*

$$\mathcal{R}_{n+2} = \mathbb{R}^{n+2}.$$

Proof. Definitions (2.5), (2.7), (2.8), the space $\mathcal{R}_{n+2}$ is a linear span of the vectors $\eta(\Phi_1(t)), \ldots, \eta(\Phi_{n+2-r}(t)), \eta(1), \ldots, \eta(t^{r-1})$. The corollary will be proven, once we show that these vectors are linearly independent.

Suppose that there exists a collection of coefficients $\{\hat{\lambda}_i\}_{i=1}^{n+2-r} \times \{\mu_i\}_{i=0}^{r-1}$, such that

$$\sum_{i=1}^{n+2-r} \hat{\lambda}_i \eta(\Phi_i(t)) + \sum_{i=0}^{r-1} \hat{\mu}_i \eta(t^i) = 0. \tag{2.26}$$

Put

$$Y(t) = \sum_{i=1}^{n+2-r} \hat{\lambda}_i \Phi_i(t) + \sum_{i=0}^{r-1} \hat{\mu}_i t^i, \qquad -1 \leq t \leq 1. \tag{2.27}$$

Clearly, $Y \in M_{n+2}$. In addition, by (2.26),

$$\eta(LY(t)) = (0, \ldots, 0), \qquad \text{for all } L \in \mathbb{R}. \tag{2.28}$$

Thus, $\eta(LY(t)) \in [-A_{n+1}, A_{n+1}]^{n+2}$, $\forall L \in \mathbb{R}$, and from Proposition 16.2.2 it follows that for all $L \in \mathbb{R}$,

$$|L\lambda_i| \leq 1, \qquad i = 1, \ldots, n + 2 - r. \tag{2.29}$$

The relations (2.29) are possible for all $L \in \mathbb{R}$, if and only if

$$\hat{\lambda}_i = 0, \qquad i = 1, \ldots, n + 2 - r. \tag{2.30}$$

Thus,

$$Y(t) = \sum_{i=0}^{r-1} \hat{\mu}_i t^i. \tag{2.31}$$

Notice that now the equation $\eta(Y) = 0$ is equivalent to the system of the linear equations

$$\sum_{i=0}^{r-1} \hat{\mu}_i \nu_j^i = 0, \qquad j = 0, \dots, n+1. \tag{2.32}$$

The determinant of the system of the first r equations in (2.32) is the Vandermonde determinant $\Pi = \prod_{1 \le l < m \le r-1} (\nu_l - \nu_m) \ne 0$. Thus, the system (2.32) has only the *trivial* solution: $\hat{\mu}_i = 0$, $i = 0, \dots, r-1$.

Thus, we have shown that the equation (2.26) has only the trivial solution. Therefore, the vectors $\{\eta(\Phi_i(t))\}_{i=1}^{n+2-r}$, $\{\eta(t^i)\}_{i=0}^{r-1}$ are linearly independent, and their span coincides with $\mathbb{R}^{n+2}$. $\qquad\qquad\Box$

16.3. Lower bounds for d_{n+1} $(W^r H^\omega[n]$, $\mathbb{C}[-1,1])$

We give two proofs of Theorem 16.1.2 – Proof I does not use Theorem 1.1.6, while Proof II employs Theorem 1.1.6 on the widths of the ball.

Proof I. Fix $n \ge r$ and an $(n+1)$-dimensional linear subspace F_{n+1} in $\mathbb{C}[-1,1]$. By Corollary 16.2.3, the set $\mathcal{R}_{n+2}$ coincides with the $(n+2)$-dimensional Euclidean space. Therefore, there exists a vector $\eta(g_*(t)) \in \mathcal{R}_{n+2}$, with $g_* \in M_{n+2}$, such that

$$\|\eta(g_*(t))\|_{l_\infty^{n+2}} = A_{n+1}, \tag{3.1}$$

and

$$\inf_{h(t) \in F_{n+1}} \|\eta(g_*(t)) - \eta(h(t))\|_{l_\infty^{n+2}} = A_{n+1}. \tag{3.2}$$

By the equation (3.1),

$$\eta(g_*) \in [-A_{n+1}, A_{n+1}]^{n+2}, \tag{3.3}$$

and, by Proposition 16.2.2 and Proposition 16.2.1, $g_* \in W^r H^\omega[n]$.

The relation (3.2) implies that

$$E\left(W^r H^\omega[n], F_{n+1}\right)_{\mathbb{C}[-1,1]} \ge \inf_{h \in F_{n+1}} \|g_* - h\|_{\mathbb{C}[-1,1]} \ge$$

$$\ge \inf_{h \in F_{n+1}} \max_{1 \le l \le n+2} |g_*(\nu_l) - h(\nu_l)| = \int_{h \in F_{n+1}} \|\eta(g_*) - \eta(h)\|_{l_\infty^{n+2}} = A_{n+1}. \tag{3.4}$$

Since the estimates (3.4) hold for an arbitrary $(n+1)$-dimensional linear subspace F_{n+1}, the inequalities (3.4) imply that

$$d_{n+1}\left(W^r H^\omega[n], \mathbb{C}[-1,1]\right) := \inf_{F_{n+1}} E\left(W^r H^\omega[n], F_{n+1}\right)_{\mathbb{C}[-1,1]} \ge A_{n+1}. \tag{3.5}$$

Proof **II**. Let

$$\mathbb{B}_r^{n+2} := \{g \in M_{n+2} \mid \|g\|_{\mathbb{C}[-1,1]} \le r\}, \qquad \text{for } r > 0. \tag{3.6}$$

By Propositions 16.2.1 and 16.2.2, for any function $g \in M_{n+2}$, the inequality $\|g\|_{\mathbb{C}[-1,1]} \le A_{n+1}$ implies that $f \in W^r H^\omega[n]$, i.e., all functions from the ball $\mathbb{B}_{A_{n+1}}^{n+2}$ of radius A_{n+1} belong to the class $W^r H^\omega[n]$. In view of Theorem 1.1.6,

$$d_{n+1}\left(W^r H^\omega[n], \mathbb{C}[-1,1]\right) \ge A_{n+1}. \tag{3.7}$$

$\square$

Appendix A

Kolmogorov Problem for Functions $f \in W^r H^\omega(\mathbb{R}_+) \ : \ \|f\|_{\mathbb{L}_p(\mathbb{R}_+)} < +\infty$

All results of this chapter on the structure of extremal functions of the problem

$$\|f^{(m)}\|_{\mathbb{L}_\infty(\mathbb{R}_+)} \to \sup, \qquad f \in W^r H^\omega(\mathbb{R}_+), \quad \|f\|_{\mathbb{L}_p(\mathbb{R}_+)} \le B, \qquad (\mathbb{K})$$

are borrowed from [12].

In the first two sections of the chapter we establish the differentiation formula for functions from $f \in W^r H^\omega[0,d]$ with a finite $\mathbb{L}_p$-norm. Then, we list sufficient conditions of extremality of a function $f \in W^r H^\omega[0,d]$ in the corresponding Kolmogorov–Landau problem

$$f^{(m)}(0) \to \sup, \qquad f \in W^r H^\omega[0,d], \quad \|f\|_{\mathbb{L}_p[0,d]} \le B. \qquad (\mathbb{K}-\mathbb{L})$$

Relying on these conditions, expressed in the form of *operator equations,* we formulate Theorem A.4.2 describing extremal functions and sharp additive inequalities of the problem $(\mathbb{K})$. In conclusion, we give the corresponding exact *multiplicative* inequalities in Hölder classes $W^r H^\alpha(\mathbb{R}_+)$.

A.1. Differentiation formulae for $f^{(m)}(0)$, $0 \le m < r$

Fix an interval $[0,d]$ and integers r, $m \in \mathbb{Z}_+ : 0 \le m \le r$.

Let $1 \le p < \infty$, and

$$p' = p'(p) = \begin{cases} \dfrac{p}{p-1}, & 1 < p < \infty, \\ \infty, & p = 1. \end{cases}$$

Fix a positive constant $B > 0$, a concave modulus of continuity ω, and a function $f \in W^r H^\omega[0,d]$ be such that $\|f\|_{\mathbb{L}_p[0,d]} \le B$.

Let a function $Z \in \mathbb{C}^r[0,d]$ be endowed with the properties

(i) $Z^{(i)}(d) = 0, \qquad i = 0, \ldots, r-1;$

(ii) $Z^{(i)}(0) = 0, \qquad i = 0, \ldots, r-2-m, r-m, \ldots, r-1;$

(iii) $(-1)^{m+1} Z^{(r-1-m)}(0) > 0;$ \hfill (1.1)

(iv) $\displaystyle\int_0^d Z(t)\, dt = 0.$

Integrating by parts and using properties (1.1) of the function Z, we arrive at the formula for $f^{(m)}(0)$:

$$f^{(m)}(0) = \frac{(-1)^{m+1}}{Z^{(r-1-m)}(0)} \left[\int_0^d f(t) Z^{(r)}(t)\, dt + (-1)^{r+1} \int_0^d f^{(r)}(t) Z(t)\, dt \right]. \quad (1.2)$$

Therefore, employing the Hölder inequality, we obtain the following estimate for the value of the mth derivative of the function f at the origin:

$$f^{(m)}(0) \le \frac{1}{|Z^{(r-1-m)}(0)|} \left[\|Z^{(r)}\|_{L_{p'}[0,d]} B + \sup_{h \in H^\omega[0,d]} \int_0^d h(t) Z(t)\, dt \right] \quad (1.3)$$

REMARK 3.1.1. The property (1.1), (iv) of the function Z assures that

$$\sup_{h \in H^\omega[0,d]} \int_0^d h(t) Z(t)\, dt = \sup_{h \in H^\omega[0,d]} \int_0^d [h(t) - h(0)]\, Z(t)\, dt = \sup_{h \in H_0^\omega[0,d]} \int_0^d h(t) Z(t)\, dt.$$
$$\quad (1.4)$$

A.2. Differentiation formula for $f^{(r)}(0)$

Let a function $Z \in \mathbb{C}^r[0,d]$ enjoy the properties

$$(i)\ \ Z^{(i)}(0) = Z^{(i)}(d) = 0, \qquad i = 0, \ldots, r-1;$$

$$(ii)\ \int_0^d Z(t)\, dt > 0. \qquad\qquad (2.1)$$

Integrating by parts and using properties (2.1) of the function Z, we derive the following formula for $f^{(r)}(0)$:

$$f^{(r)}(0) = \left(\int_0^d Z(t)\, dt \right)^{-1} \left[(-1)^r \int_0^d f(t) Z^{(r)}(t)\, dt - \int_0^d [f^{(r)}(t) - f^{(r)}(0)] Z(t)\, dt \right]$$
$$\quad (2.2)$$

Therefore, employing the Hölder inequality, we obtain the estimate

$$|f^{(r)}(0)| \le \left| \int_0^d Z(t)\, dt \right| \left[\|Z^{(r)}\|_{L_{p'}[0,d]} B + \sup_{h \in H_0^\omega[0,d]} \int_0^d h(t) Z(t)\, dt \right] \quad (2.3)$$

A.3.   Sufficient conditions of extremality in the problem $(\mathbb{K} - \mathbb{L})$

A.3.1. Corollaries of differentiation formulas

From the identities (1.2) for $m \in \{0, \ldots, r - 1\}$ and (2.2) for $m = r$ we derive the following *sufficient conditions of extremality* of functions $X \in W^r H^\omega[0, d]$ and $Z \in \mathbb{C}^r[0, d]$ in the inequalities (1.3) for $0 \leq m < r$ and (2.3) for $m = r$ (i.e. conditions for the transformation of these inequalities into the equalities), respectively:

$$(\alpha) \quad Z^{(r)}(t) = |X(t)|^{p-1} \operatorname{sign} X(t), \qquad t \in [0, d];$$

$$(\beta) \quad \int_0^d [X^{(r)}(t) - X^{(r)}(0)] Z(t)\, dt = \sup_{h \in H_0^\omega[0,d]} \int_0^d h(t) Z(t)\, dt; \qquad (3.1)$$

$$(\gamma) \quad \|X\|_{\mathbb{L}_p[0,d]} = B.$$

REMARK A.3.1. The equation (3.1), (α) represents the case of the equality sign in the Hölder inequality

$$\int_0^d f(t) Z^{(r)}(t)\, dt \leq \|f\|_{\mathbb{L}_p[0,d]} \|Z^{(r)}\|_{\mathbb{L}_{p'}[0,d]}.$$

A.3.2. Extremality conditions in the form of an operator equation

We introduce *the differentiation operator* $D_r : \mathbb{C}^r[0, d] \to \mathbb{C}[0, d]$,

$$(D_r\, x)\,(t) := x^{(r)}(t) - x^{(r)}(0), \qquad x \in \mathbb{C}^r[0, d], \qquad (3.2)$$

the Hölder operator $\mathcal{H} : \mathbb{C}[0, d] \to \mathbb{C}[0, d]$,

$$(\mathcal{H}\, x)\,(t) := |x(t)|^{p-1} \operatorname{sign} x(t), \qquad x \in \mathbb{C}[0, d], \qquad (3.3)$$

the integration operator $I_r : \mathbb{C}[0, d] \to \mathbb{C}^r[0, d]$,

$$(I_r\, x)\,(t) := \frac{(-1)^r}{r!} \int_0^d (y - t)_+^{r-1} x(y)\, dy, \qquad x \in \mathbb{C}[0, d], \qquad (3.4)$$

and the operator M from the set $\mathbb{M}_n[0, d]$, $n \in \mathbb{N} \cup \{+\infty\}$ (see (2.2.2) of Definition 2.2.1) of integrable functions on $[0, d]$ with a finite or countable ordered set of sign changes to the functional class $H_0^\omega[0, d]$:

$$(M\, x)\,(t) := z(t), \qquad x \in \mathbb{M}_n[0, d], \qquad (3.5)$$

where the function $z \in H_0^\omega[0, d]$ is uniquely defined by the property

$$\int_0^d z(t)x(t)\, dt = \sup_{h \in H_0^\omega[0,d]} \int_0^d h(t)x(t)\, dt. \qquad (3.6)$$

The structure of the extremal function $z(t)$ in (3.6) is described in Theorem X. We also introduce the mapping $\Gamma : \mathbb{C}^r[0, d] \to \mathbb{R}^{r-1}$:

$$\Gamma x = \left(x'(0), x''(0), \dots, x^{(r-m-2)}(0), x^{(r-m)}(0), \dots, x^{(r)}(0) \right), \qquad x \in \mathbb{C}^r[0, d]. \qquad (3.7)$$

Then, the conditions (3.1) of the extremality of a function $X(\cdot)$ in the Kolmogorov-Landau problem $(\mathbb{K} - \mathbb{L})$ can be reformulated as follows:
the function $X \in W^r H^\omega[0, d]$ is extremal in the problem $(\mathbb{K} - \mathbb{L})$, if $\|X\|_{L_p[0,d]} = B$, and $X(\cdot)$ satisfies the operator equation

$$[D_r - M \circ I_r \circ \mathcal{H}] X = 0 \qquad (3.8)$$

with the zero boundary conditions

$$\Gamma \left((I_r \circ \mathcal{H}) X \right) = 0. \qquad (3.9)$$

A.4.   Sharp inequalities in problems (K) and (K − L)

A.4.1. Kolmogorov–Landau inequalities in $W^r H^\omega(\mathbb{R}_+)$

The following result describes functions $X_n \in W^r H^\omega[0, d_n]$ with n simple zeroes on $[0, d_n]$, extremal in the problem $(\mathbb{K} - \mathbb{L})$ for $d = d_n$.

THEOREM A.4.1. *Let $r, m \in \mathbb{N} : 0 \le m \le r; n \in \mathbb{N} : n \ge r, B > 0, 1 \le p < +\infty$, and ω be a modulus of continuity. There exists a positive number $d_n = d_n(\omega, B, r, m, p)$ and a function $X_n = X_{n,\omega,B,r,m,p} \in W^r H^\omega[0, d_n]$ endowed with the following properties:*
  (1) *X_n has precisely n simple zeroes $\{t_i = t_i(\omega, B, r, m, n)\}_{i=1}^n$ on $[0, d_n]$;*
  (2) *$X_n^{(r)}$ exhibits exactly $n - r$ simple zeroes on $[0, d_n]$, if $0 \le m < r$, and $n - r + 1$ simple zeroes on $[0, d_n]$, if $m = r$;*
  (3) *$\|X_n\|_{L_p[0, d_n]} = B$;*
  (4) *X_n is a solution of the equation (3.3.8) satisfying the boundary conditions (3.3.9).*

Theorem A.4.1 plays the same role in the theory of inequalities for intermediate derivatives of functions from $W^r H^\omega(\mathbb{R}_+)$ with the finite L_p-norm as Theorem 6.0.1 in the Kolmogorov–Landau inequalities for functions from $W^r H^\omega(\mathbb{R}_+)$ with $\|f\|_{L_\infty(I)} \le \infty$.

In [12] we apply the limiting procedure to the sequence $\{X_n\}_{n \in \mathbb{N}}$ to construct the extremal function in the problem $(\mathbb{K})$. The major results are listed in the following subsection.

A.4.2. Solution of the problem (K)

Theorem A.4.2 characterizes the extremal function $X = X_{\omega,r,m,B,p}$ of the problem (K).

THEOREM A.4.2. *There exists a function $Z(t) = Z_{\omega,r,m,B,p}(t)$, $t \in \mathbb{R}_+$, satisfying conditions (1.1) in the case $0 \le m < r$ and (2.1) in the case of $m = r$ for $d = +\infty$, and the function $X = X_{\omega,r,m,B,p} \in W^r H^\omega(\mathbb{R}_+)$ endowed with the properties (3.1) for $d = +\infty$.*

REMARK A.4.1. If $d = +\infty$ the equalities $Z^{(i)}(d) = 0$, $i = 0,\dots,r-1$, are understood in the sense that $\lim\limits_{t \to +\infty} Z^{(i)}(t) = 0$, $i = 0,\dots,r-1$.

In [12] we also established analogs of the differention formulae

$$f^{(m)}(0) = \frac{(-1)^{m+1}}{Z^{(r-1-m)}(0)}\left[\int\limits_{\mathbb{R}_+} f(t)Z^{(r)}(t)\,dt + (-1)^{r+1}\int\limits_{\mathbb{R}_+} f^{(r)}(t)Z(t)\,dt\right]. \quad (4.1)$$

for $0 \le m < r$ and

$$f^{(r)}(0) = \left(\int\limits_{\mathbb{R}_+} Z(t)dt\right)^{-1}\left[\int\limits_{\mathbb{R}_+} f(t)Z^{(r)}(t)dt + (-1)^r\int\limits_{\mathbb{R}_+}[f^{(r)}(t) - f^{(r)}(0)]Z(t)dt\right] \quad (4.2)$$

Therefore, we have the following estimates for the norm $\|f^{(m)}\|_{L_\infty(\mathbb{R}_+)}$ of the intermediate derivative:

$$f^{(m)}(0) \le \frac{1}{|Z^{(r-1-m)}(0)|}\left[\|Z^{(r)}\|_{L_{p'}(\mathbb{R}_+)}B + \sup_{h \in H^\omega(\mathbb{R}_+)}\int\limits_{\mathbb{R}_+} h(t)Z(t)\,dt\right], \quad (4.3)$$

if $0 \le m < r$ and

$$\|f^{(r)}\|_{L_\infty(\mathbb{R}_+)} \le \left|\int\limits_{\mathbb{R}_+} Z(t)\,dt\right|\left[\|Z^{(r)}\|_{L_{p'}(\mathbb{R}_+)}B + \sup_{h \in H_0^\omega(\mathbb{R}_+)}\int\limits_{\mathbb{R}_+} h(t)Z(t)\,dt\right] \quad (4.4)$$

Properties of functions $X = X_{\omega,r,m,B,p}$ and $Z = Z_{\omega,r,m,B,p}$ guarantee that inequalities (4.3) and (4.4) are sharp.

A.4.3. Problem (K) in the Hölder classes

Fix $r \in \mathbb{N}$, $m \in \mathbb{Z}_+ : 0 \le m < r$, and $\alpha \in (0,1]$, $p \in [1,+\infty)$.

Let $\widehat{X} := X_{\omega_\alpha, r, m, 1, p}$ be the extremal function of the problem $(\mathbb{K})$ for $\omega(t) = \omega_\alpha(t) = t^\alpha$ and $B = 1$. Let also $\widehat{Z}$ be the corresponding function $Z_{\omega_\alpha, r, m, 1}$. We introduce the constants $P = P_{\alpha, r, m, p}$, $Q = Q_{\alpha, r, m, p}$ and $S = S_{\alpha, r, m, p}$:

$$
P = \begin{cases} \|Z^{(r)}\|_{L_{p'}(\mathbb{R}_+)} \cdot |\widehat{Z}^{r-1-m}(0)|^{-1}; \\[2ex] \|Z^{(r)}\|_{L_{p'}(\mathbb{R}_+)} \cdot \left| \displaystyle\int_{\mathbb{R}_+} \widehat{Z}(t)\, dt \right|; \end{cases} \qquad Q = \int_{\mathbb{R}_+} \Re_{\omega_\alpha}(\widehat{Z}; t) \omega_\alpha'(t)\, dt; \qquad (4.5)
$$

$$
S := (r + \alpha + p^{-1}) \left(\frac{P}{r + \alpha - m} \right)^{\frac{r+\alpha-m}{r+\alpha+p^{-1}}} \left(\frac{Q}{m + p^{-1}} \right)^{\frac{m+p^{-1}}{r+\alpha+p^{-1}}}. \qquad (4.6)
$$

The sharp *multiplicative* inequalities are characterized in the following corollary of Theorem A.4.2.

COROLLARY A.4.3. *If* $f \in W^r H^\alpha(\mathbb{R}_+)$, *then*

$$
\|f^{(m)}\|_{L_\infty(\mathbb{R}_+)} \le S \|f\|_{L_p(\mathbb{R}_+)}^{\frac{r+\alpha-m}{r+\alpha+p^{-1}}},
$$

where the constant $S = S_{\alpha, r, m, p}$ *is introduced in* (4.5), (4.6).

Appendix B

Kolmogorov Problems in $W^1H^\omega(\mathbb{R}_+)$ and $W^1H^\omega(\mathbb{R})$

In this chapter we offer a detailed description of the structure of extremal functions of the Kolmogorov problem in $W^1H^\omega(\mathbb{R})$ and $W^1H^\omega(\mathbb{R}_+)$:

$$\|f^{(m)}\|_{\mathbb{L}_\infty(\mathbb{I})} \to \sup, \qquad f \in W^1H^\omega(\mathbb{I}), \quad \|f\|_{\mathbb{L}_p(\mathbb{I})} \le B, \tag{1.1}$$

where $m = 0, 1, \ 1 \le p < +\infty$, and $\mathbb{I} = \mathbb{R} \vee \mathbb{R}_+$.

B.1. Preliminary remarks

Let a function $\mathcal{Z}(t) = \mathcal{Z}_{\omega,m,B,p,\mathbb{I}}$ be endowed with the properties

$$\begin{aligned} &(i) \quad \mathcal{Z}^{(m)}(0) = 0, \qquad (-1)^{m+1}\mathcal{Z}^{1-m}(0) > 0; \\ &(ii) \quad \lim_{t\to\infty} \mathcal{Z}(t) = \lim_{t\to\infty} \mathcal{Z}'(t) = 0. \end{aligned} \tag{1.2}$$

In [12] we established the existence of functions $X \in W^1H^\omega(\mathbb{I})$ and $\mathcal{Z}$ satisfying conditions (1.2) and exhibiting the following properties:

$$\begin{aligned} &(\alpha) \quad \mathcal{Z}''(t) = |X(t)|^{p-1}\operatorname{sign} X(t), \qquad t \in \mathbb{I}; \\ &(\beta) \quad \int_{\mathbb{I}} [X'(t) - X'(0)]\mathcal{Z}'(t)\,dt = \sup_{h \in H_0^\omega(\mathbb{I})} \int_{\mathbb{I}} h(t)\mathcal{Z}'(t)\,dt; \\ &(\gamma) \quad \|X\|_{\mathbb{L}_p(\mathbb{I})} = B. \end{aligned} \tag{1.3}$$

REMARK $B.1.1$. In the case $\mathbb{I} = \mathbb{R}_+$ the function $\mathcal{Z}(x)$ is the indefinite integral $\int_{+\infty}^{x} Z(t)\,dt$ of the function $Z(t)$ from Theorem A.4.2 for $r = 1$.

In the following four sections of the chapter we examine specific features of the functions $\mathcal{Z}$ and X in each of the possible cases $m \in \{0, 1\}$ and $\mathbb{I} = \{\mathbb{R}, \mathbb{R}_+\}$.

B.2. Maximization of the norm $\|f\|_{\mathbb{L}_\infty(\mathbb{R}_+)}$

First, we describe the structure of extremal functions of the problem

$$\|f\|_{\mathbb{L}_\infty(\mathbb{R}_+)} \to \sup, \qquad f \in W^1H^\omega(\mathbb{R}_+), \quad \|f\|_{\mathbb{L}_p(\mathbb{R}_+)} \le B. \tag{2.1}$$

B.2.1. Differentiation formulae and inequalities

In addition to enjoying the properties (1.2), (1.3), the function $\mathcal{Z}(t)$ satisfies the following conditions:

(i) $\mathcal{Z}$ has a countable number of zeroes $\{\xi_i\}_{i=0}^{\infty} : 0 = \xi_0 < \xi_1 < \xi_2 < \dots$;

(ii) $\mathcal{Z}'$ has precisely one zero η_i on $[\xi_i, \xi_{i+1}]$ for each $i \in \mathbb{Z}_+$. $\qquad$ (2.2)

By (2.2), each of the restrictions $\mathcal{Z}\big|_{[\xi_i, \xi_{i+1}]}$ is *a simple kernel* on $[\xi_i, \xi_{i+1}]$ for $i \in \mathbb{Z}_+$.

The following differentiation formula holds for any function $f \in W^1 H^\omega(\mathbb{R}_+)$:
$\|f\|_{\mathbb{L}_p(\mathbb{R}_+)} \le B$:

$$f(0) = -\mathcal{Z}'(0)^{-1} \left(\int_0^\infty f(x) \mathcal{Z}''(x)\, dx + \int_0^\infty f'(x) \mathcal{Z}'(x)\, dx \right). \qquad (2.3)$$

Consequently,

$$|f(0)| \le |\mathcal{Z}'(0)|^{-1} \left(\|\mathcal{Z}''\|_{\mathbb{L}_{p'}(\mathbb{R}_+)} B + \int_{\mathbb{R}_+} \Re_\omega(\mathcal{Z}; t) \omega'(t)\, dt \right). \qquad (2.4)$$

B.2.2. Rearrangements $\Re_\omega(\mathcal{Z}; t)$ and shifts $X_j(\cdot) := X(\cdot + \xi_j)$

We showed in [12] that the lengths of the intervals $\{[\xi_{i-1}, \xi_i]\}_{i \in \mathbb{N}}$ decrease:

$$\xi_i - \xi_{i-1} > \xi_{i+1} - \xi_i, \qquad i \in \mathbb{N}. \qquad (2.5)$$

Therefore, as we explained in Subsection 2.2.6 (consult (2.2.30), $\bigcup_{i=1}^{\infty} [\xi_{i-1}, \xi_i]$ is a trivial partition of the half-line $\mathbb{R}_+$ in the sense of Definition 2.2.6. Thus, for each of the natural i's the restriction $X'\big|_{[\xi_{i-1}, \xi_i]}$ is extremal in the problem

$$\int_{\xi_{i-1}}^{\xi_i} h(t) \mathcal{Z}'(t)\, dt \to \sup, \qquad h \in H^\omega[\xi_{i-1}, \xi_i], \qquad (2.6)$$

and the extremal rearrangement $\Re_\omega(\mathcal{Z}; t)$ is the sum of rearrangements of simple kernels $\mathcal{Z}\big|_{[\xi_{i-1}, \xi_i]}$:

$$\Re_\omega(\mathcal{Z}; t) := \sum_{i=1}^{\infty} \Re\left(\mathcal{Z}\big|_{[\xi_{i-1}, \xi_i]}; t \right). \qquad (2.7)$$

The Korneichuk's Lemma 2.1.1 gives the formulas for the derivative of the extremal function X:

$$X'(t) = \left\{ \begin{array}{ll} (-1)^{j+1} \omega'(\rho_i(t) - t), & \xi_j \le t < \eta_j \\ (-1)^{j+1} \omega'(t - \rho_i^{-1}(t)), & \eta_j < t \le \xi_{j+1} \end{array} \right\} \quad i \in \mathbb{Z}_+, \qquad (2.8)$$

where $\rho_j : [\xi_j, \eta_j] \to [\eta_j, \xi_{j+1}]$ is defined by the equations

$$\mathcal{Z}(x) = \mathcal{Z}(\rho_j(x)), \qquad x \in [\xi_j, \eta_j], \quad \rho_j \in [\eta_j, \xi_{j+1}]. \tag{2.9}$$

For any $j \in \mathbb{N}$ we introduce functions $X_j(t)$ and $\mathcal{Z}_j(t)$ by the formulae

$$X_j(t) = X(t + \xi_j), \qquad \mathcal{Z}_j(t) = \mathcal{Z}(t + \xi_j), \qquad t \in \mathbb{R}_+. \tag{2.10}$$

For $j \in \mathbb{N}$, put also

$$B_j := \|X_j\|_{L_p(\mathbb{R}_+)}; \qquad \eta_i(j) = \eta_{i+j}, \quad \xi_i(j) = \xi_{i+j}, \quad i \in \mathbb{Z}_+, \tag{2.11}$$

where $\{\eta_i\}_{i\in\mathbb{N}}$ and $\{\xi_i\}_{i\in\mathbb{N}}$ are the points of sign change of the functions $\mathcal{Z}'$ and $\mathcal{Z}$, respectively.

The refterence to the properties (1.2), (1.3) and (2.2), (2.3) of X and $\mathcal{Z}$ reveals that functions X_j and $\mathcal{Z}_j$ enjoy all properties of the functions X and $\mathcal{Z}$ *with respect to the collections of points* $\{\eta_i(j)\}_{i\in\mathbb{N}}$ *and* $\{\xi_i(j)\}_{i\in\mathbb{N}}$. Therefore, for each $j \in \mathbb{N}$, the function X_j solves the problem

$$\|f\|_{L_\infty(\mathbb{R}_+)} \to \sup, \qquad f \in W^r H^\omega(\mathbb{R}_+) : \|f\|_{L_p(\mathbb{R}_+)} \leq B_j. \tag{2.11}$$

B.2.3. Special properties of extremal functions in Hölder classes $W^1 H^\alpha(\mathbb{R}_+)$

Let us consider the variant of the problem (2.1) in the particular case of the Hölder modulus of continuity $\omega_\alpha(t) = t^\alpha$ for $\alpha \in (0, 1]$:

$$\|f\|_{L_\infty(\mathbb{R}_+)} \to \sup, \qquad f \in W^1 H^\alpha(\mathbb{R}_+), \quad \|f\|_{L_p(\mathbb{R}_+)} \leq B. \tag{2.12}$$

Let $\mathcal{R}$ be the extremal function in (2.12) for $B = 1$, and $\mathcal{R}_j(\cdot)$ be the shifts $\mathcal{R}(\xi_j + \cdot)$. Then, by Corollary 2.2.4, the solution R_B of the problem (2.12) is a dilated and rescaled function $\mathcal{R}$:

$$R_B(t) = B^{\frac{1+\alpha}{2+\alpha}} \mathcal{R}\left(B^{-\frac{1}{2+\alpha}} t\right), \qquad t \in \mathbb{R}_+. \tag{2.13}$$

Moreover, we showed in [12] that

$$\mathcal{R}(t) = (-1)^j B^{\frac{1+\alpha}{2+\alpha}} \mathcal{R}\left(B_j^{-\frac{1}{2+\alpha}}(t + \xi_j)\right), \qquad t \in \mathbb{R}_+. \tag{2.14}$$

The property (2.14) of the function $\mathcal{R}$ immediately implies that the points $\{\xi_i\}_{i=1}^\infty$ constitute *a geometric mesh*:

$$\frac{\xi_{j+1} - \xi_j}{\xi_j - \xi_{j-1}} = \mathcal{A}(\alpha), \qquad \forall j \in \mathbb{N}, \tag{2.15}$$

i.e. $\xi_j = \xi_1 \cdot \gamma^{j-1}$ for some constant $\gamma = \gamma(\alpha) : \gamma \in (0, 1]$.

Consequently, in the Hölder class $W^1 H^\alpha(\mathbb{R}_+)$ we have the familiar variant of *the self-similarity property*: for all $i \in \mathbb{N}$,

$$\mathcal{R}\big|_{[\xi_i, \xi_{i+1}]}(t + \xi_i) = (-1)^i \left(\frac{\xi_{i+1} - \xi_i}{\xi_1}\right)^{1+\alpha} \mathcal{R}\big|_{[0, \xi_1]}\left(\frac{\xi_1}{\xi_{i+1} - \xi_i} t\right), \quad 0 \leq t \leq \xi_{i+1} - \xi_i. \tag{2.16}$$

In particular, (2.15) implies that $\mathcal{R}$ has a compact support.

B.3. Maximization of the norm $\|f'\|_{\mathbb{L}_\infty(\mathbb{R}_+)}$

In this section we examine the structure of extremal functions of the problem

$$\|f'\|_{\mathbb{L}_\infty(\mathbb{R}_+)} \to \sup, \qquad f \in W^1 H^\omega(\mathbb{R}_+), \quad \|f\|_{\mathbb{L}_p(\mathbb{R}_+)} \le B. \qquad (3.1)$$

In addition to (1.2) and (1.3), the function $\mathcal{Z}(t)$ exhibits the following properties:

(i) $\mathcal{Z}$ has a countable number of zeroes $\{\xi_i\}_{i=1}^\infty : 0 = \xi_0 < \xi_1 < \xi_2 < \dots$;

(ii) $\mathcal{Z}'$ has precisely one zero η_i on $[\xi_i, \xi_{i+1}]$ for each $i \in \mathbb{N}$. $\qquad (3.2)$

By (3.2), each of the restrictions $\mathcal{Z}\big|_{[\xi_i, \xi_{i+1}]}$ is *a simple kernel* on $[\xi_i, \xi_{i+1}]$ for $i \in \mathbb{N}$, while $\mathcal{Z}$ is monotone and positive on $[0, \xi_1)$.

The following differentiation formula holds for any function $f \in W^1 H^\omega(\mathbb{R}_+)$: $\|f\|_{\mathbb{L}_p(\mathbb{R}_+)} \le B$:

$$f'(0) = \mathcal{Z}(0)^{-1} \left(\int_0^\infty f(x) \mathcal{Z}''(x)\, dx + \int_0^\infty [f'(x) - f'(0)] \mathcal{Z}'(x)\, dx \right). \qquad (3.3)$$

Consequently,

$$\|f'\|_{\mathbb{L}_\infty(\mathbb{R}_+)} \le |\mathcal{Z}(0)|^{-1} \left(\|\mathcal{Z}''\|_{\mathbb{L}_{p'}(\mathbb{R}_+)} B + \int_{\mathbb{R}_+} \Re_\omega(\mathcal{Z}; t) \omega'(t)\, dt \right). \qquad (3.4)$$

We also showed that the lengths of the intervals $\{[\xi_{i-1}, \xi_i]\}_{i \in \mathbb{N}}$ decrease:

$$\xi_i - \xi_{i-1} > \xi_{i+1} - \xi_i, \qquad i \in \mathbb{N}. \qquad (3.5)$$

Therefore, for $i \in \mathbb{N}$ the restriction $X'\big|_{[\xi_i, \xi_{i+1}]}$ is extremal in the problem

$$\int_{\xi_i}^{\xi_{i+1}} h(t) \mathcal{Z}'(t)\, dt \to \sup, \qquad h \in H^\omega[\xi_i, \xi_{i+1}], \qquad (3.6)$$

while the restriction $X'\big|_{[0,\xi_1]}$ is given by the formula

$$X'(t) = X'(0) - \omega(t), \qquad t \in [0, \xi_1]. \qquad (3.7)$$

In accordance with (3.5), the extremal rearrangement $\Re_\omega(\mathcal{Z}; t)$ of the kernel $\mathcal{Z}$ has the support $[0, \xi_1]$ and is given by the formula

$$\Re_\omega(\mathcal{Z}; t) := \mathcal{Z}(t) \mathcal{X}\big|_{[0,\xi_1]} + \sum_{i=1}^\infty \Re\left(\mathcal{Z}\big|_{[\xi_{i-1}, \xi_i]}; t \right), \qquad (3.8)$$

where $\mathcal{X}\big|_{[0,\xi_1]}$ is the indicator of the interval $[0, \xi_1]$.

Let us introduce the functions $\hat{X}(t)$ and $\hat{Z}(t)$ on the half-line $\mathbb{R}_+$ by the formulae

$$\hat{X}(t) = X(t+\xi_1), \qquad \hat{Z}(t) = \mathcal{Z}(t+\xi_1), \qquad t \in \mathbb{R}_+. \tag{3.9}$$

Put also

$$\hat{B} := \|\hat{X}\|_{L_p(\mathbb{R}_+)}; \qquad \hat{\xi}_i = \xi_{i+1}, \quad \hat{\eta}_i = \eta_{i+1}, \qquad i \in \mathbb{Z}_+, \tag{3.10}$$

where $\{\xi_i\}_{i\in\mathbb{N}}$ and $\{\eta_i\}_{i\in\mathbb{N}}$ are the points of sign change of $\mathcal{Z}$ and $\mathcal{Z}'$, respectively.

The reference to the properties (1.2), (1.3) and (3.2), (3.3) of the extremal functions X and $\mathcal{Z}$ demonstrates that the functions $\hat{X}$ and $\hat{Z}$ enjoy all properties of extremal functions X and $\mathcal{Z}$ from §B.2 with respect to the collections $\{\hat{\xi}_i\}_{i\in\mathbb{N}}$ and $\{\hat{\eta}_i\}_{i\in\mathbb{N}}$. Therefore, the function $\hat{X}$ is extremal in the problem

$$\|f'\|_{L_\infty(\mathbb{R}_+)} \to \sup, \qquad f \in W^r H^\omega(\mathbb{R}_+), \quad \|f\|_{L_p(\mathbb{R}_+)} \le \hat{B}. \tag{3.11}$$

In particular, in the case of the Hölder class $W^1 H^\alpha(\mathbb{R}_+)$ the lengths $\triangle_i := \xi_{i+1} - \xi_i$ of the interval $[\xi_i, \xi_{i+1}]$ constitute a geometric mesh:

$$\frac{\triangle_{i+1}}{\triangle_i} = \mathcal{A}(\alpha), \qquad \forall i \in \mathbb{N}, \tag{3.12}$$

with *the same constant* as in (2.15). In addition, the function $X\big|_{[\xi_1,+\infty)}$ is self-similar: for $i \in \mathbb{N}$,

$$X\big|_{[\xi_i,\xi_{i+1}]}(t+\xi_i) = (-1)^i \left(\frac{\xi_{i+1}-\xi_i}{\xi_2-\xi_1}\right)^{1+\alpha} X\big|_{[\xi_1,\xi_2]}\left(\frac{\xi_2-\xi_1}{\xi_{i+1}-\xi_i}t\right), \; 0 \le t \le \xi_{i+1}-\xi_i. \tag{3.13}$$

B.4.   Maximization of the norm $\|f\|_{L_\infty(\mathbb{R})}$

The following problem is under our consideration:

$$\|f\|_{L_\infty(\mathbb{R})} \to \sup, \qquad f \in W^1 H^\omega(\mathbb{R}), \quad \|f\|_{L_p(\mathbb{R})} \le B. \tag{4.1}$$

In this case the function $\mathcal{Z}(t)$ is *even*, and its restriction to the positive half-line enjoys the following properties:

(i) $\mathcal{Z}\big|_{\mathbb{R}_+}$ has a countable number of zeroes $\{\xi_i\}_{i=1}^\infty : 0 < \xi_1 < \xi_2 < \cdots$;

(ii) $\mathcal{Z}'$ has precisely one zero η_i on $[\xi_i, \xi_{i+1}]$ for each $i \in \mathbb{N}$. $\qquad\qquad$ (4.2)

Let f be a function from $W^1 H^\omega(\mathbb{R})$ such that $\|f\|_{L_p(\mathbb{R})} \le B$. Integration by parts gives us the following formula for the value of the function f at zero:

$$f(0) = -\frac{1}{2}\mathcal{Z}'(0)^{-1}\left(\int_\mathbb{R} f(x)\mathcal{Z}''(x)\,dx + \int_\mathbb{R} f'(x)\mathcal{Z}'(x)\,dx\right). \tag{4.3}$$

Let ξ_1 be the first zero of $\mathcal{Z}$ on $(0, +\infty)$. Then,

$$\int\limits_{\mathbb{R}} f'(x)\mathcal{Z}'(x)\,dx = \int\limits_{-\xi_1}^{\xi_1} f'(x)\mathcal{Z}'(x)\,dx + \int\limits_{\mathbb{R}\setminus[-\xi_1,\xi_1]} f'(x)\mathcal{Z}'(x)\,dx. \qquad (4.4)$$

We showed that the lengths of the intervals $\{[\xi_i, \xi_{i+1}]\}_{i \in \mathbb{N}}$ monotonely decrease. Thus, taking into account the fact that $\mathcal{Z}'$ is odd and applying the Korneichuk Lemma 2.2.1 on each of the intervals $[\xi_i, \xi_{i+1}]$, we estimate the second summand in (4.4) as follows:

$$\sup_{h \in H^\omega(\mathbb{R})} \int\limits_{\mathbb{R}\setminus[-\xi_1,\xi_1]} h(x)\mathcal{Z}'(x)\,dx \leq \sum_{i=1}^{\infty} \int\limits_{0}^{\xi_{i+1}-\xi_i} \left[2\sum_{i=1}^{\infty} \Re(\mathcal{Z}\big|_{[\xi_{i-1},\xi_i]}; t)\right] \omega'(t)dt.$$
$$(4.5)$$

The finction $\mathcal{Z}'\big|_{[-\xi_1,\xi_1]}$ is odd and has 0 as its only zero. Therefore, $\mathcal{Z}\big|_{[-\xi_1,\xi_1]}$ is a simple kernel, and we can apply Corollary 2.2.2 to obtain the following formula for the function X' on the interval $[-\xi_1, \xi_1]$:

$$\mathcal{Z}'(x) = \left\{ \begin{array}{ll} -\dfrac{1}{2}\omega(2x), & x \in [0, \xi_1]; \\[2ex] \dfrac{1}{2}\omega(-2x), & x \in [-\xi_1, 0]. \end{array} \right\}. \qquad (4.6)$$

Letting

$$\Re_\omega(\mathcal{Z}; t) := \Re\left(\mathcal{Z}\big|_{[\xi_{-1},\xi_1]}; t\right) + 2\sum_{i=1}^{\infty} \Re\left(\mathcal{Z}\big|_{[\xi_i,\xi_{i+1}]}; t\right), \qquad t \in \mathbb{R}_+. \qquad (4.7)$$

we arrive at the estimate

$$\|f\|_{L_\infty(\mathbb{R})} \leq \frac{1}{2}|\mathcal{Z}'(0)|^{-1}\left(\|\mathcal{Z}'\|_{L_{p'}(\mathbb{R})}B + \int\limits_{\mathbb{R}_+} \omega'(t)\Re_\omega(\mathcal{Z}; t)\,dt\right). \qquad (4.8)$$

As in the previous cases, for each $j \in \mathbb{N}$ the shift $\widetilde{X}(t) := X(t + t_j)$, $t \in \mathbb{R}_+$, is extremal in the problem

$$\|f\|_{L_\infty(\mathbb{R}_+)} \to \sup, \qquad f \in W^r H^\omega(\mathbb{R}_+), \quad \|f\|_{L_p(\mathbb{R}_+)} \leq B_j, \qquad (4.9)$$

where $B_j := \|X\|_{L_p[\xi_j, +\infty)}$.

B.5.   Maximization of the norm $\|f'\|_{L_\infty}(\mathbb{R})$

The following problem is under our consideration:

$$\|f'\|_{L_\infty(\mathbb{R})} \to \sup, \qquad f \in W^1 H^\omega(\mathbb{R}), \quad \|f\|_{L_p(\mathbb{R})} \le B. \qquad (5.1)$$

In this case the function $\mathcal{Z}(t)$ is *odd,* and its restriction to the positive half-line enjoys the following properties:

(i) $\mathcal{Z}\big|_{\mathbb{R}_+}$ has a countable number of zeroes $\{\xi_i\}_{i=0}^\infty : 0 = \xi_0 < \xi_1 < \dots;$

(ii) $\mathcal{Z}'\big|_{\mathbb{R}_+}$ has precisely one zero η_i on $[\xi_i, \xi_{i+1}]$ for each $i \in \mathbb{N}$.

$$(5.2)$$

Let f be a function from $W^1 H^\omega(\mathbb{R})$ such that $\|f\|_{L_p(\mathbb{R})} \le B$. The differentiation formula for the value of f' at zero is as follows:

$$f'(0) = \frac{1}{2}\mathcal{Z}(0)^{-1}\left(\int_{\mathbb{R}} f(x)\mathcal{Z}''(x)\,dx + \int_{\mathbb{R}} [f'(x) - f'(0)]\mathcal{Z}'(x)\,dx \right). \qquad (5.3)$$

Put

$$\mathfrak{R}_\omega(\mathcal{Z};t) := 2|\mathcal{Z}(t)|\chi_{[0,\xi_1]}(t) + 2\sum_{i=1}^\infty \mathfrak{R}\left(\mathcal{Z}\big|_{[\xi_i,\xi_{i+1}]};t\right), \qquad t \in \mathbb{R}_+, \qquad (5.4)$$

where, as usual, $\chi_{[a,b]}(\cdot)$ is the indicator function of the interval $[a,b]$. Repeating the technicalities of Sections 2 and 4, we arrive at the estimate

$$\|f'\| \le \frac{1}{2}|\mathcal{Z}(0)|^{-1}\left(\|f\|_{L_p(\mathbb{R})} B + \int_{\mathbb{R}_+} \omega'(t)\mathfrak{R}_\omega(\mathcal{Z};t)\,dt \right). \qquad (5.4)$$

Each of the kernels $\mathcal{Z}\big|_{[\xi_i,\xi_{i+1}]}$ is simple for $i \in \mathbb{N}$, and $\{\triangleright_i := \xi_{i+1}-\xi_i\}_{i\in\mathbb{N}}$ constitute a decreasing sequence, so one can use the Korneichuk's formulas (2.8), (2.9) for the characterization of the derivative X' on $\mathbb{R} \setminus [-\xi_1, \xi_1]$. The reasoning of Section B.3 enables us to conclude that

$$X'(t) = X'(0) - \omega(|t|), \qquad t \in [-\xi_1, \xi_1]. \qquad (5.6)$$

Finally, the shift $\widetilde{X}_j(t) := X(t + t_j)$ is extremal in the problem

$$\|f\|_{L_\infty(\mathbb{R}_+)} \to \sup, \qquad f \in W^r H^\omega(\mathbb{R}_+), \quad \|f\|_{L_p(\mathbb{R}_+)} \le B_j, \qquad (5.7)$$

where $B_j := \|X\|_{L_p[\xi_j, +\infty)}, \; j \in \mathbb{N}$.

Bibliography

[1] N. I. Akhiezer, *Theory of Approximation*, F. Ungar Pub. Co., New York, N. Y., 1956.

[2] N. I. Akhiezer, M. G. Krein, *On the best approximation of differentiable periodic functions by trigonometric sums*, DAN SSSR **15** (1937), 107–112. (Russian)

[3] V. V. Arestov, *On sharp inequalities between the norms of functions and their derivatives*, Acta Sci. Math. **33** (1972), 243–267.

[4] V. V. Arestov, *Approximation of linear operators and related extremal problems*, Trudy Steklov Math. Inst. **138** (1975), 29–42.

[5] S. K. Bagdasarov, *Zolotarev ω-splines* (1996), (submitted).

[6] S. K. Bagdasarov, *Maximization of functionals in $H^\omega[a, b]$*, Matem. Sbornik **189** (1998), no. 2. (Russian)

[7] S. K. Bagdasarov, *Extremal functions of integral functionals in $H^\omega[a, b]$* (1996), Izvestiya RAN, (to appear). (Russian)

[8] S. K. Bagdasarov, *Zolotarev ω-polynomials in $W^r H^\omega[0, 1]$*, J. Approx. Theory **90** (1997), no. 3, 340–378, (to appear).

[9] S. K. Bagdasarov, *The general construction of Chebyshev ω-splines of the given norm* (1996), Algebra and Analysis (St.-Petersburg Math. Journal) (to appear). (Russian)

[10] S. K. Bagdasarov, *Chebyshev ω-splines extremal in the Kolmogorov–Landau problem* (1996), Journal of Approximation Theory (submitted).

[11] S. K. Bagdasarov, *Sharp inequalities between the upper bounds of intermediate derivatives in the functional classes $W^r H^\omega(\mathbb{R})$ and $W^r H^\omega(\mathbb{R}_+)$* (1996), Journal of Approximation Theory (submitted).

[12] S. K. Bagdasarov, *Kolmogorov problem for intermediate derivatives and optimal control* (1996), in preparation.

[13] S. N. Bernstein, *On V. A. Markov theorem*, Trudy Leningrad. Indust. Inst., Razdel Fiz.-Mat. Nauk **5** (1938), 8–13, reprinted in Collected Works, Vol. 2, Izdat. Akad. Nauk SSSR, Moscow, 1954, pp. 281–286.

[14] S. N. Bernstein, *Selected Works*, vol. I, Izdat. Akad. Nauk SSSR, Moscow, 1952.

[15] K. Borsuk, *Drei Sätze über die n-dimensionale euklidische Sphäre*, Fund. Math. **20** (1933), 177–191.

[16] Yu. G. Bosse (G. E. Shilov), *Inequalities between derivatives*, Sbornik rabot Nauch. Stud. Kruzhkov Mosk. Un-ta (1937), 17–27. (Russian)

[17] H. Cartan, *Sur les classes de fonctions définies par des inégalités portant sur leurs dérivées successives*, Act. Sci. Ind. **867** (1940), Hermann, Paris.

[18] A. S. Cavaretta, *An elementary proof of Kolmogorov's theorem*, Amer. Math. Monthly **81** (1974), 480–486.

[19] A. S. Cavaretta, I. J. Schoenberg, *Solution of Landau's problem concerning higher derivatives on the half-line*, MRC T.S.R. **1050** (1970), Madison, Wisconsin; Also in *Proc. of the Intern. Conf. on Constructive Function Theory, Golden Sands (Varna) May 19–25, 1970, Publ. House Bulg. Acad. Sci., Sofia 1972, 297–308.*

[20] A. S. Cavaretta, *A refinement of Kolmogorov's inequality*, Journal of Approximation Theory **27** (1979), 45–60.

[21] P. L. Chebyshev, *Problems for the smallest quantities connected with approximate representation of functions*, Zap. Akad. Nauk (1859), (see [22]).

[22] P. L. Chebyshev, *Selected Papers*, Akad Nauk SSSR, Moscow, 1955.

[23] R. A. DeVore, G. G. Lorentz, *Constructive Approximation*, Springer-Verlag, Berlin Heidelberg, 1993.

[24] R. A. DeVore, H. Kierstead, G. G. Lorentz, *A proof of Borsuk's theorem*, in Functional Analysis (E. Odell, H. Rosenthal, eds.), vol. 1332, Springer, Berlin, pp. 195–202.

[25] J. Favard, *Sur les meilleures procédés d'approximation de certaines classes des fonctions par des polynômes trigonométriques*, Bull. Sci. Math. **61** (1937), 209–224, 243–256.

[26] V. N. Gabushin, *Inequalities for norms of functions and their derivatives in the L_p metrics*, Matem. Zametki **1** (1967), no. 3, 291–298; transl. in English in Math. Notes **1** (1967), no. 3, 194–198.

[27] V. N. Gabushin, *Exact constants in inequalities between norms of the derivatives of a function*, Math. Notes **4** (1968), no. 2, 630–634.

[28] V. N. Gabushin, *Best approximation of a differentiation operator on the half-line*, Matem. Zametki **6** (1969), 573–582; English transl. in Math. Notes **6** (1969), 573–582.

[29] Yu. I. Grigoryan, *N-widths of certain sets in functional spaces*, Math. Notes **13** (1973), no. 5, 637–644.

[30] V. A. Gusev, *Derivative functionals of an algebraic polynomial and V. A. Markov's theorem*, Izv. Akad. Nauk SSSR, Ser. Mat. **25** (1961), 367–384, (see translation in [90]).

[31] J. Hadamard, *Sur le module maximum d'une fonction et de ses dérivées*, Soc. math. France, Comptes rendus, des Séances **41** (1914), 68–72.

[32] G. H. Hardy, J. Littlewood, and G. Pólya, *Inequalities*, Cambridge Univ. Press, New York, 1934.

[33] D. Jackson, *Über die Genauigkeit der Annäherung stetiger Funktionen durch ganze rationale Funktionen gegebenen Grades und trigonometrischen Summen gegebener Ordnung*, Diss., Göttingen (1911).

[34] H. Kallioniemi, *On representation formulas for intermediate derivatives*, Mathematica Scandivica (1976), 315–326.

[35] R. R. Kallman, G.-C. Rota, *The inequality $\|f'\|^2 \leq 4\|f\|\|f''\|$*, Inequalities II, Academic Press, New York, 1970, pp. 187–191.

[36] S. Karlin, *Interpolation properties of generalized perfect splines and the solutions of certain extremal problems*, Trans. AMS **206** (1975), 25–66.

[37] S. Karlin, *Oscillatory perfect splines and related extremal problems*, in "Studies in Spline Functions and Approximation Theory" (S. Karlin, C. A. Micchelli, A. Pinkus, and I. J. Schoenberg, eds.), Academic Press, New York, N. Y., 1976, pp. 371–460.

[38] S. Karlin, *Some one-sided numerical differentiation formulae and applications*, in "Studies in Spline Functions and Approximation Theory" (S. Karlin, C. A. Micchelli,

A. Pinkus, and I. J. Schoenberg, eds.), Academic Press, New York, N. Y., 1976, pp. 485–500.

[39] S. Karlin, *Total Positivity*, Stanford Univ. Press, 1968.

[40] A. N. Kolmogorov, *Über die beste Annäherung von Functionen einer gegebenen Funktionenklasse*, Ann. Math. **37** (1936), 107–110.

[41] A. N. Kolmogorov, *On inequalities between upper bounds of successive derivatives of an arbitrary function defined on an infinite interval*, Uch. zap. MGU, Matematika **3** (1939), 3–16, see translation in Amer. Math. Soc. Transl. Ser. 1, Vol. 2 (1962), 233–243. (Russian)

[42] A. N. Kolmogorov, *Izbrannye trudy. Matematika i mehanika (Selected works. Mathematics and mechanics)*, Nauka, Moscow, 1985. (Russian)

[43] Kong-Ming Shong, *Some extensions of a theorem of Hardy, Littlewood and Polya and their applications*, Can. J. Math. **26** (1974), 1321–1340.

[44] N. P. Korneichuk, *Exact value of best approximations and widths of some classes of functions*, Doklady Akad. Nauk SSSR **150** (1963), 1218–1220.

[45] N. P. Korneichuk, *Upper bounds of best approximations on the classes of differentiable periodic functions in the metrics of C and L*, Doklady Akad. Nauk SSSR **190** (1970), 269–271.

[46] N. P. Korneichuk, *Extremal values of functionals and the best approximation on classes of periodic functions*, Izv. AN SSSR, Ser. Mat. **35** (1971), 423–434.

[47] N. P. Korneichuk, *Inequalities for differentiable periodic functions and the best approximation of one class by another*, Izv. Akad. Nauk, Ser. Mat. **36** (1972), 423–434.

[48] N. P. Korneichuk, *Extremal Problems of Approximation Theory*, Nauka, Moscow, 1976. (Russian)

[49] N. P. Korneichuk, *Splines in Approximation Theory*, Nauka, Moscow, 1984. (Russian)

[50] N. P. Korneichuk, *Exact Constants in Approximation Theory*, Cambridge University Press, Cambridge, New York, 1990, Series: Encyclopedia of Mathematics and Its Applications, vol. 38.

[51] N. P. Korneichuk, *S. M. Nikol'skii and the development of research on approximation theory in the USSR*, Russian mathematical surveys **40** (1985), 83–156.

[52] M. G. Krein, M. A. Krasnosel'skii, D. P. Mil'man, *On deficiency numbers of linear operators in Banach spaces and some geometric problems*, Sb. Trudov Inst. Mat. Akad. Nauk Ukr. SSR **11** (1948), 97–112.

[53] N. P. Kupcov, *Kolmogorov estimates for derivatives in $L_2[0, \infty)$*, Proc. Steklov Institute of Mathematics **138** (1975), 101–125.

[54] E. Landau, *Einige Ungleichungen für zweimal differentierbare Funktionen*, Proc. London Math. Soc. **13** (1913), 43–49.

[55] A. A. Ligun, *Diameters of certain classes of differentiable periodic functions*, Math. Notes **27** (1980), no. 1, 34–41.

[56] G. G. Lorentz, *Approximation of Functions*, Holt, Rinehalt and Winston, New York, 1966.

[57] G. G. Lorentz, M. von Golitschek, Yu. Makovoz, *Constructive Approximation. Advanced Problems.*, Springer, Berlin, 1996, A Series of Comprehensive Studies in Mathematics 304.

[58] Yu. I. Lyubich, *On inequalities between powers of linear operators*, Izv. AN SSSR **24** (1960), 825–864.

[59] G. G. Magaril-Il'yaev, *On Kolmogorov inequalities on a half-line*, Vestnik Moskovskogo Universiteta (Moscow University Bulletin), Matematika **31** (1976), no. 5, 33–41.

[60] G. G. Magaril-Il'yaev, *Inequalities for the derivatives and the duality*, Trudy Mat. Inst. Steklov **161** (1983), 183–194. (Russian)

[61] V. N. Malozemov, A. B. Pevny, *Polynomial'nye splainy (Polynomial splines)*, Izd. Len. Univ., Leningrad, 1986. (Russian)

[62] A. A. Markov, *On a problem of D. I. Mendeleev*, Zap. Akad. Nauk St. Petersburg **62** (1889), 1–24. (Russian)

[63] V. A. Markov, *On functions deviating least from zero in a given interval*, Izdat. Akad. Nauk, St. Petersburg (1892).

[64] A. P. Matorin, *On inequalities between the maxima of absolute values of a function and its derivatives on a half-line*, Amer. Math. Soc. Transl., Series 2, 8 (1958), 13–17. (Russian)

[65] G. Meinardus, *Approximation of Functions: Theory and Numerical Methods*, Springer-Verlag Inc., N. Y., 1967.

[66] C. A. Micchelli, A. Pinkus, *Some problems on approximation of functions of two variables and n-widths of integral operators*, J. Approx. Theory **24** (1978), 51–77.

[67] C. A. Micchelli, T. J. Rivlin, S. Winograd, *The optimal recovery of smooth smooth functions*, Numer. Math. **26** (1976), no. 2, 191–200.

[68] V. P. Motornyi, *On the best quadrature formula of the form $\sum_{k=1}^{n} p_k f(x_k)$ for some classes of periodic differentiable functions*, Izv. AN SSSR **38** (1974), no. 3, 583–614.

[69] V. P. Motornyi, V. I. Ruban, *Diameters of some classes of differentiable periodic functions in the space L*, Math. Notes **17** (1975), no. 4, 313–320.

[70] S. M. Nikol'skii, *Approximation of periodic functions by trigonometric polynomials*, Trudy MIAN **15** (1945), 1–76.

[71] S. M. Nikol'skii, *La série de Fourier d'une fonction dont le module de continuité est donné*, Dokl. Akad. Nauk SSSR **52** (1946), 191–194.

[72] A. Pinkus, *Some extremal properties of perfect splines and the pointwise Landau problem on the finite interval*, Journal of Approximation Theory **23** (1978), no. 2, 37–64.

[73] A. Pinkus, *N-Widths of Sobolev spaces in L_p*, CA **1** (1985), 15–62.

[74] A. Pinkus, *N-widths in Approximation Theory*, Springer Verlag, Berlin, 1985.

[75] V. I. Ruban, *Even diameters of the classes $W^{(r)}H_\omega$ in the space $C_{2\pi}$*, Math. Notes **15** (1974), no. 3, 222–225.

[76] M. Sato, *The Landau inequality for bounded intervals with $\left\| f^{(3)} \right\|$ finite*, Journal of Approximation Theory **34** (1982), 159–166.

[77] A. C. Schaeffer, R. J. Duffin, *On some inequalities of S. Bernstein and W. Markoff for derivatives of polynomials*, Bull. Amer. Math. Soc. **44** (1938), 289–297.

[78] I. J. Schoenberg, *The elementary case of Landau's problem on inequalities between derivatives*, Amer. Math. Monthly **80** (1973), 121–158.

[79] I. J. Schoenberg, A. Whitney, *On Polya frequency functions*, Trans. Amer. Math. Soc. **74** (1953), no. 2, 246–259.

[80] S. B. Stechkin, *Inequalities between norms of derivatives of an arbitrary function*, Acta Sci. Math. **26** (1965), Szeged, 225–230. (Russian)

[81] S. B. Stechkin, Math. Zametki **1** (1967), no. 2, 137–148 Best approximation of linear operators.

[82] E. M. Stein, *Functions of exponential type*, Ann. of Math. **65** (1957), no. 3, 582–592.

[83] B. Szekëfalvi-Nagy, *Über Integralungleichungen zwischen einer Funktion und ihrer Ableitung*, Acta Sci. Math. **10** (1941), 64–74.

[84] L. V. Taikov, *Inequalities of Kolmogorov type and best formulas for numerical differentiation*, Math. Zametki **4** (1968), 233–238; English transl. in Math. Notes **4** (1968), 631–634.

[85] V. M. Tihomirov, *Widths of sets in functional spaces and theory of best approximations*, Usp. Mat. Nauk **15** (1960), no. 3, 81–120.

[86] V. M. Tihomirov, S. B. Babadjanov, *On the widths of a functional class in the space $L_p (p \geq 1)$*, Izv. Akad. Nauk UzSSR Ser. Fiz.-Mat. Nauk **2** (1967), 24–30.

[87] V. M. Tihomirov, *Best methods of approximation and interpolation of differentiable functions in the space $C[-1, 1]$*, Math USSR Sbornik **9** (1969), 277–289.

[88] V. M. Tihomirov, *Some Problems of Approximation Theory*, Izd. MGU, Moscow, 1976.

[89] E. V. Voronovskaja, *The functional of the first derivative and improvement of a theorem of A. A. Markov*, Izv. Akad, Nauk SSSR, Ser. Mat. **23** (1959), 951–962, (see translation in [90]).

[90] E. V. Voronovskaja, *The functional method and its application*, vol. 28, AMS, Providence, 1970.

[91] E. I. Zolotarev, *Application of elliptic functions to problems on functions deviating least or most from zero*, Zapiski St.-Peterburg Akad. Nauk **30** (1877), no. 5, 1–59, reprinted in Collected works, Vol.II, Izdat. Akad. Nauk SSSR, Leningrad, 1932. (Russian)

[92] A. Zvjagincev, A. Lepin, *On Kolmogorov's inequalities between the upper bounds of the derivatives of a function for $n = 3$*, Latviiskii matematicheskii ezhegodnik (Latvian annual mathematical journal) **26** (1982), Zinatne, Riga, 176–181. (Russian)

[93] A. Zygmund, *Trigonometric Series*, vol. I, Mir, Moscow, 1965.

Index

OPERATOR THEORY: ADVANCES AND APPLICATIONS

BIRKHÄUSER VERLAG

Edited by

I. Gohberg,

School of Mathematical Sciences, Tel-Aviv University, Ramat Aviv, Israel

This series is devoted to the publication of current research in operator theory, with particular emphasis on applications to classical analysis and the theory of integral equations, as well as to numerical analysis, mathematical physics and mathematical methods in electrical engineering.

———

91   A.L. Skubachevskii
     Elliptic Functional Differential Equations and Applications.
     1997, ISBN 3-7643-5404-6

92   A.Ya. Shklyar
     Complete Second Order Linear Differential Equations in Hilbert Spaces
     1997. ISBN 3-7643-5377-5

93   Y. Egorov / B.-W. Schulze
     Pseudo-Differential Operators, Singularities, Applications
     1997. ISBN 3-7643-5484-4

94   M.I. Kadets / V.M. Kadets
     Series in Banach Spaces. Conditional and Unconditional Convergence.
     1997. ISBN 3-7643-5401-1

95   H. Dym / V. Katsnelson / B. Fritzsche / B. Kirstein (Eds)
     Topics in Interpolation Theory
     1997. ISBN 3-7643-5723-1

96   D. Alpay / A. Dijksma / J. Rovnyak / H. de Snoo
     Schur Functions, Operator Colligations, and Reproducing Kernel Pontryagin Spaces
     1997. ISBN 3-7643-5763-0

97   M.L. Gorbachuk / V.I. Gorbachuk
     M.G. Krein's Lectures on Entire Operators
     1997. ISBN 3-7643-5704-5

98   I. Gohberg / Yu. Lyubich (Eds)
     New Results in Operator Theory and Its Applications
     The Israel M. Glazman Memorial Volume
     1997. ISBN 3-7643-5775-4

99   T. Ayerbe Toledano / T. Dominguez Benavides / G. López Acedo
     Measures of Noncompactness in Metric Fixed Point Theory
     1997. ISBN 3-7643-5794-0

100 C. Foias / A.E. Frazho / I. Gohberg / M.A. Kaashoek
     Metric Constrained Interpolation, Commutant Lifting and System
     1998. ISBN 3-7643-5889-0

101 S.D. Eidelman / N.V. Zhitarashu
     Parabolic Boundary Value Problems
     1998. ISBN 3-7643-2972-6

102 I. Gohberg / R. Mennicken / C.Tretter (Eds)
     Differential and Integral Operators. International Workshop on Operator Theory and Applications, IWOTA 95, in Regensburg, July 31–August 4, 1995.
     1998. ISBN 3-7643-5890-4

103. I. Gohberg / R. Mennicken / C. Tretter (Eds)
     Recent Progress in Operator Theory. International Workshop on Operator Theory and Applications, IWOTA 95, in Regensburg, July 31–August 4, 1995.
     1998. ISBN 3-7643-5891-2

104. Bercovici, H. / Foias, C. (Eds)
     Nonselfadjoint Operator Algebras, Operator Theory, and Related Topics. The Carl M. Pearcy Anniversary Volume.
     1998. ISBN 3-7643-5954-4